STUDENT'S SOLUTION MANUAL

RAYMOND A. BARNETT and MICHAEL R. ZIEGLER's

ANALYTIC
TRIGONOMETRY
WITH APPLICATIONS

Sixth Edition

prepared by

FRED SAFIER
City College of San Francisco

PWS PUBLISHING COMPANY
I(T)P *An International Thomson Publishing Company*

Boston • Albany • Bonn • Cincinnati • Detroit • London • Madrid • Melbourne
Mexico City • New York • Paris • San Francisco • Singapore • Tokyo • Toronto • Washington

PWS PUBLISHING COMPANY
20 Park Plaza, Boston, MA 02116-4324

International Thomson Publishing
The trademark ITP is used under license.

Copyright © 1995 by PWS Publishing Company, a division of International Thomson Publishing Inc.

ISBN 0-534-94347-0

Printed and bound in the United States of America by Malloy Lithographing.

3 4 5 6 7 8 9—99 98 97

Chapter 1   Right Triangle Ratios

EXERCISE 1.1 Angles, Degrees, and Arcs

1. Since one complete revolution has measure $360°$, $\frac{1}{2}$ revolution has measure $\frac{1}{2}(360°) = 180°$.

3. Since one complete revolution has measure $360°$, $\frac{1}{8}$ revolution has measure $\frac{1}{8}(360°) = 45°$.

5. Since one complete revolution has measure $360°$, $\frac{2}{3}$ revolution has measure $\frac{2}{3}(360°) = 240°$.

7. Since $123°$ is between $90°$ and $180°$, this is an obtuse angle.

9. A $180°$ angle is called a straight angle.

11. Since $45°$ is between $0°$ and $90°$, this is an acute angle.

13. None of these.

15. $10°33' = (10 \cdot 60 + 33)' = 633'$

17. $1°10'12'' = (1 \cdot 60 \cdot 60 + 10 \cdot 60 + 12)'' = 4{,}212''$

19. $72' = (60 + 12)' = 60' + 12' = 1°12'$

21. Since $21' = \frac{21}{60}^\circ$ and $4'' = \frac{4}{3{,}600}^\circ$, then
$$43°21'4'' = \left(43 + \frac{21}{60} + \frac{4}{3{,}600}\right)^\circ \approx 43.351° \text{ to three decimal places}$$

23. Since $12' = \frac{12}{60}^\circ$ and $47'' = \frac{47}{3{,}600}^\circ$, then
$$2°12'47'' = \left(2 + \frac{12}{60} + \frac{47}{3{,}600}\right)^\circ \approx 2.213° \text{ to three decimal places}$$

25. Since $17' = \frac{17}{60}^\circ$ and $41'' = \frac{41}{3{,}600}^\circ$, then
$$103°17'41'' = \left(103 + \frac{17}{60} + \frac{41}{3{,}600}\right)^\circ \approx 103.295° \text{ to three decimal places}$$

27. $13.633° = 13°(0.633 \times 60)' = 13°37.98' = = 13°37'(0.98 \times 60)'' \approx 13°37'59''$

29. $83.017° = 83°(0.017 \times 60)' = 83°1.02' = 83°1'(0.02 \times 60)'' = 83°1'1''$

31. $187.204° = 187°(0.204 \times 60)' = 187°12.24' = 187°12'(0.24 \times 60)' = 187°12'14''$

33. To compare α and β, we convert α to decimal form. Since $9' = \frac{9}{60}^\circ$ and $17'' = \frac{17}{3{,}600}^\circ$, then
$$27°9'17'' = \left(27 + \frac{9}{60} + \frac{17}{3{,}600}\right)^\circ \approx 27.155°. \text{ Since } 27.155° < 27.163°, \text{ we conclude that } \alpha < \beta.$$

Chapter 1   Right Triangle Ratios

35. To compare α and β, we convert β to decimal form. Since $47' = \dfrac{47°}{60}$ and $13'' = \dfrac{13°}{3{,}600}$, then

$$12°47'13'' = \left(12 + \frac{47}{60} + \frac{13}{3{,}600}\right)° \approx 12.787°. \text{ Since } 12.807° > 12.787°, \text{ we conclude that } \alpha > \beta.$$

37. To compare α and β, we convert α to decimal form. Since $7' = \dfrac{7°}{60}$ and $3'' = \dfrac{3°}{3{,}600}$, then

$$248°7'3'' = \left(248 + \frac{7}{60} + \frac{3}{3{,}600}\right)° \approx 248.118°. \text{ Since } 248.118° < 248.132°, \text{ we conclude}$$

that $\alpha < \beta$.

39. Since $\dfrac{s}{C} = \dfrac{\theta°}{360°}$, then

$$\frac{s}{1000 \text{ cm}} = \frac{36°}{360°}$$

$$s = \frac{36}{360}(1000 \text{ cm}) = 100 \text{ cm}$$

41. Since $\dfrac{s}{C} = \dfrac{\theta°}{360°}$, then

$$\frac{25 \text{ km}}{C} = \frac{20°}{360°}$$

$$C = \frac{360}{20}(25 \text{ km}) = 450 \text{ km}$$

43. Since $\dfrac{s}{C} = \dfrac{\theta°}{360°}$ and $C = 2\pi R$, then $\dfrac{s}{2\pi R} = \dfrac{\theta°}{360°}$

$$\frac{s}{2(3.14)(5{,}400{,}000 \text{ mi})} \approx \frac{2.6°}{360°}$$

$$s \approx \frac{2(3.14)(5{,}400{,}000 \text{ mi})(2.6)}{360}$$

$$\approx 240{,}000 \text{ mi (to nearest 10,000 mi)}$$

45. Since $\dfrac{s}{C} = \dfrac{\theta°}{360°}$ and $\theta = 12°31'4'' = \left(12 + \dfrac{31}{60} + \dfrac{4}{3{,}600}\right)° = 12.517°$, then

$$\frac{50.2 \text{ cm}}{C} \approx \frac{12.517°}{360°}$$

$$C \approx \frac{360}{12.517}(50.2 \text{ cm}) \approx 1440 \text{ cm (to nearest 10 cm)}$$

47. Since $\dfrac{A}{\pi R^2} = \dfrac{\theta}{360°}$, then $\dfrac{A}{\pi(25.2 \text{ cm})^2} = \dfrac{47.3°}{360°}$

$$A \approx \frac{47.3}{360}(3.14)(25.2 \text{ cm})^2 \approx 262 \text{ cm}^2$$

49. Since $\dfrac{A}{\pi R^2} = \dfrac{\theta}{360°}$, then $\dfrac{98.4 \text{ m}^2}{\pi(12.6 \text{ m})^2} = \dfrac{\theta}{360°}$

$$\theta = \frac{98.4}{(3.14)(12.6)^2} \cdot 360° = 71.1°$$

51. Since $\dfrac{s}{C} = \dfrac{\theta}{360°}$ and $C = 2\pi R$, then $\dfrac{s}{2\pi R} = \dfrac{\theta}{360°}$

$$R = \frac{s}{2\pi} \cdot \frac{360°}{\theta}$$

$$\approx \frac{11.5 \text{ mm}}{2(3.14)} \cdot \frac{360}{118.2} \approx 5.58 \text{ mm}$$

53. Since $\dfrac{s}{C} = \dfrac{\theta}{360°}$ and $C = 2\pi R$, then $\dfrac{s}{2\pi R} = \dfrac{\theta}{360°}$

$$s = 2\pi R \cdot \dfrac{\theta}{360°}$$

$$\approx 2(3.14)(5.49 \text{ mm}) \cdot \dfrac{119.7}{360} \approx 11.5 \text{ mm}$$

In Problems 55–61 we use the diagram and reason as follows: Since the cities have the same longitude, θ is given by their difference in latitude.

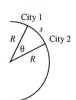

55. Since $\dfrac{s}{C} = \dfrac{\theta}{360°}$ and $C = 2\pi R$, then $\dfrac{s}{2\pi R} = \dfrac{\theta}{360°}$; $\theta = 47°40' - 37°50' = 9°50' = \left(9 + \dfrac{50}{60}\right)^{\circ}$

$$s = 2\pi R \cdot \dfrac{\theta}{360°} \approx 2(3.14)(3960 \text{ mi}) \cdot \dfrac{9 + \dfrac{50}{60}}{360} \approx 679 \text{ mi}$$

57. Since $\dfrac{s}{C} = \dfrac{\theta}{360°}$ and $C = 2\pi R$, then $\dfrac{s}{2\pi R} = \dfrac{\theta}{360°}$; $\theta = 40°50' - 32°50' = 8°$

$$s = 2\pi R \cdot \dfrac{\theta}{360°} \approx 2(3.14)(3960) \cdot \dfrac{8}{360} \approx 553 \text{ mi}$$

59. To find the length of s in nautical miles, since 1 nautical mile is the length of $1'$ on the circle shown in the diagram, we need only find how many minutes are in the angle θ. Since

$$\theta = 47°40' - 37°50' = 9°50' = (9 \times 60 + 50)', \ \theta = 590'.$$

Therefore, s = 590 nautical miles.

60. To find the length of s in nautical miles, since 1 nautical mile is the length of $1'$ on the circle shown in the diagram, we need only find how many minutes are in the angle θ. Since

$$\theta = 40°50' - 32°50' = 8° = (8 \times 60)', \ \theta = 480'.$$

Therefore, s = 480 nautical miles.

63. We use the intercepted arc (2°) to approximate its chord (the diameter of the object). Since

$$\dfrac{s}{2\pi R} = \dfrac{\theta^{\circ}}{360°}$$

$$s = \dfrac{2\pi R \theta}{360} \approx \dfrac{2(3.14)(1,000 \text{ m})2}{360} \approx 34.9 \text{ m}$$

65. We use the intercepted arc $(32')$ to approximate its chord (the diameter of the sun). Since

$$\dfrac{s}{2\pi R} = \dfrac{\theta^{\circ}}{360°} \text{ and } \theta = 32' = \dfrac{32^{\circ}}{60} \approx 0.533^{\circ}$$

$$s = \dfrac{2\pi R \theta}{360} \approx \dfrac{2(3.142)(92,956,000 \text{ mi})(0.533)}{360} \approx 865,000 \text{ mi}$$

Chapter 1   Right Triangle Ratios

EXERCISE 1.2 Similar Triangles

1. Since $\dfrac{a}{a'} = \dfrac{b}{b'}$ by Euclid's Theorem, then $\dfrac{5}{2} = \dfrac{15}{b'}$, $b' = \dfrac{2(15)}{5} = 6$

3. Since $\dfrac{a}{a'} = \dfrac{c}{c'}$ by Euclid's Theorem, then $\dfrac{12}{2.0} = \dfrac{c}{18}$, $c = \dfrac{(18)(12)}{2.0} = 110$

5. Since $\dfrac{b}{b'} = \dfrac{c}{c'}$ by Euclid's Theorem, then $\dfrac{52,000}{8.0} = \dfrac{18,000}{c'}$, $c' = \dfrac{(8.0)(18,000)}{52,000} = 2.8$

7. Since the triangles are similar, the sides are proportional and we can write

$$\dfrac{b}{1.0} = \dfrac{10.1 \text{ m}}{0.47} \qquad\qquad \dfrac{c}{1.1} = \dfrac{10.1 \text{ m}}{0.47}$$

$$b = 21 \text{ m} \qquad\qquad c = \dfrac{(10.1 \text{ m})(1.1)}{0.47} = 24 \text{ m}$$

9. Since the triangles are similar, the sides are proportional and we can write

$$\dfrac{a}{0.47} = \dfrac{51 \text{ in.}}{1.0} \qquad\qquad \dfrac{c}{1.1} = \dfrac{51 \text{ in.}}{1.0}$$

$$a = \dfrac{(0.47)(51 \text{ in.})}{1.0} = 24 \text{ in.} \qquad c = \dfrac{(1.1)(51 \text{ in.})}{1.0} = 56 \text{ in.}$$

11. Since the triangles are similar, the sides are proportional and we can write

$$\dfrac{a}{0.47} = \dfrac{2.0 \times 10^5 \text{ km}}{1.1} \qquad\qquad \dfrac{b}{1.0} = \dfrac{2.0 \times 10^5 \text{ km}}{1.1}$$

$$a = \dfrac{(0.47)(2.0 \times 10^5 \text{ km})}{1.1} \qquad b = \dfrac{2.0 \times 10^5 \text{ km}}{1.1}$$

$$= 8.5 \times 10^4 \text{ km} \qquad\qquad = 1.8 \times 10^5 \text{ km}$$

13. Since the triangles are similar, the sides are proportional and we can write

$$\dfrac{b}{1.0} = \dfrac{23.4 \text{ m}}{0.47} \qquad\qquad \dfrac{c}{1.1} = \dfrac{23.4 \text{ m}}{0.47}$$

$$b = \dfrac{23.4 \text{ m}}{0.47} = 50 \text{ m} \qquad c = \dfrac{(1.1)(23.4 \text{ m})}{0.47} = 55 \text{ m}$$

15. Since the triangles are similar, the sides are proportional and we can write

$$\dfrac{a}{0.47} = \dfrac{2.478 \times 10^9 \text{ yd}}{1.0} \qquad\qquad \dfrac{c}{1.1} = \dfrac{2.489 \times 10^9 \text{ yd}}{1.0}$$

$$a = (0.47)(2.489 \times 10^9 \text{ yd}) \qquad c = (1.1)(2.489 \times 10^9 \text{ yd})$$

$$= 1.2 \times 10^9 \text{ yd} \qquad\qquad = 2.7 \times 10^9 \text{ yd}$$

17. Since the triangles are similar, the sides are proportional and we can write

$$\dfrac{a}{0.47} = \dfrac{8.39 \times 10^{-5} \text{ mm}}{1.1} \qquad\qquad \dfrac{b}{1.0} = \dfrac{8.39 \times 10^{-5} \text{ mm}}{1.1}$$

$$a = \dfrac{(0.47)(8.39 \times 10^{-5} \text{ mm})}{1.1} \qquad b = \dfrac{8.39 \times 10^{-5} \text{ mm}}{1.1}$$

$$= 3.6 \times 10^{-5} \text{ mm} \qquad\qquad = 7.6 \times 10^{-5} \text{ mm}$$

19. We make a scale drawing of the triangle, choosing
    a' to be 2.00 in., $\angle A' = 70°$, $\angle C' = 90°$. Now
    measure c' (approximately 2.13 in.) and set up a
    proportion. Thus,

    $$\frac{c}{2.13 \text{ in.}} = \frac{101 \text{ ft}}{2.00 \text{ in.}}$$

    $$c \approx \frac{2.13}{2.00}(101 \text{ ft}) \approx 108 \text{ ft}$$

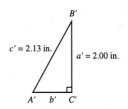

Drawing not to scale

21. In the drawing,
    we note that
    triangles LBT and
    LNM are similar.
    MN = 9 ft.

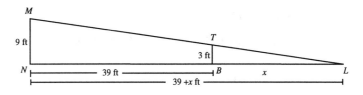

    $NB = \frac{1}{2}$ (length of court) $= \frac{1}{2}$ (78 ft) = 39 ft. TB = 3 ft. BL is to be found. Let BL = x.

    Then, $\frac{BL}{NL} = \frac{TB}{MN}$. NL = NB + BL. Thus, $\frac{x}{39 + x} = \frac{3}{9}$.

    $$9(39 + x)\frac{x}{39 + x} = 9(39 + x)\frac{3}{9} \text{ (clear of fractions)}$$

    $$9x = (39 + x)3 = 117 + 3x$$
    $$6x = 117$$
    $$x = 19.5 \text{ ft}$$

23. Since the triangles ABC and DEC in the figure are similar, we can write $\frac{AB}{DE} = \frac{AC}{CD}$. Then,

    $$\frac{AB}{5.5 \text{ ft}} = \frac{24 \text{ ft}}{2.1 \text{ ft}}$$

    $$AB = \frac{5.5}{2.1}(24 \text{ ft}) = 63 \text{ ft}$$

25. Let us make a scale drawing of the figure in the text as follows: pick any convenient length, say
    2 in., for A′C′. Copy the 15° angle CAB and the 90° angle ACB using a protractor. Now, measure
    B′C′ (approximately 0.55 in.) and set up a proportion. Thus,

    $$\frac{x}{0.55 \text{ in.}} = \frac{4.0 \text{ km}}{2 \text{ in.}}$$

    $$x \approx \frac{0.55}{2}(4.0 \text{ km}) \approx 1.1 \text{ km}$$

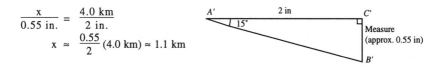

EXERCISE 1.3   Trigonometric Ratios and Right Triangles

1. $\frac{a}{c}$ 3. $\frac{b}{a}$ 5. $\frac{c}{a}$ 7. $\sin \theta$ 9. $\tan \theta$ 11. $\csc \theta$

13. Set calculator in degree mode and use sin key.

    $\sin 25.6° = 0.432$.

Chapter 1   Right Triangle Ratios

15. Set calculator in degree mode, convert to decimal degrees, and use tan key.
$$35°20' = \left(35 + \frac{20}{60}\right)^° = (35.333333...)^°$$

$$\tan 35°20' = \tan(35.333333...)^° = 0.709$$

17. Use the reciprocal relationship sec θ = 1/cos θ. Set calculator in degree mode, use cos key, then take reciprocal.
$$\sec 44.8° = 1.41$$

19. Set calculator in degree mode and use cos key.
$$\cos 72.9° = 0.294$$

21. Use the reciprocal relationship cot θ = 1/tan θ. Set calculator in degree mode, use tan key, then take reciprocal.
$$\cot 54.9° = 0.703$$

23. Use the reciprocal relationship csc θ = 1/sin θ. Set calculator in degree mode, convert to decimal degrees, use sin key, then take reciprocal.
$$67°30' = \left(67 + \frac{30}{60}\right)^° = 67.5°$$

$$\csc 67°30' = \csc 67.5° = 1.08$$

25. If sin θ = 0.8032, then

   θ = $\sin^{-1}$ 0.8032 = 53.44°

27. θ = arccos 0.7153

   = 44.332° = 44°20'

29. θ = $\tan^{-1}$ 1.948

   = 62.826° = 62°50'

31. θ = $\sin^{-1}(0.3772)$

   = 22.160° = 22°9'37"

33. *Solve for the complementary angle*: 90° − θ = 90° − 58°40' = 31°20'

   Solve for b: Since θ = 58°40' = $\left(58 + \frac{40}{60}\right)^°$ = (58.666...)$^°$ and c = 15.0 mm, we look for a trigonometric ratio that involves θ and c (the known quantities) and b (the unknown quantity). We choose the sine. sin θ = $\frac{b}{c}$

$$b = c \sin \theta = (15.0 \text{ mm})(\sin 58.666...°) = 12.8 \text{ mm}$$

   Solve for a: We choose the cosine to find a. Thus,
$$\cos \theta = \frac{a}{c}$$
$$a = c \cos \theta = (15.0)(\cos 58.666...°) = 7.80 \text{ mm}$$

35. *Solve for the complementary angle*: 90° − θ = 90° − 83.7° = 6.3°

   Solve for a: Since θ = 83.7° and b = 3.21 km, we look for a trigonometric ratio that involves θ and b (the known quantities) and a (the unknown quantity). We choose the tangent.
$$\tan \theta = \frac{b}{a}$$
$$a = \frac{b}{\tan \theta} = \frac{3.21 \text{ km}}{\tan 83.7°} = 0.354 \text{ km}$$

Solve for c: We choose the sine to find c. Thus, $\sin \theta = \dfrac{b}{c}$

$$c = \frac{b}{\sin \theta} = \frac{3.21 \text{ km}}{\sin 83.7°} = 3.23 \text{ km}$$

37. *Solve for the complementary angle:* $90° - \theta = 90° - 71.5° = 18.5°$

   Solve for a: Since $\theta = 71.5°$ and b = 12.8 in., we look for a trigonometric ratio that involves

   θ and b (the known quantities) and a (the unknown quantity). We choose the tangent.

$$\tan \theta = \frac{b}{a}$$

$$a = \frac{b}{\tan \theta} = \frac{12.8 \text{ in.}}{\tan 71.5°} = 4.28 \text{ in.}$$

   Solve for c: We choose the sine to find c. Thus, $\sin \theta = \dfrac{b}{c}$

$$c = \frac{b}{\sin \theta} = \frac{12.8 \text{ in.}}{\sin 71.5°} = 13.5 \text{ in.}$$

39. *Solve for θ:* $\theta = 90° - (90° - \theta) = 90° - 33°40' = 56°20'$

   Solve for b: Since $\theta = 56°20'$ and a = 22.4 cm, we look for a trigonometric ratio that involves

   θ and a (the known quantities) and b (the unknown quantity). We choose the tangent.

$$\tan \theta = \frac{b}{a}$$

$$b = a \tan \theta = (22.4 \text{ cm})(\tan 56°20') = 33.6 \text{ cm}$$

   Solve for c: We choose the cosine to find c. Thus, $\cos \theta = \dfrac{a}{c}$

$$c = \frac{a}{\cos \theta} = \frac{22.4 \text{ cm}}{\cos 56°20'} = 40.4 \text{ cm}$$

41. *Solve for θ:* $\sin \theta = \dfrac{b}{c} = \dfrac{63.8 \text{ ft}}{134 \text{ ft}} = 0.476$

$$\theta = \sin^{-1} 0.476 = 28.4° = 28°30' \text{ to nearest } 10'$$

   Solve for the complementary angle: $90° - \theta = 90° - 28°30' = 61°30'$

   Solve for a: We will use the tangent. Thus, $\tan \theta = \dfrac{b}{a}$

$$a = \frac{b}{\tan \theta} = \frac{63.8 \text{ ft}}{\tan 28°30'} = 118 \text{ ft}$$

43. *Solve for θ:* $\tan \theta = \dfrac{b}{a} = \dfrac{132 \text{ mi}}{108 \text{ mi}} = 1.22$

$$\theta = \tan^{-1} 1.22 = 50.7° \text{ to nearest } 0.1°$$

   Solve for the complementary angle: $90° - \theta = 90° - 50.7° = 39.3°$

   Solve for c: We will use the sine. Thus, $\sin \theta = \dfrac{b}{c}$

$$c = \frac{b}{\sin \theta} = \frac{132 \text{ mi}}{\sin 50.7°} = 171 \text{ mi}$$

45. (A) If $\theta = 11°$, $(\sin \theta)^2 + (\cos \theta)^2 = (\sin 11°)^2 + (\cos 11°)^2 = (0.1908...)^2 + (0.9816...)^2 = 1$

   (B) If $\theta = 6.09°$, $(\sin \theta)^2 + (\cos \theta)^2 = (\sin 6.09°)^2 + (\cos 6.09°)^2 = (0.106...)^2 + (0.994...)^2 = 1$

Chapter 1   Right Triangle Ratios

(C) If $\theta = 43°24'47''$, $(\sin \theta)^2 + (\cos \theta)^2 = (\sin 43°24'47'')^2 + (\cos 43°24'47'')^2$

$$= (0.687...)^2 + (0.726...)^2 = 1$$

47. (A) If $\theta = 19°$, $\sin \theta - \cos(90° - \theta) = \sin 19° - \cos(90° - 19°)$

$$= \sin 19° - \cos 71° = 0.3256 - 0.3256 = 0$$

(B) If $\theta = 49.06°$, $\sin \theta - \cos(90° - \theta) = \sin 49.06° - \cos(90° - 49.06°)$

$$= \sin 49.06° - \cos 40.94° = 0.7554 - 0.7554 = 0$$

(C) If $\theta = 72°51'12''$, $\sin \theta - \cos(90° - \theta) = \sin 72°51'12'' - \cos(90° - 72°51'12'')$

$$= \sin 72°51'12'' - \cos(17°8'48'')$$

$$= 0.9556 - 0.9556 = 0$$

49. *Solve for the complementary angle:* $90° - \theta = 90° - 37.46° = 52.54°$

Solve for a: We choose the tangent to find a. Thus, $\tan \theta = \dfrac{b}{a}$

$$a = \frac{b}{\tan \theta} = \frac{5.317 \text{ cm}}{\tan 37.46°} = 6.939 \text{ cm}$$

Solve for c: We choose the sine to find c. Thus, $\sin \theta = \dfrac{b}{c}$

$$c = \frac{b}{\sin \theta} = \frac{5.317 \text{ cm}}{\sin 37.46°} = 8.742 \text{ cm}$$

51. *Solve for the complementary angle:* $90° - \theta = 90° - 83°12' = 6°48'$

Solve for b: We choose the tangent to find b. Thus, $\tan \theta = \dfrac{b}{a}$

$$b = a \tan \theta$$

$$= (23.82 \text{ mi})(\tan 83°12')$$

$$= 199.8 \text{ mi}$$

Solve for c: We choose the cosine to find c. Thus, $\cos \theta = \dfrac{a}{c}$

$$c = \frac{a}{\cos \theta} = \frac{23.82 \text{ mi}}{(\cos 83°12')} = 201.2 \text{ mi}$$

53. *Solve for θ:* $\tan \theta = \dfrac{b}{a} = \dfrac{42.39 \text{ cm}}{56.04 \text{ cm}}$; $\tan \theta = 0.7564$; $\theta = \tan^{-1} 0.7564 = 37.105° = 37°6'$

Solve for the complementary angle: $90° - \theta = 90° - 37°6' = 52°54'$

Solve for c: We will use the sine. Thus, $\sin \theta = \dfrac{b}{c}$

$$c = \frac{b}{\sin \theta} = \frac{42.39 \text{ cm}}{\sin 37°6'} = 70.27 \text{ cm}$$

55. *Solve for θ:* $\sin \theta = \dfrac{b}{c} = \dfrac{35.06 \text{ cm}}{50.37 \text{ cm}} = 0.6960$

$$\theta = \sin^{-1}(0.6960) = 44.11°$$

Solve for the complementary angle: $90° - \theta = 90° - 44.11° = 45.89°$

Solve for a: We choose the cosine to find a. Thus, $\cos\theta = \dfrac{a}{c}$

$$a = c\cos\theta$$
$$= (50.37\text{ cm})(\cos 44.11°) = 36.17\text{ cm}$$

57. According to the Pythagorean theorem, $a^2 + b^2 = c^2$. Then, using Definition 1, we have

$$(\sin\theta)^2 + (\cos\theta)^2 = \left(\frac{b}{c}\right)^2 + \left(\frac{a}{c}\right)^2 = \frac{b^2 + a^2}{c^2} = \frac{a^2 + b^2}{c^2} = \frac{c^2}{c^2} = 1$$

59. (A) According to Definition 1, $\cot\theta = \dfrac{a}{b} = \dfrac{1}{b/a} = \dfrac{1}{\tan\theta}$

    (B) According to Definition 1, $\csc(90° - \theta) = \dfrac{c}{a} = \sec\theta$

61. (A) In right triangle OAD, $\sin\theta = \dfrac{\text{Opp}}{\text{Hyp}} = \dfrac{AD}{OD} = \dfrac{AD}{1} = AD$

    (B) In tright triangle OCD, $\tan\theta = \dfrac{\text{Opp}}{\text{Adj}} = \dfrac{DC}{OD} = \dfrac{DC}{1} = DC$

    (C) In right triangle ODE, $\csc\theta = \csc OED = \dfrac{\text{Hyp}}{\text{Opp}} = \dfrac{OE}{OD} = \dfrac{OE}{1} = OE$

63. (A) As θ approaches 90°, AD approaches OD, which has measure 1. Thus, $\sin\theta\,(= AD)$ approaches 1.

    (B) As θ approaches 90°, EC approaches being parallel to the x axis. Thus, DC increases without bound, so $\tan\theta\,(= DC)$ increases without bound.

    (C) As θ approaches 90°, OE approaches OF, which has measure 1. Thus, $\csc\theta\,(= OE)$ approaches 1.

65. (A) As θ approaches 0°, OA approaches OB, which has measure 1. Thus, $\cos\theta\,(= OA)$ approaches 1.

    (B) As θ approaches 0°, EC approaches being parallel to the y axis. Thus, ED increases without bound, so $\cot\theta\,(= ED)$ increases without bound.

    (C) As θ approaches 0°, OC approaches OB, which has measure 1. Thus, $\sec\theta\,(= OC)$ approaches 1.

EXERCISE 1.4   Right Triangle Applications

1. $\sin 61° = \dfrac{\text{Opp}}{\text{Hyp}} = \dfrac{x}{8.0\text{ m}}$

    $x = (8.0\text{ m})(\sin 61°) = 7.0\text{ m}$

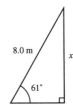

Chapter 1   Right Triangle Ratios

3.

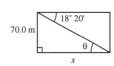

$$\cot \theta = \frac{\text{Adj}}{\text{Opp}}$$

$$\cot 18°20' = \frac{x}{70.0 \text{ m}}$$

$$x = (70.0 \text{ m})(\cot 18°20') = 211 \text{ m}$$

(If line p crosses parallel lines m and
n, then angles α and β have the same
measure. Thus, $\theta = 18°20'$.)

5. We first sketch a figure and label the known parts.

$$\tan 15° = \frac{\text{Opp}}{\text{Adj}}$$

$$= \frac{x}{4.0 \text{ km}}$$

$x = (4.0 \text{ km})(\tan 15°)$

$= 1.1 \text{ km}$

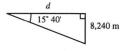

7. We first sketch a figure and label the known parts.

$$\tan 15°40' = \frac{\text{Opp}}{\text{Adj}} = \frac{8,240 \text{ m}}{d}$$

$$d = \frac{8,240 \text{ m}}{\tan 15°40'} = 29,400 \text{ m or } 29.4 \text{ km}$$

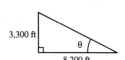

9. We first sketch a figure and label the known parts.

$$\tan \theta = \frac{\text{Opp}}{\text{Adj}} = \frac{3,300 \text{ ft}}{8,200 \text{ ft}} = 0.40...$$

$$\theta = \tan^{-1}(0.40...) = 22°$$

11. (A) In triangle ABC, $\angle \theta$ is complementary
to 75°,

thus $\theta = 15°$.

BC $=$ x $=$ roof overhang

AC $=$ 19 ft

$$\tan \theta = \frac{\text{Opp}}{\text{Adj}} = \frac{x}{19 \text{ ft}}$$

$$x = (19 \text{ ft})(\tan 15°) = 5.1 \text{ ft}$$

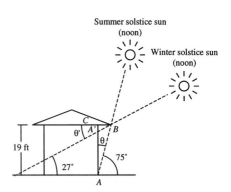

(B) In triangle A'BC, $\theta' = 17°$

A'C $=$ y $=$ how far down shadow will reach

$$\tan 27° = \frac{\text{Opp}}{\text{Adj}} = \frac{y}{5.1 \text{ ft}}$$

$$y = (5.1 \text{ ft})(\tan 27°) = 2.6 \text{ ft}$$

13. We first sketch a figure and label the known parts.

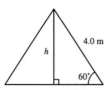

From geometry we know that each angle of an equilateral triangle has measure 60°.

$$\sin 60° = \frac{h}{4.0 \text{ m}}$$

$$h = (4.0 \text{ m})(\sin 60°) = 3.5 \text{ m}$$

15. We note that since AB is a side of a nine-sided

regular polygon,

$$\angle BCA = \frac{1}{9}(\text{circumference}) = \frac{1}{9}(360°) = 40°.$$

Since ABC is an isosceles triangle,

$$\angle FCA = \frac{1}{2}\angle BCA = \frac{1}{2}(40°) = 20°,$$

and also,

$$AF = \frac{1}{2}AB = \frac{1}{2}x.$$

Therefore, in right triangle AFC,

$$\sin 20° = \frac{(1/2)x}{8.32 \text{ cm}}$$

$$x = 2(8.32 \text{ cm})(\sin 20°) = 5.69 \text{ cm}$$

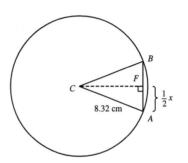

17. We note: by the symmetry of the cone, ATC is

an isosceles right triangle, hence

$$\angle TAC = \angle TCA = 45°$$

Since the mast is perpendicular to the deck, TBA

and TBC are also right triangles, and since each

has a 45° angle, these are also isosceles right triangles.

Then it follows that TB = AB and TB = AC. Since

the length TB is given as 67.0 feet, the diameter

$$AC = AB + BC = 67.0 \text{ feet} + 67.0 \text{ feet}$$

$$= 134.0 \text{ feet}.$$

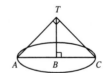

19. Label the required sides AD and AB. Then

$$AD = AC + CD = AC + 18.$$

In right triangle ABC,

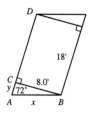

$$\sin A = \frac{BC}{AB} \qquad\qquad \tan A = \frac{BC}{AC}$$

$$\sin 72° = \frac{8.0 \text{ feet}}{x} \qquad \tan 72° = \frac{8.0 \text{ feet}}{y}$$

$$x = \frac{8.0 \text{ feet}}{\sin 72°} \qquad y = \frac{8.0 \text{ feet}}{\tan 72°}$$

$$= 8.4 \text{ feet} \qquad\qquad = 2.6 \text{ feet}$$

Thus the sides of the parallelogram are AB = 8.4 feet and

AD = 18 feet + 2.6 feet ≈ 21 feet

21. (A) We note that ∠TSC = 90° − α, hence ∠C = α.

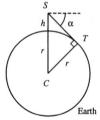

Thus in triangle CST,

$$\cos\alpha = \frac{\text{Adj}}{\text{Hyp}} = \frac{r}{r+h}$$

(B) $(r + h)\cos\alpha = (r+h)\cdot\dfrac{r}{r+h}$

$$r\cos\alpha + h\cos\alpha = r$$

$$h\cos\alpha = r - r\cos\alpha$$

$$h\cos\alpha = r(1 - \cos\alpha)$$

$$r = \frac{h\cos\alpha}{1 - \cos\alpha}$$

(C) $r = \dfrac{(335 \text{ miles})\cos 22°47'}{1 - \cos 22°47'} = 3960 \text{ miles}$

23. We note that since ABC is an isosceles triangle,

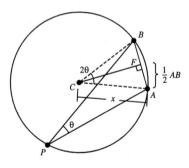

$$\angle FCA = \frac{1}{2}\angle BCA = \frac{1}{2}(2\theta) = \theta,$$

and also,

$$AF = \frac{1}{2}AB = \frac{1}{2}(6.0 \text{ mi}) = 3.0 \text{ mi.}$$

We are to find AC, the radius of the circle.

In right triangle AFC,

$$\sin\angle FCA = \frac{AF}{AC} \qquad \sin 21° = \frac{3.0 \text{ mi}}{x}$$

$$\sin\theta = \frac{3.0 \text{ mi}}{x} \qquad x = \frac{3.0 \text{ mi}}{\sin 21°} = 8.4 \text{ mi}$$

25.   In the figure, r denotes the radius of the parallel of latitude,

R the radius of the earth, i.e., the radius of the equator.

Clearly, $\cos\theta = \dfrac{\text{Adj}}{\text{Hyp}} = \dfrac{r}{R}$. Since $L = 2\pi r$ and $E = 2\pi R$,

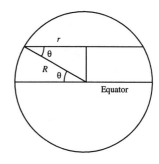

we have $r = R\cos\theta$

$2\pi r = 2\pi R\cos\theta$

$L = E\cos\theta$

E is given as 24,900 miles. In the particular case of

San Francisco, $\theta = 38°$. Hence,

$L = (24{,}900 \text{ mi})\cos 38° = 19{,}600 \text{ mi}$

27.   (A)   Let PB = y and PS = x. In right triangle SPB,

$$\csc\theta = \frac{x}{c} \qquad\qquad \cot\theta = \frac{y}{c}$$
$$x = c\csc\theta \qquad\qquad y = c\cot\theta$$

Thus the lifeguard runs a distance $d - y = d - c$

$\cot\theta$ at speed p.

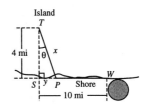

This requires a time $t_1 = $ \F(distance run,rate run)

$= \dfrac{d - c\cot\theta}{p}$

The lifeguard then swims a distance $x = c\csc\theta$ at

speed q.

This requires a time $t_2 = $ \F(distance swum,rate

swum) $= \dfrac{c\csc\theta}{q}$

Hence total time $T = t_1 + t_2 = $ \F(d − c cot θ,p) +

$\dfrac{c\csc\theta}{q}$

(B) $T = \dfrac{(380 \text{ m}) - (76 \text{ m})\cot 77°}{6.5 \text{ m/sec}} + \dfrac{(76 \text{ m})\csc 77°}{1.4 \text{ m/sec}} = 110 \text{ sec}$

(C)   We have already found that the distance run LP = d − c cot θ. Hence,

$LP = 380 \text{ m} - (76 \text{ m})\cot 77° = 360 \text{ m}$

29.   We note that the pipeline consists of ocean section TP, and
shore section PW = 10 mi − SP.

Let x = TP and y = SP. In right triangle SPT,

$$\cos\theta = \frac{4 \text{ mi}}{x}, \ \tan\theta = \frac{y}{4 \text{ mi}}, \text{ and} \theta = 30°$$
$$x = \frac{4 \text{ mi}}{\cos 30°} = 4.6188 \text{ mi}$$
$$y = (4 \text{ mi})(\tan 30°) = 2.3094 \text{ mi}$$

Chapter 1   Right Triangle Ratios

Thus,

$$\text{the cost of the pipeline} = \left(\begin{array}{c}\text{cost of ocean}\\\text{section per mile}\end{array}\right)\left(\begin{array}{c}\text{number of}\\\text{ocean miles} = x\end{array}\right)$$

$$+ \left(\begin{array}{c}\text{cost of shore}\\\text{section per mile}\end{array}\right)\left(\begin{array}{c}\text{number of shore}\\\text{miles} = 10-y\end{array}\right)$$

$$= \left(20{,}000\,\frac{\$}{\text{mi}}\right)(4.6188\text{ mi}) + \left(10{,}000\,\frac{\$}{\text{mi}}\right)(7.6906\text{ mi})$$

$$= \$92{,}376 + \$76{,}906 = \$169{,}282$$

To the nearest thousand dollars, the cost of the pipeline is $169,000.

31. A simple way to solve the system of equations

$$\tan 42° = \frac{y}{x} \qquad\qquad \tan 25° = \frac{y}{1.0 + x} \quad\text{for } y \text{ is to clear of fractions,}$$

then eliminate x from the resulting equivalent system of equations.

$$x \tan 42° = y \qquad (1.0 + x)(\tan 25°) = y$$

$$x = \frac{y}{\tan 42°}$$

Therefore, $\left(1.0 + \dfrac{y}{\tan 42°}\right)(\tan 25°) = y$

$$\tan 25° + \frac{\tan 25°}{\tan 42°}y = y \qquad \text{(Distributive property)}$$

$$\tan 25° = y - \frac{\tan 25°}{\tan 42°}y = \left(1 - \frac{\tan 25°}{\tan 42°}\right)y$$

$$y = \frac{\tan 25°}{1 - \dfrac{\tan 25°}{\tan 42°}} = 0.97\text{ km}$$

33. Labeling the diagram with the information given
in the problem, we note: We are asked to find

d = how far apart the two buildings are, and

h = the height of the apartment building.

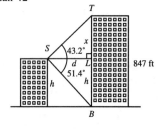

Note also that x + h = 847, so that x = 847 – h.

In right triangle SLT, $\tan 43.2° = \dfrac{x}{d} = \dfrac{847 - h}{d}$

In right triangle SLB, $\tan 51.4° = \dfrac{h}{d}$

We solve the system of equations $\tan 51.4° = \dfrac{h}{d}$ and $\tan 43.2° = \dfrac{847 - h}{d}$ by clearing of fractions,
then eliminating h.

(1) $d \tan 51.4° = h$ and $d \tan 43.2° = 847 - h$

Adding, $d \tan 51.4° + d \tan 43.2° = 847$

$$d(\tan 51.4° + \tan 43.2°) = 847$$

$$d = \frac{847}{\tan 51.4° + \tan 43.2°} = 386\text{ ft apart}$$

Substituting in (1), h = d tan 51.4° = (386 ft) tan 51.4° = 484 ft high.

14

35. We are given t = 2 sec, v = 11.1 ft/sec, θ = 10.0°. Thus,

$$g = \frac{v}{(\sin\theta)t} = \frac{11.1 \text{ ft/sec}}{(\sin 10.0°)(2 \text{ sec})} = 32.0 \text{ ft/sec}^2$$

37. From the Pythagorean theorem:

Since   AB = s and MB = $\frac{s}{2}$ Since   CN = CM = $\frac{s}{2}$

$$AM^2 = AB^2 + MB^2 \qquad\qquad NM^2 = CN^2 + CM^2$$
$$= s^2 + \left(\frac{s}{2}\right)^2 = \frac{5s^2}{4} \qquad\qquad = \left(\frac{s}{2}\right)^2 + \left(\frac{s}{2}\right)^2 = \frac{2s^2}{4}$$
$$AM = s\frac{\sqrt{5}}{2} \qquad\qquad\qquad NM = s\frac{\sqrt{2}}{2}$$

From the fact that NA = MA, thus triangle AMN is isosceles: AE bisects NM, hence

$$ME = \frac{1}{2}NM = \frac{1}{2} \cdot s\frac{\sqrt{2}}{2} = s\frac{\sqrt{2}}{4}$$

From the Pythagorean theorem, once again:

$$AE^2 + EM^2 = MA^2$$
$$AE^2 + \left(s\frac{\sqrt{2}}{4}\right)^2 = \left(s\frac{\sqrt{5}}{2}\right)^2$$
$$AE^2 + \frac{2s^2}{16} = \frac{5s^2}{4}$$
$$AE^2 = \frac{18s^2}{16}$$
$$AE = s\frac{3\sqrt{2}}{4}$$

From the fact that triangle AMN is isosceles, once again:

$$\angle NMA = \angle MNA = \frac{1}{2}(180° - \theta)$$

Since $\angle NFM = \angle AEN = 90°$ and $\angle NMF = \angle ENA = \frac{1}{2}(180° - \theta)$, triangles NMF and ANE are similar. Hence,

$$\frac{NF}{NM} = \frac{AE}{AN} = \frac{AE}{MA} = s\frac{3\sqrt{2}}{4} \div s\frac{\sqrt{5}}{2} = \frac{3\sqrt{2}}{2\sqrt{5}}.$$

Finally, $\sin \theta = \dfrac{NF}{NA} = \dfrac{NF}{NM} \cdot NM \div NA = \dfrac{3\sqrt{2}}{2\sqrt{5}} \cdot s\dfrac{\sqrt{2}}{2} \div s\dfrac{\sqrt{5}}{2} = \dfrac{3\sqrt{2}\,\sqrt{2}}{4\sqrt{5}} \cdot \dfrac{2}{\sqrt{5}} = \dfrac{3}{5}$

CHAPTER 1   REVIEW EXERCISE

1. 2°1′20″ = (2 · 60 · 60 + 1 · 60 + 20)″ = 7,280″

2. Since a circumference has degree measure 360, $\frac{1}{6}$ a circumference has degree measure $\frac{1}{6}$ of 360; that

is, $\frac{1}{6}$ (360) = 60 degrees, written 60°.

3. Since $\dfrac{a}{a'} = \dfrac{c}{c'}$ by Euclid's Theorem, then $\dfrac{a}{2} = \dfrac{20,000}{5}$; $a = \dfrac{(2)(20,000)}{5} = 8,000$

Chapter 1   Right Triangle Ratios

4. Since $23' = \dfrac{23°}{60}$, then $36°23' = \left(36 + \dfrac{23}{60}\right)° \approx 36.38°$ to two decimal places

5. We draw a figure (the one shown is not drawn to
   scale) and label the known parts. Since triangles
   BTA and LMY are similar, we have

   $$\dfrac{BT}{LM} = \dfrac{BA}{LY}$$

   $$\dfrac{x}{1\ yd} = \dfrac{31\ ft}{2.0\ in.}$$

   $$\dfrac{x}{36\ in.} = \dfrac{31\ ft}{2.0\ in.}$$

   $$x = \dfrac{(31\ ft)(36\ in.)}{2.0\ in.}$$

   $$= 560\ ft\ \text{(to two significant digits)}$$

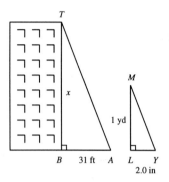

6. (A) $\dfrac{b}{c}$ (B) $\dfrac{c}{a}$ (C) $\dfrac{b}{a}$ (D) $\dfrac{c}{b}$ (E) $\dfrac{a}{c}$ (F) $\dfrac{a}{b}$

7. *Solve for the complementary angle*: $90° - \theta = 90° - 35.2° = 54.8°$

   Solve for a: We choose the cosine to find a. Thus, $\cos \theta = \dfrac{a}{c}$

   $$a = c \cos \theta$$

   $$= (20.2\ cm)(\cos 35.2°) = 16.5\ cm$$

   Solve for b: We choose the sine to solve for b. Thus, $\sin \theta = \dfrac{b}{c}$

   $$b = c \sin \theta$$

   $$= (20.2\ cm)(\sin 35.2°) = 11.6\ cm$$

8. Since $\dfrac{s}{C} = \dfrac{\theta}{360°}$, then $\dfrac{8.00\ cm}{20.0\ cm} = \dfrac{\theta}{360°}$ and $\theta = \dfrac{8.00}{20.0}(360°) = 144°$

9. In 20 minutes the tip of the hand travels through 1/3 of the circumference of a circle of radius 2 in.
   (since it travels through an entire circumference in 60 minutes). Thus, $s = \dfrac{1}{3}C$ and $C = 2\pi R$.

   That is, $s = \dfrac{1}{3}(2\pi R) = \dfrac{1}{3}(2\pi)(2\ in.) = \dfrac{4}{3}\pi\ in. \approx 4.19\ in.$

10. $74.273° = 74°(0.273 \times 60)' = 74°16.38' = 74°16'(0.38 \times 60)'' = 74°16'23''$

11. To compare α and β, we convert α to decimal form. Since $47' = \dfrac{47°}{60}$ and $18'' = \dfrac{18°}{3600}$, then

    $$32°47'18'' = \left(32 + \dfrac{47}{60} + \dfrac{18}{3600}\right)° \approx 32.788°$$

    Since $32.788° > 32.783°$, we conclude that $\alpha > \beta$.

12. Since $\dfrac{a}{a'} = \dfrac{b}{b'}$ by Euclid's Theorem, then

$$\frac{4.1 \times 10^{-6} \text{ mm}}{1.5 \times 10^{-4} \text{ mm}} = \frac{b}{2.6 \times 10^{-4} \text{ mm}}$$

$$b = (2.6 \times 10^{-4} \text{ mm})\frac{4.1 \times 10^{-6}}{1.5 \times 10^{-4}} = 7.1 \times 10^{-6} \text{ mm}$$

13. (A) $\cos \theta$ (B) $\tan \theta$ (C) $\sin \theta$ (D) $\sec \theta$ (E) $\csc \theta$ (F) $\cot \theta$

14. *Solve for the complementary angle:* $90° - \theta = 90° - 62°20' = 27°40'$

   Solve for b: We choose the tangent to find b. Thus, $\tan \theta = \dfrac{b}{a}$

$$b = a \tan \theta$$
$$= (4.00 \times 10^{-8} \text{ m})(\tan 62°20')$$
$$= 7.63 \times 10^{-8} \text{ m}$$

   Solve for c: We choose the cosine to find c. Thus, $\cos \theta = \dfrac{a}{c}$

$$c = \frac{a}{\cos \theta}$$
$$= \frac{4.00 \times 10^{-8} \text{ m}}{\cos 62°20'} = 8.61 \times 10^{-8} \text{ m}$$

15. (A) If $\tan \theta = 1.662$, then $\theta = \tan^{-1} 1.662 = 58.97°$

   (B) $\theta = \arccos 0.5607 = 55.896° = 55°50'$

   (C) $\theta = \sin^{-1} 0.0138 = 0.7907° = 0°47'27''$

16. *Solve for θ:* We choose the tangent to solve for θ. Thus, $\tan \theta = \dfrac{b}{a} = \dfrac{13.3 \text{ mm}}{15.7 \text{ mm}} = 0.8471$

$$\theta = \tan^{-1} 0.8471 = 40.3°$$

   Solve for the complementary angle: $90° - \theta = 90° - 40.3° = 49.7°$

   Solve for c: We choose the sine to solve for c. Thus, $\sin \theta = \dfrac{b}{c}$

$$c = \frac{b}{\sin \theta} = \frac{13.3 \text{ mm}}{\sin 40.3} = 20.6 \text{ mm}$$

17. $40.3° = 40°(0.3 \times 60)' = 40°20'$ (to the nearest 10'); $90° - \theta = 90° - 40°20' = 49°40'$

18. We first sketch a figure and label the known parts.

   From geometry we know that each angle of an
   equilateral triangle has measure 60°.

   $\sin 60° = \dfrac{h}{10 \text{ ft}}$

   $h = (10 \text{ ft})(\sin 60°) = 8.7 \text{ ft}$

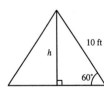

19. Since $\dfrac{s}{C} = \dfrac{\theta}{360°}$ and $C = 2\pi R$ then

$$\dfrac{s}{2\pi R} = \dfrac{\theta°}{360°}$$

$$\dfrac{s}{2(3.14)(1500\ \text{ft})} \approx \dfrac{36°}{360°}$$

$$s \approx \dfrac{2(3.14)(1500\ \text{ft})(36)}{360} \approx 940\ \text{ft}$$

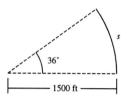

1500 ft

20. Since $\dfrac{A}{\pi r^2} = \dfrac{\theta}{360°}$, then $\dfrac{A}{\pi(18.3\ \text{ft})^2} = \dfrac{36.5°}{360°}$

$$A = \dfrac{36.5}{360}\ \pi\ (18.3\ \text{ft})^2 = 107\ \text{ft}^2$$

21. *Solve for θ*: $\theta = 90° - (90° - \theta) = 90° - 23°43' = 66°17'$

Solve for a: We choose the cosine to find a. Thus, $\cos \theta = \dfrac{a}{c}$

$$a = c \cos \theta$$

$$= (232.6\ \text{km})(\cos 66°17') = 93.56\ \text{km}$$

Solve for b: We choose the sine to find b. Thus, $\sin \theta = \dfrac{b}{c}$

$$b = c \sin \theta$$

$$= (232.6\ \text{km})(\sin 66°17') = 213.0\ \text{km}$$

22. *Solve for θ*: We choose the cosine to find θ. Thus, $\cos \theta = \dfrac{a}{c} = \dfrac{2{,}421\ \text{m}}{4{,}883\ \text{m}} = 60.28°$

Solve for the complementary angle: $90° - \theta = 90° - 60.28° = 29.72°$

Solve for b: We choose the sine to find b. Thus, $\sin \theta = \dfrac{b}{c}$

$$b = c \sin \theta$$

$$= (4{,}883\ \text{m})(\sin 60.28°) = 4{,}241\ \text{m}$$

23. Use the reciprocal relationship $\csc \theta = 1/\sin \theta$. Set calculator in degree mode, use sin key, then take reciprocal. $\csc 72.3142° = 1.0496$

24. We note: triangles PFH and PBT are similar, hence

$$\dfrac{PF}{FH} = \dfrac{PB}{BT}$$

$$\dfrac{s}{5.5\ \text{ft}} = \dfrac{20\ \text{ft} + s}{18\ \text{ft}}$$

$$5.5(18\ \text{ft}) \cdot \dfrac{s}{5.5\ \text{ft}} = 5.5(18\ \text{ft}) \cdot \dfrac{20\ \text{ft} + s}{18\ \text{ft}}$$

$$18\ s = 5.5(20\ \text{ft} + s) = 110\ \text{ft} + 5.5\ s$$

$$12.5\ s = 110\ \text{ft}$$

$$s = \dfrac{110\ \text{ft}}{12.5} = 8.8\ \text{ft}$$

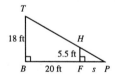

25. In right triangle ABC, the length of the ramp = AB.

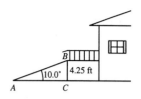

$$\sin A = \frac{BC}{AB}$$

$$AB = \frac{BC}{\sin A}$$

$$= \frac{4.25 \text{ ft}}{\sin 10.0°} = 24.5 \text{ ft.}$$

The distance of the end of the ramp from the porch = AC.

$$\tan A = \frac{BC}{AC}$$

$$AC = \frac{BC}{\tan A}$$

$$= \frac{4.25 \text{ ft}}{\tan 10.0°} = 24.1 \text{ ft}$$

26. From the figure, it is clear that $\tan \theta = \frac{a}{b}$.

Given: percentage of inclination $\frac{a}{b} = 4\% = 0.04$, then $\tan \theta = 0.04$; $\theta = 2.3°$

Given: angle of inclination $\theta = 4°$, then $\frac{a}{b} = \tan 4° = 0.07$ or 7%

27. Since $\frac{s}{c} = \frac{\theta}{360°}$ and $C = 2\pi r$, then $\frac{s}{2\pi r} = \frac{\theta}{360°}$; $s = 2\pi r \cdot \frac{\theta}{360°}$.

Since the cities have the same longitude, θ is given by their difference in latitude.

$$\theta = 44°31' - 30°42' = 13°49' = \left(13 + \frac{49}{60}\right)° .$$

Thus, we have

$$s \approx 2(3.14)(3960)\frac{13 + \frac{49}{60}}{360} \approx 954 \text{ miles}$$

28. We first sketch a figure and label the known parts.

From the figure we note:

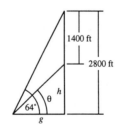

$$h + 1400 \text{ ft} = 2800 \text{ ft}$$

$$h = 1400 \text{ ft}$$

$$\tan 64° = \frac{2800 \text{ ft}}{g}$$

$$\tan \theta = \frac{h}{g} = \frac{1400 \text{ ft}}{g}$$

Thus, $g = \frac{2800 \text{ ft}}{\tan 64°}$

and $\tan \theta = \frac{1400 \text{ ft}}{\left(\frac{2800 \text{ ft}}{\tan 64°}\right)} .$

Then, $\tan \theta = \frac{1400 \tan 64°}{2800}$

$$= \frac{1}{2} \tan 64° = 1.025$$

$$\theta = 46°$$

Chapter 1   Right Triangle Ratios

29.

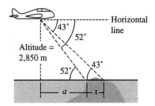

We note: $\cot 52° = \dfrac{a}{2{,}850 \text{ m}}$, $\cot 43° = \dfrac{a + x}{2{,}850 \text{ m}}$.

Then, $\cot 43° - \cot 52° = \dfrac{a + x}{2{,}850 \text{ m}} - \dfrac{a}{2{,}850 \text{ m}} = \dfrac{a + x - a}{2{,}850 \text{ m}} = \dfrac{x}{2{,}850 \text{ m}}$

$x = 2{,}850 \text{ m}(\cot 43° - \cot 52°) = 830 \text{ m}$

30. We note:

In right triangle BCP_1, $\cot 73.5° = \dfrac{x}{h}$

In right triangle BCP_2, $\cot 54.2° = \dfrac{x + 525 \text{ m}}{h}$

Then, $\cot 54.2° - \cot 73.5° = \dfrac{x + 525 \text{ m}}{h} - \dfrac{x}{h}$

$= \dfrac{x + 525 \text{ m} - x}{h}$

$= \dfrac{525 \text{ m}}{h}$

$h = \dfrac{525 \text{ m}}{\cot 54.2° - \cot 73.5°} = 1{,}240 \text{ m}$

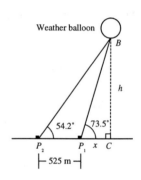

31. We use the figure and the notation of Problem 25, Ex. 1–4.

$\quad$ r = radius of the parallel of latitude

$\quad$ R = radius of the earth

$\quad$ θ = latitude

$\quad$ $\cos \theta = \dfrac{r}{R}$, $r = R \cos \theta$

$\quad$ L = length of the parallel of latitude

$\quad$ L = $2\pi r$

$\quad$ L = $2\pi R \cos \theta$

To keep the sun in the same position, the plane must fly at a rate v sufficient to fly a distance L in

24 hours, thus $\quad v = \dfrac{L}{24} = \dfrac{2\pi R \cos \theta}{24}$

$\quad v \approx \dfrac{2\pi(3960 \text{ mi}) \cos 42°50'}{24 \text{ hr}} \approx 760 \text{ mi/hr}$

32. (A) Since α and β are complementary (see diagram) $\beta = 90° - \alpha$.

$\quad$ (B) In the right triangle containing r, h, and α, $\tan \alpha = \dfrac{r}{h}$. Thus, $r = h \tan \alpha$.

20

(C) Using similar triangles, we can write $\dfrac{H}{h} = \dfrac{R}{r}$. Then

$$H = \frac{R}{r}\,h,$$

$$H - h = \frac{R}{r}\,h - h = h\left(\frac{R}{r} - 1\right) = h\left(\frac{R-r}{r}\right) = \frac{h}{r}(R-r)$$

Since $\dfrac{r}{h} = \tan\alpha,\ \dfrac{h}{r} = 1 \div \dfrac{r}{h} = 1 \div \tan\alpha = \cot\alpha.$ Hence, we can write $H - h = (R - r)\cot\alpha.$

Chapter 2   Trigonometric Functions

EXERCISE 2.1 Degrees and Radians

1. Since 90° is one-half of 180°, the corresponding radian measure must be one-half of π, or $\pi/2$ rad.

3. Since 60° is one-third of 180°, the corresponding radian measure must be one-third of π, or $\pi/3$ rad.

5. Since 120° is twice 60° (see Problem 3), the corresponding radian measure must be twice $\pi/3$, or $2\pi/3$ rad.

7. Since $\pi/4$ is one-fourth of π, the corresponding degree measure must be one-fourth of 180°, or 45°.

9. Since $\pi/6$ is one-sixth of π, the corresponding degree measure must be one-sixth of 180°, or 30°.

11. Since $5\pi/6$ is 5 times $\pi/6$ (see Problem 9), the corresponding degree measure must be 5 times 30°, or 150°.

13. Since 30° corresponds to a radian measure of $\pi/6$ rad, we have:

$$60° = 2 \cdot 30° \text{ corresponds to } 2 \cdot \pi/6 \text{ or } \pi/3 \text{ rad.}$$

$$90° = 3 \cdot 30° \text{ corresponds to } 3 \cdot \pi/6 \text{ or } \pi/2 \text{ rad.}$$

$$120° = 4 \cdot 30° \text{ corresponds to } 4 \cdot \pi/6 \text{ or } 2\pi/3 \text{ rad.}$$

$$150° = 5 \cdot 30° \text{ corresponds to } 5 \cdot \pi/6 \text{ or } 5\pi/6 \text{ rad.}$$
$$180° = 6 \cdot 30° \text{ corresponds to } 6 \cdot \pi/6 \text{ or } \pi \text{ rad.}$$

15.

17.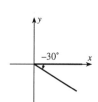

19.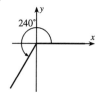

Coterminal angles
have measures

$$60° + 360° = 420°$$
and
$$60° - 360° = -300°$$

Coterminal angles
have measures

$$-30° + 360° = 330°$$
and
$$-30° - 360° = -390°$$

Coterminal angles
have measures

$$240° + 360° = 600°$$
and
$$240° - 360° = -120°$$

21. Since $\theta = \dfrac{s}{R}$, we have   (A) $\theta = \dfrac{6 \text{ cm}}{3.0 \text{ cm}} = 2$ rad   (B) $\theta = \dfrac{4.5 \text{ cm}}{3.0 \text{ cm}} = 1.5$ rad

23. $\dfrac{\theta_{deg}}{180°} = \dfrac{\theta_{rad}}{\pi \text{ rad}}$

$\theta_{rad} = \dfrac{\pi \text{ rad}}{180°} \theta_{deg} = \dfrac{\pi}{180}(18)$

$= \dfrac{\pi}{10}$ rad   Exact form

≈ 0.3142 rad   Approximation

25. $\dfrac{\theta_{deg}}{180°} = \dfrac{\theta_{rad}}{\pi \text{ rad}}$

$\theta_{rad} = \dfrac{\pi \text{ rad}}{180°} \theta_{deg} = \dfrac{\pi}{180}(27)$

$= \dfrac{3\pi}{20}$ rad   Exact form

$= 0.4712$ rad   Approximation

27. $\dfrac{\theta_{\text{deg}}}{180°} = \dfrac{\theta_{\text{rad}}}{\pi \text{ rad}}$

$\theta_{\text{rad}} = \dfrac{\pi \text{rad}}{180°}\,\theta_{\text{deg}} = \dfrac{\pi}{180}(130)$

$\quad = \dfrac{13\pi}{18}\text{ rad} \qquad$ Exact form

$\quad \approx 2.269 \text{ rad} \qquad$ Approximation

29. $\dfrac{\theta_{\text{deg}}}{180°} = \dfrac{\theta_{\text{rad}}}{\pi \text{ rad}}$

$\theta_{\text{deg}} = \dfrac{180°}{\pi \text{ rad}}\,\theta_{\text{rad}} = \dfrac{180}{\pi}(1.6)$

$\quad = \left(\dfrac{288}{\pi}\right)^{\circ} \qquad$ Exact form

$\quad \approx 91.67° \qquad$ Approximation

31. $\dfrac{\theta_{\text{deg}}}{180°} = \dfrac{\theta_{\text{rad}}}{\pi \text{ rad}}$

$\theta_{\text{deg}} = \dfrac{180°}{\pi \text{ rad}}\,\theta_{\text{rad}} = \dfrac{180}{\pi}\cdot\dfrac{\pi}{12}$

$\quad = 15° \qquad$ Exact form

33. $\dfrac{\theta_{\text{deg}}}{180°} = \dfrac{\theta_{\text{rad}}}{\pi \text{ rad}}$

$\theta_{\text{deg}} = \dfrac{180°}{\pi \text{ rad}}\,\theta_{\text{rad}} = \dfrac{180}{\pi}\cdot\dfrac{\pi}{60}$

$\quad = 3° \qquad$ Exact form

35.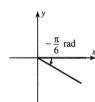

37.

39.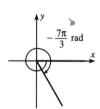

41. (A) $\dfrac{\theta_{\text{deg}}}{180°} = \dfrac{\theta_{\text{rad}}}{\pi \text{ rad}}$

$\theta_{\text{deg}} = \dfrac{180°}{\pi \text{ rad}}\,\theta_{\text{rad}}$

If $\theta_{\text{rad}} = 8.30$

$\theta_{\text{deg}} = \dfrac{180}{\pi}(8.30) = 476°$

If $\theta_{\text{rad}} = -11.6$

$\theta_{\text{deg}} = \dfrac{180}{\pi}(-11.6) = -665°$

(B) $\dfrac{\theta_{\text{deg}}}{180°} = \dfrac{\theta_{\text{rad}}}{\pi \text{ rad}}$

$\theta_{\text{rad}} = \dfrac{\pi \text{ rad}}{180°}\,\theta_{\text{deg}}$

If $\theta_{\text{deg}} = 563°$

$\theta_{\text{rad}} = \dfrac{\pi}{180}(563) = 9.83 \text{ rad}$

If $\theta_{\text{deg}} = -1{,}230°$

$\theta_{\text{rad}} = \dfrac{\pi}{180}(-1{,}230) = -21.5 \text{ rad}$

43. (A) $s = R\theta$

$\quad = (25.0)(2.33) \approx 58.3 \text{ m}$

(C) $s = R\theta$

$\quad = (25.0)(0.821) \approx 20.5 \text{ m}$

(B) $s = \dfrac{\pi}{180}R\theta$

$\quad = \dfrac{\pi}{180}(25.0)(19.0) \approx 8.29 \text{ m}$

(D) $s = \dfrac{\pi}{180}R\theta$

$\quad = \dfrac{\pi}{180}(25.0)(108) \approx 47.1 \text{ m}$

45. (A) $A = \dfrac{1}{2}R^2\theta$

$\quad = \dfrac{1}{2}(14.0)^2(0.473) \approx 46.4 \text{ cm}^2$

(C) $A = \dfrac{1}{2}R^2\theta$

$\quad = \dfrac{1}{2}(14.0)^2(1.02) \approx 1.0 \times 10^2 \text{ cm}^2$

(B) $A = \dfrac{\pi}{360}R^2\theta$

$\quad = \dfrac{\pi}{360}(14.0)^2(25.0) = 42.8 \text{ cm}^2$

(D) $A = \dfrac{\pi}{360}R^2\theta$

$\quad = \dfrac{\pi}{360}(14.0)^2(112) = 192 \text{ cm}^2$

47. 432° is coterminal with (432 − 360)° or 72°. Since 72° is between 0° and 90°, its terminal side lies in quadrant I.

49. $-\dfrac{14\pi}{3}$ is coterminal with $-\dfrac{14\pi}{3} + 2\pi = -\dfrac{8\pi}{3}$, and thus with $-\dfrac{8\pi}{3} + 2\pi = -\dfrac{2\pi}{3}$. Since $-\dfrac{2\pi}{3}$ is between $-\dfrac{\pi}{2}$ and $-\pi$, its terminal side lies in quadrant III.

51. 1,243° is coterminal with (1,243 − 360)° or 883°, and thus with (883 − 360)° or 523°, and thus with (523 − 360)° = 163°. Since 163° is between 90° and 180°, its terminal side lies in quadrant II.

53. $\dfrac{\theta_{\text{deg}}}{180°} = \dfrac{\theta_{\text{rad}}}{\pi\,\text{rad}}$, $\theta_{\text{rad}} = \dfrac{\pi\,\text{rad}}{180°}(\theta_{\text{deg}}) = \dfrac{\pi}{180}(57.3421) = 1.0008\,\text{rad}$

55. $\dfrac{\theta_{\text{deg}}}{180°} = \dfrac{\theta_{\text{rad}}}{\pi\,\text{rad}}$, $\theta_{\text{deg}} = \dfrac{180°}{\pi\,\text{rad}}(\theta_{\text{rad}}) = \dfrac{180}{\pi}(0.3184) = 18.2430°$

57. $\dfrac{\theta_{\text{deg}}}{180°} = \dfrac{\theta_{\text{rad}}}{\pi\,\text{rad}}$, $\theta_{\text{rad}} = \dfrac{\pi\,\text{rad}}{180°}(\theta_{\text{deg}}) = \dfrac{\pi}{180}\left(26 + \dfrac{23}{60} + \dfrac{14}{3,600}\right) = 0.4605\,\text{rad}$

59. At 2:30, the minute hand has moved $\dfrac{1}{2}$ of a circumference from its position at the top of the clock. The hour hand has moved $2\dfrac{1}{2}$ twelfths of a circumference from the same position. Therefore, they form an angle of

$$\frac{1}{2}C - \frac{2\frac{1}{2}}{12}C \text{ radians,}$$

where C = a total circumference or 2π radians. Thus, the desired angle is

$$\frac{1}{2}(2\pi) - \frac{2\frac{1}{2}}{12}(2\pi) = \pi - \frac{5\pi}{12} = \frac{7\pi}{12}\,\text{rad} \approx 1.83\,\text{rad}$$

61. We are to find the arc length subtended by a central angle of 32°, in a circle of radius 22 cm.
$$s = \frac{\pi}{180}R\theta = \frac{\pi}{180}(22)(32) = 12\,\text{cm}$$

63. We are to find the angle, in degrees, that subtends an arc of 24 inches, in a circle of radius 72 inches.
$$s = \frac{\pi}{180}R\theta; \quad \theta = \frac{180s}{\pi R} = \frac{180(24)}{\pi(72)} = 19°$$

65. Since $s = R\theta$, we have
$$\text{diameter} \approx s = (1.5 \times 10^8\,\text{km})(9.3 \times 10^{-3}\,\text{rad})$$
$$\approx 1.4 \times 10^6\,\text{km}$$

67. Since $s = \dfrac{\pi}{180}R\theta$, we have
$$\text{width of field} \approx s = \frac{\pi}{180}(1,250\,\text{ft})(8) = 175\,\text{ft}$$

In Problems 69, 71, 73, we use the diagram and reason as follows: Since the cities have the same longitude, θ is given by their difference in latitude.

69. $s = \dfrac{\pi}{180} R\theta$, $\theta = 42°20' - 40°0' = 2°20' = \left(2 + \dfrac{20}{60}\right)^{\circ}$

  $s = \dfrac{\pi}{180} (3,964 \text{ miles})\left(2 + \dfrac{20}{60}\right) = 161 \text{ miles}$

71. $s = \dfrac{\pi}{180} R\theta$, $\theta = 49°54' - 37°43' = 12°11' = \left(12 + \dfrac{11}{60}\right)^{\circ}$

  $s = \dfrac{\pi}{180} (3,964 \text{ miles})\left(12 + \dfrac{11}{60}\right) = 843 \text{ miles}$

73. $s = \dfrac{\pi}{180} R\theta$, $\theta = 27°51' - 23°8' = 4°43' = \left(4 + \dfrac{43}{60}\right)^{\circ}$

  $s = \dfrac{\pi}{180} (3,964 \text{ miles})\left(4 + \dfrac{43}{60}\right) = 326 \text{ miles}$

75. Assuming that an angle corresponding to an entire circumference is swept out in 1 year (52 weeks), and that the amount swept out in one week is proportional to the time, we can write

$$\frac{\text{angle}}{\text{time}} = \frac{\text{angle}}{\text{time}}$$
$$\frac{\theta}{1} = \frac{2\pi}{52}$$
$$\theta = \frac{2\pi}{52} = \frac{\pi}{26} \text{ rad} \approx 0.12 \text{ rad}$$

77. We use the proportion $\dfrac{\text{error in distance}}{\text{error in time}} = \dfrac{\text{actual distance}}{\text{actual time}}$. Let x = error in distance. The actual time,

$$1 \text{ year} = (365 \text{ days})\left(24 \,\frac{\text{hours}}{\text{day}}\right)\left(3,600 \,\frac{\text{seconds}}{\text{hour}}\right).$$

Thus,
$$\frac{x}{365 \text{ seconds}} = \frac{2\pi R}{365 \cdot 24 \cdot 3,600 \text{ seconds}}$$
$$x = 365 \cdot \frac{2\pi(9.3 \times 10^7 \text{ miles})}{365 \cdot 24 \cdot 3,600}$$
$$= \frac{2\pi(9.3 \times 10^7 \text{ miles})}{24 \cdot 3,600}$$
$$= 6,800 \text{ miles}$$

79. Since $A = \dfrac{1}{2} R^2\theta$ and $P = s + 2R = R\theta + 2R$, we can eliminate θ between the two equations and write $2A = R^2\theta$, $\theta = \dfrac{2A}{R^2}$.

$$P = R\left(\frac{2A}{R^2}\right) + 2R = \frac{2A}{R} + 2R.$$

Thus, $P = \dfrac{2(52.39)}{10.5} + 2(10.5) \approx 31 \text{ ft}$

81. Since there are 2π rad in 1 revolution, there are $5(2\pi) = 10\pi$ rad in 5 revolutions, $3.6(2\pi) = 7.2\pi$ rad in 3.6 revolutions, and $2\pi n$ rad in n revolutions.

83. Since the two wheels are coupled together, the distance (arc length) that the drive wheel turns is equal to the distance that the shaft turns. Thus,

$$s = R_1\theta_1 \qquad s = R_2\theta_2$$

$$R_1\theta_1 = R_2\theta_2$$

$$\theta_1 = \frac{R_2}{R_1}\,\theta_2 = \frac{26}{12}\,(3 \text{ revolutions})$$

$$= 6.5 \text{ revolutions}$$

In Problem 81, we noted that there are $2\pi n$ rad in n revolutions, hence there are $2\pi(6.5)$ ≈ 40.8 rad in 6.5 revolutions.

85. Since $s = \dfrac{\pi}{180}\,R\theta$, we have $\dfrac{s}{R} = \dfrac{\pi}{180}\,\theta$, $\theta = \dfrac{180}{\pi}\left(\dfrac{s}{R}\right)$. Here,

$$R = \frac{1}{2}\,(32) = 16 \text{ in. and } s = 20 \text{ ft} = 20 \text{ ft}\left(12\,\frac{\text{in.}}{\text{ft}}\right) = 240 \text{ in.}$$

Thus, $\theta = \dfrac{180°}{\pi}\left(\dfrac{240}{16}\right) = 859°.$

EXERCISE 2.2 Linear and Angular Velocity

1. $V = R\omega = 6(0.5) = 3$ mm/sec

3. $\omega = \dfrac{V}{R} = \dfrac{102}{6.0} = 17$ rad/sec

5. $\omega = \dfrac{\theta}{t} = \dfrac{2\pi}{1.7} = 3.7$ rad/hr

7. $\omega = \dfrac{\theta}{t} = \dfrac{8.07}{13.6} = 0.593$ rad/sec

9. 1,500 revolutions per second $= 1,500 \cdot 2\pi$ rad/sec $= 3,000\pi$ rad/sec

$$V = R\omega = \left(\frac{1}{2}\cdot 16\right)(3,000\pi) \text{ mm/sec} = 24,000\pi \text{ mm/sec}$$

$$= 24,000\pi\,\frac{\text{mm}}{\text{sec}} \cdot \frac{1}{1000}\,\frac{\text{m}}{\text{mm}} = 75\,\frac{\text{m}}{\text{sec}}$$

11. $\omega = \dfrac{V}{R} = \dfrac{20,000 \text{ mph}}{4,300 \text{ mi}} = 4.65$ rad/hr

13. $\omega = \dfrac{V}{R} = \dfrac{335.3 \text{ m/sec}}{\frac{1}{2}\,(3.000 \text{ m})} = 223.5 \text{ rad/sec} = (223.5 \text{ rad/sec})\left(\dfrac{1}{2\pi}\,\dfrac{\text{revolution}}{\text{radian}}\right) = 35.6$ rev/sec

15. The earth travels 1 revolution, or 2π radian, in 1 year, or $24 \cdot 365$ hours. Thus,

$$\omega = \frac{\theta}{t} = \frac{2\pi}{24 \cdot 365} = \frac{\pi}{4,380}\,\frac{\text{rad}}{\text{hr}}$$

$$V = R\omega = \left(\frac{\pi}{4,380}\,\frac{\text{rad}}{\text{hr}}\right)(9.3 \times 10^7 \text{ mi})$$

$$= 6.67 \times 10^4 \text{ mi/hr or } 66,700 \text{ mi/hr}$$

17. (A) Jupiter travels 1 revolution, or 2π radian, in 9 hr 55 min, or $\left(9 + \dfrac{55}{60}\right)$ hr. Thus,

$$\omega = \frac{\theta}{t} = \frac{2\pi}{9 + \dfrac{55}{60}} = 0.633 \, \frac{\text{rad}}{\text{hr}}$$

(B) Note that $R = \dfrac{1}{2} \times$ diameter. Thus,

$$V = R\omega = \left(\frac{1}{2} \times 88{,}700 \text{ miles}\right)\left(0.633 \, \frac{\text{rad}}{\text{hr}}\right) = 28{,}100 \text{ mi/hr}$$

19. The satellite travels 1 revolution, or 2π radian, in 23.93 hr. So,

$$\omega = \frac{\theta}{t} = \frac{2\pi}{23.93} = 0.2626 \, \frac{\text{rad}}{\text{hr}}$$

The radius, R, of the satellite's orbit is given by adding the satellite's distance above the earth's

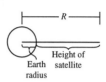

Height of satellite
Earth radius

surface to the radius of the earth (see sketch). Thus,

$$V = R\omega = (22{,}300 + 3{,}964 \text{ miles})\left(0.2626 \, \frac{\text{rad}}{\text{hr}}\right) = 6{,}900 \text{ mi/hr}$$

21. Using subscript E to denote quantities associated with the earth, and subscript S to denote quantities associated with the satellite, we can write:

$$\omega_E = \frac{\theta_E}{t} \qquad \omega_S = \frac{\theta_S}{t} \qquad \theta_E = \omega_E t \qquad \theta_S = \omega_S t$$

Using the hint, $2\pi = \theta_S - \theta_E$, hence,

$$2\pi = \omega_S t - \omega_E t = (\omega_S - \omega_E)\, t$$
$$t = \frac{2\pi}{\omega_S - \omega_E}$$

Thus, $t = \dfrac{2\pi}{\dfrac{2\pi}{1.51} - \dfrac{2\pi}{23.93}} = 1.61 \text{ hr}$

23. In triangle ABC, we can write $\tan \theta = \dfrac{a}{b} = \dfrac{a}{15}$. Thus, $a = 15 \tan \theta$. But, $\theta = \omega t$, where ω is given by 1 revolution per second $= 2\pi \, \dfrac{\text{rad}}{\text{sec}}$. Thus, $\theta = 2\pi t$, and $a = 15 \tan 2\pi t$

EXERCISE 2.3 Trigonometric Functions

1. $P(a, b) = (3, 4),$ $R = \sqrt{a^2 + b^2} = \sqrt{3^2 + 4^2} = 5$

$$\sin \theta = \frac{b}{R} = \frac{4}{5} \qquad \csc \theta = \frac{R}{b} = \frac{5}{4}$$
$$\cos \theta = \frac{a}{R} = \frac{3}{5} \qquad \sec \theta = \frac{R}{a} = \frac{5}{3}$$
$$\tan \theta = \frac{b}{a} = \frac{4}{3} \qquad \cot \theta = \frac{a}{b} = \frac{3}{4}$$

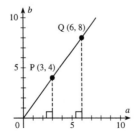

Chapter 2   Trigonometric Functions

$$Q(a, b) = (6, 8), \quad R = \sqrt{a^2 + b^2} = \sqrt{6^2 + 8^2} = 10$$

$$\sin \theta = \frac{b}{R} = \frac{8}{10} = \frac{4}{5} \qquad \csc \theta = \frac{R}{b} = \frac{10}{8} = \frac{5}{4}$$

$$\cos \theta = \frac{a}{R} = \frac{6}{10} = \frac{3}{5} \qquad \sec \theta = \frac{R}{a} = \frac{10}{6} = \frac{5}{3}$$

$$\tan \theta = \frac{b}{a} = \frac{8}{6} = \frac{4}{3} \qquad \cot \theta = \frac{a}{b} = \frac{6}{8} = \frac{3}{4}$$

3. $P(a, b) = (4, -3), \quad R = \sqrt{a^2 + b^2} = \sqrt{4^2 + (-3)^2} = 5$

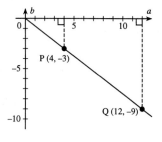

$$\sin \theta = \frac{b}{R} = \frac{-3}{5} = -\frac{3}{5} \qquad \csc \theta = \frac{R}{b} = \frac{5}{-3} = -\frac{5}{3}$$

$$\cos \theta = \frac{a}{R} = \frac{4}{5} \qquad \sec \theta = \frac{R}{a} = \frac{5}{4}$$

$$\tan \theta = \frac{b}{a} = \frac{-3}{4} = -\frac{3}{4} \qquad \cot \theta = \frac{a}{b} = \frac{4}{-3} = -\frac{4}{3}$$

$$Q(a, b) = (12, -9), \quad R = \sqrt{a^2 + b^2} = \sqrt{12^2 + (-9)^2} = 15$$

$$\sin \theta = \frac{b}{R} = \frac{-9}{15} = -\frac{3}{5} \qquad \csc \theta = \frac{R}{b} = \frac{15}{-9} = -\frac{5}{3}$$

$$\cos \theta = \frac{a}{R} = \frac{12}{15} = \frac{4}{5} \qquad \sec \theta = \frac{R}{a} = \frac{15}{12} = \frac{5}{4}$$

$$\tan \theta = \frac{b}{a} = \frac{-9}{12} = -\frac{3}{4} \qquad \cot \theta = \frac{a}{b} = \frac{12}{-9} = -\frac{4}{3}$$

5.   We sketch a reference triangle and label what we know.

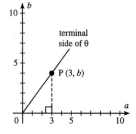

Since $\cos \theta = \dfrac{a}{R} = \dfrac{3}{5}$, we know that a = 3 and R = 5.

Use the Pythagorean theorem to find b:

$$3^2 + b^2 = 5^2$$

$$b^2 = 25 - 9 = 16$$

$$b = 4$$

b is positive since P(a, b) is in quadrant I.

We can now find the other five functions using Definition 1:

$$\sin \theta = \frac{b}{R} = \frac{4}{5} \qquad \csc \theta = \frac{R}{b} = \frac{5}{4}$$

$$\tan \theta = \frac{b}{a} = \frac{4}{3} \qquad \cot \theta = \frac{a}{b} = \frac{3}{4}$$

$$\sec \theta = \frac{R}{a} = \frac{5}{3}$$

7.   We sketch a reference triangle and label what we know.

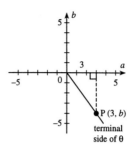

Since $\cos\theta = \dfrac{a}{R} = \dfrac{3}{5}$, we know that a = 3 and R = 5.

Use the Pythagorean theorem to find b:

$$3^2 + b^2 = 5^2$$

$$b^2 = 25 - 9 = 16$$

$$b = -4$$

b is negative since P(a, b) is in quadrant IV.

We can now find the other five functions using Definition 1:

$$\sin\theta = \frac{b}{R} = \frac{-4}{5} = -\frac{4}{5} \qquad \csc\theta = \frac{R}{b} = \frac{5}{-4} = -\frac{5}{4}$$

$$\tan\theta = \frac{b}{a} = \frac{-4}{3} = -\frac{4}{3} \qquad \cot\theta = \frac{a}{b} = \frac{3}{-4} = -\frac{3}{4}$$

$$\sec\theta = \frac{R}{a} = \frac{5}{3}$$

9.   We sketch a reference triangle and label what we know.

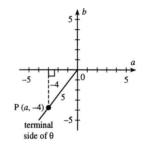

Since $\csc\theta = \dfrac{R}{b} = -\dfrac{5}{4} = \dfrac{5}{-4}$, we know that b = −4

and R = 5 (R is never negative). Use the Pythagorean

theorem to find a:

$$a^2 + (-4)^2 = 5^2$$

$$a^2 = 25 - 16 = 9$$

$$a = -3$$

a is negative since P(a, b) is in quadrant III.

We can now find the other five functions using Definition 1:

$$\sin\theta = \frac{b}{R} = \frac{-4}{5} = -\frac{4}{5} \qquad \sec\theta = \frac{R}{a} = \frac{5}{-3} = -\frac{5}{3}$$

$$\cos\theta = \frac{a}{R} = \frac{-3}{5} = -\frac{3}{5} \qquad \cot\theta = \frac{a}{b} = \frac{-3}{-4} = \frac{3}{4}$$

$$\tan\theta = \frac{b}{a} = \frac{-4}{-3} = \frac{4}{3}$$

11. We sketch a reference triangle and label what we know.

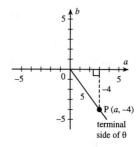

Since $\csc \theta = \dfrac{R}{b} = -\dfrac{5}{4} = \dfrac{5}{-4}$, we know that $b = -4$
and $R = 5$ (R is never negative). Use the Pythagorean

theorem to find a:

$$a^2 + (-4)^2 = 5^2$$

$$a^2 = 25 - 16 = 9$$

$$a = 3$$

a is positive since $P(a, b)$ is in quadrant IV.

We can now find the other five functions using Definition 1:

$$\sin \theta = \frac{b}{R} = \frac{-4}{5} = -\frac{4}{5} \qquad \sec \theta = \frac{R}{a} = \frac{5}{3}$$

$$\cos \theta = \frac{a}{R} = \frac{3}{5} \qquad \cot \theta = \frac{a}{b} = \frac{3}{-4} = -\frac{3}{4} \qquad \tan \theta = \frac{b}{a} = \frac{-4}{3} = -\frac{4}{3}$$

13. Degree mode: $\tan 89° = 57.29$ 15. Radian mode: $\cos (3 \text{ rad}) = -0.9900$

17. Use the reciprocal relationship $\csc \theta = \dfrac{1}{\sin \theta}$. Degree mode: $\csc 162° = \dfrac{1}{\sin 162°} = 3.236$

19. Use the reciprocal relationship $\cot \theta = \dfrac{1}{\tan \theta}$. Degree mode: $\cot 341° = \dfrac{1}{\tan 341°} = -2.904$

21. Radian mode: $\sin 13 = 0.4202$

23. Use the reciprocal relationship $\cot \theta = \dfrac{1}{\tan \theta}$. Radian mode: $\cot 2 = \dfrac{1}{\tan 2} = -0.4577$

25. Use the reciprocal relationship $\sec \theta = \dfrac{1}{\cos \theta}$. Radian mode: $\sec 74 = \dfrac{1}{\cos 74} = 5.824$

27. Degree mode: $\sin 428° = 0.9272$ 29. Radian mode: $\cos (-12) = 0.8439$

31. Use the reciprocal relationship $\cot \theta = \dfrac{1}{\tan \theta}$. Degree mode: $\cot (-167°) = \dfrac{1}{\tan (-167°)} = 4.331$

33. $(a, b) = \left(\sqrt{3}, \ 1\right), \quad R = \sqrt{a^2 + b^2} = \sqrt{\left(\sqrt{3}\right)^2 + 1^2} = \sqrt{4} = 2$

$$\sin \theta = \frac{b}{R} = \frac{1}{2} \qquad \cos \theta = \frac{a}{R} = \frac{\sqrt{3}}{2} \qquad \tan \theta = \frac{b}{a} = \frac{1}{\sqrt{3}}$$

$$\csc \theta = \frac{R}{b} = \frac{2}{1} = 2 \qquad \sec \theta = \frac{R}{a} = \frac{2}{\sqrt{3}} \qquad \cot \theta = \frac{a}{b} = \frac{\sqrt{3}}{1} = \sqrt{3}$$

35. $(a, b) = \left(1, -\sqrt{3}\right)$, $R = \sqrt{a^2 + b^2} = \sqrt{1^2 + \left(-\sqrt{3}\right)^2} = \sqrt{4} = 2$

$\sin \theta = \dfrac{b}{R} = \dfrac{-\sqrt{3}}{2} = -\dfrac{\sqrt{3}}{2}$ $\cos \theta = \dfrac{a}{R} = \dfrac{1}{2}$ $\tan \theta = \dfrac{b}{a} = \dfrac{-\sqrt{3}}{1} = -\sqrt{3}$

$\csc \theta = \dfrac{R}{b} = \dfrac{2}{-\sqrt{3}} = -\dfrac{2}{\sqrt{3}}$ $\sec \theta = \dfrac{R}{a} = \dfrac{2}{1} = 2$ $\cot \theta = \dfrac{a}{b} = \dfrac{1}{-\sqrt{3}} = -\dfrac{1}{\sqrt{3}}$

37. $(a, b) = \left(\sqrt{2}, -\sqrt{2}\right)$, $R = \sqrt{a^2 + b^2} = \sqrt{\left(\sqrt{2}\right)^2 + \left(-\sqrt{2}\right)^2} = \sqrt{4} = 2$

$\sin \theta = \dfrac{b}{R} = \dfrac{-\sqrt{2}}{2} = -\dfrac{\sqrt{2}}{2}$ $\cos \theta = \dfrac{a}{R} = \dfrac{\sqrt{2}}{2}$ $\tan \theta = \dfrac{b}{a} = \dfrac{-\sqrt{2}}{\sqrt{2}} = -1$

$\csc \theta = \dfrac{R}{b} = \dfrac{2}{-\sqrt{2}} = -\sqrt{2}$ $\sec \theta = \dfrac{R}{a} = \dfrac{2}{\sqrt{2}} = \sqrt{2}$ $\cot \theta = \dfrac{a}{b} = \dfrac{\sqrt{2}}{-\sqrt{2}} = -1$

39. I, IV   41. I, III   43. I, IV   45. III, IV   47. II, IV   49. III, IV

51. Since $\sin \theta < 0$ and $\cot \theta > 0$, the terminal side of θ

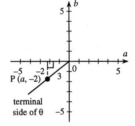

lies in quadrant III. We sketch a reference triangle and label what we know. Since $\sin \theta = \dfrac{b}{R} = -\dfrac{2}{3} = \dfrac{-2}{3}$, we know that $b = -2$ and $R = 3$ (R is never negative).

Use the Pythagorean theorem to find a:

$a^2 + (-2)^2 = 3^2$

$a^2 = 9 - 4 = 5$

$a = -\sqrt{5}$

a is negative since the terminal side of θ lies in quadrant III.

We can now find the other five functions using Definition 1:

$\csc \theta = \dfrac{R}{b} = \dfrac{3}{-2} = -\dfrac{3}{2}$ $\sec \theta = \dfrac{R}{a} = \dfrac{3}{-\sqrt{5}} = -\dfrac{3}{\sqrt{5}}$

$\cos \theta = \dfrac{a}{R} = \dfrac{-\sqrt{5}}{3} = -\dfrac{\sqrt{5}}{3}$ $\cot \theta = \dfrac{a}{b} = \dfrac{-\sqrt{5}}{-2} = \dfrac{\sqrt{5}}{2}$

$\tan \theta = \dfrac{b}{a} = \dfrac{-2}{-\sqrt{5}} = \dfrac{2}{\sqrt{5}}$

Chapter 2   Trigonometric Functions

53. Since $\sin \theta < 0$ and $\tan \theta < 0$, the terminal side of θ

lies in quadrant IV. We sketch a reference triangle and

label what we know. Since $\sin \theta = \dfrac{b}{R} = -\dfrac{2}{3} = \dfrac{-2}{3}$,

we know that $b = -2$ and $R = 3$ (R is never negative).

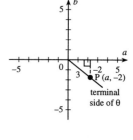

Use the Pythagorean theorem to find a:

$$a^2 + (-2)^2 = 3^2$$
$$a^2 = 9 - 4 = 5$$
$$a = \sqrt{5}$$

a is positive since the terminal side of θ lies in quadrant IV.

We can now find the other functions using Definition 1:

$\csc \theta = \dfrac{R}{b} = \dfrac{3}{-2} = -\dfrac{3}{2}$ $\qquad$ $\sec \theta = \dfrac{R}{a} = \dfrac{3}{\sqrt{5}}$

$\cos \theta = \dfrac{a}{R} = \dfrac{\sqrt{5}}{3}$ $\qquad$ $\cot \theta = \dfrac{a}{b} = \dfrac{\sqrt{5}}{-2} = -\dfrac{\sqrt{5}}{2}$ $\qquad$ $\tan \theta = \dfrac{b}{a} = \dfrac{-2}{\sqrt{5}} = -\dfrac{2}{\sqrt{5}}$

55. Since $\sec \theta > 0$ and $\sin \theta < 0$, the terminal side of θ

lies in quadrant IV. We sketch a reference triangle and

label what we know. Since $\sec \theta = \dfrac{R}{a} = \sqrt{3} = \dfrac{\sqrt{3}}{1}$,

we know that $R = \sqrt{3}$ and $a = 1$. Use the Pythagorean

theorem to find b:

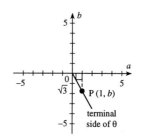

$$1^2 + b^2 = \left(\sqrt{3}\right)^2$$
$$b^2 = 3 - 1 = 2$$
$$b = -\sqrt{2}$$

b is negative since the terminal side of θ lies in quadrant IV.
We can now find the other five functions using Definition 1:

$\sin \theta = \dfrac{b}{R} = \dfrac{-\sqrt{2}}{\sqrt{3}} = -\dfrac{\sqrt{2}}{\sqrt{3}}$ $\qquad$ $\csc \theta = \dfrac{R}{b} = \dfrac{\sqrt{3}}{-\sqrt{2}} = -\dfrac{\sqrt{3}}{\sqrt{2}}$

$\cos \theta = \dfrac{a}{R} = \dfrac{1}{\sqrt{3}}$ $\qquad$ $\cot \theta = \dfrac{a}{b} = \dfrac{1}{-\sqrt{2}} =$

$-\dfrac{1}{\sqrt{2}}\tan \theta = \dfrac{b}{a} = \dfrac{-\sqrt{2}}{1} = -\sqrt{2}$

57. Degree mode: $\cos 308.25° = 0.6191$ $\qquad\qquad$ 59. Radian mode: $\tan 1.371 = 4.938$

61. Use the reciprocal relationship $\cot \theta = \dfrac{1}{\tan \theta}$.

Degree mode: $\cot(-265.33°) = \dfrac{1}{\tan(-265.33°)} = -0.08169$

63. Use the reciprocal relationship $\sec \theta = \dfrac{1}{\cos \theta}$.

Radian mode: $\sec(-4.013) = \dfrac{1}{\cos(-4.013)} = -1.553$

65. Degree mode: $\cos 208°12'55'' = \cos(208.21528...°)$ Convert to decimal degrees.

$$= -0.8812$$

67. Use the reciprocal relationship $\csc \theta = \dfrac{1}{\sin \theta}$

Degree mode: $\csc 112°5'38'' = \csc(112.09389...°)$ Convert to decimal degrees.

$$= \dfrac{1}{\sin(112.09389...°)} = 1.079$$

69. Use the reciprocal relationship $\sec \theta = \dfrac{1}{\cos \theta}$

Radian mode: $\sec(-1,000) = \dfrac{1}{\cos(-1,000)} = 1.778$

71. Degree mode: $\sin(405.33°) = 0.7112$

73. Degree mode: $\cos(-168°32'5'') = \cos(-168.53472...°) = -0.9800$

75. When the terminal side of an angle lies along the vertical axis, the coordinates of any point on the terminal side have the form (0, b), that is, a = 0. Therefore, $\tan \theta = \dfrac{b}{a}$ and $\sec \theta = \dfrac{R}{a}$ are not defined.

77. (A) Since $\theta = \dfrac{s}{R}$, and s = 6, and R, the measure of CA, is 5, we have $\theta = \dfrac{6}{5} = 1.2$ rad

(B) Since $\cos \theta = \dfrac{a}{R}$ and $\sin \theta = \dfrac{b}{R}$, we have

$a = R \cos \theta = 5 \cos 1.2$ $b = R \sin \theta = 5 \sin 1.2$

Thus, (a, b) = (5 cos 1.2, 5 sin 1.2) = (1.81, 4.66)

79. (A) Since $\theta = \dfrac{s}{R}$, and s = 2, and R, the measure of CA, is 1, we have $\theta = \dfrac{2}{1} = 2$ rad

(B) Since $\cos \theta = \dfrac{a}{R}$ and $\sin \theta = \dfrac{b}{R}$, we have

$a = R \cos \theta = 1 \cos 2$ $b = R \sin \theta = 1 \sin 2$

Thus, (a, b) = (1 cos 2, 1 sin 2) = (-0.416, 0.909)

81. From the figure, we note $s = R\theta$

$$R = \sqrt{a^2 + b^2} = \sqrt{4^2 + 3^2} = \sqrt{25} = 5$$
$$\tan \theta = \dfrac{b}{a} = \dfrac{3}{4}$$
$$\theta = \tan^{-1} \dfrac{3}{4}$$

Thus, $s = 5 \tan^{-1} \dfrac{3}{4} \approx 3.22$ units (calculator in radian mode)

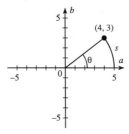

83. If $\theta = 0°$, $I = k \cos 0° = k \cdot 1 = k$

If $\theta = 20°$, $I = k \cos 20° = k(0.94) = 0.94k$

If $\theta = 40°$, $I = k \cos 40° = k(0.77) = 0.77k$

Chapter 2   Trigonometric Functions

If $\theta = 60°$, $I = k \cos 60° = k(0.50) = 0.50k$

If $\theta = 80°$, $I = k \cos 80° = k(0.17) = 0.17k$

85.   If $\theta = 15°$ (summer solstice), $I = k \cos 15° = k(0.97) = 0.97k$

     If $\theta = 63°$ (winter solstice), $I = k \cos 63° = k(0.45) = 0.45k$

87.   (A) If $n = 6$, $A = \dfrac{n}{2} \sin \left(\dfrac{360}{n}\right)^° = \dfrac{6}{2} \sin \left(\dfrac{360}{6}\right)^° = 3 \sin 60° = 2.59808$

       If $n = 10$, $A = \dfrac{10}{2} \sin \left(\dfrac{360}{10}\right)^° = 5 \sin 36° = 2.93893$

       If $n = 100$, $A = \dfrac{100}{2} \sin \left(\dfrac{360}{100}\right)^° = 50 \sin 3.6° = 3.13953$

       If $n = 1000$, $A = \dfrac{1000}{2} \sin \left(\dfrac{360}{1000}\right)^° = 500 \sin(0.36)° = 3.14157$

       If $n = 10{,}000$, $A = \dfrac{10{,}000}{2} \sin \left(\dfrac{360}{10{,}000}\right) = 5{,}000 \sin(0.036)° = 3.14159$

     (B) As n gets larger and larger, the area of the polygon gets closer and closer to the area of the circle,

         which is $A = \pi r^2 = \pi(1)^2 = \pi \approx 3.14159265 \ldots$ units.

     (C) No.

89.   From the diagram, we can see that $x = a + \ell$.

     To determine a, we note that $\cos \theta = \dfrac{a}{R}$,

     $R = 1$, and $\theta = 20\pi t$. Thus,

       $\cos 20\pi t = \dfrac{a}{1}$ and $a = \cos 20\pi t$

     To determine ℓ, we note that triangle PFL is

     a right triangle. Thus, from the Pythagorean

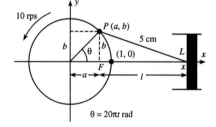

$\theta = 20\pi t$ rad

     theorem,

$$b^2 + \ell^2 = 5^2$$

$$\ell^2 = 25 - b^2$$

     Since $\sin \theta = \dfrac{b}{R}$, $R = 1$, and $\theta = 20\pi t$, we

     have $\sin 20\pi t = \dfrac{b}{1}$ and $b = \sin 20\pi t$.

     Thus, $\ell^2 = 25 - (\sin 20\pi t)^2$,

     $\ell = \sqrt{25 - (\sin 20\pi t)^2}$

     $x = a + \ell = \cos 20\pi t + \sqrt{25 - (\sin 20\pi t)^2}$

91.   Since $I = 35 \sin(48\pi t - 12\pi)$ and $t = 0.13$,

           $= 35 \sin(48\pi(0.13) - 12\pi)$

           (calculator in radian mode)

         $\approx 35 \sin(-18.095574\ldots) \approx 24$ amperes

93. In the figure, note that ABC and CDE are

    right triangles, and that L = a + b. Then,

    $\sec \theta = \dfrac{a}{8}$ from triangle ABC, so a = 8 sec θ

    $\csc \theta = \dfrac{b}{12}$ from triangle CDE, so b = 12 csc θ

    Thus, L = a + b = 8 sec θ + 12 csc θ

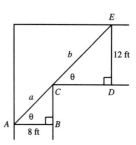

95. (A) If the angle of inclination θ = 63.5°, then m = tan θ = tan 63.5° = 2.01

    If the angle of inclination θ = 172°, then m = tan θ = tan 172° = −0.14

    (B) If the angle of inclination θ = 143°, then m = tan θ = tan 143°. Then the equation of the line

    is given by y − 6 = tan 143°(x − (−3))

    $$y = \tan 143°(x + 3) + 6 = x \tan 143° + 3 \tan 143° + 6$$

    $$y = -0.75x + 3.74$$

EXERCISE 2.4 Additional Applications

1. Use $\dfrac{n_2}{n_1} = \dfrac{\sin \alpha}{\sin \beta}$,

   where n_2 = 1.33, n_1 = 1.00, and α = 40.6°

   Solve for β: $\dfrac{1.33}{1.00} = \dfrac{\sin 40.6°}{\sin \beta}$

   $\sin \beta = \dfrac{\sin 40.6°}{1.33}$

   $\beta = \sin^{-1}\left(\dfrac{\sin 40.6°}{1.33}\right)$

   = 29.3°

3. Use $\dfrac{n_2}{n_1} = \dfrac{\sin \alpha}{\sin \beta}$,

   where n_2 = 1.66, n_1 = 1.33, and α = 32.0°

   Solve for β: $\dfrac{1.66}{1.33} = \dfrac{\sin 32.0°}{\sin \beta}$

   $\sin \beta = \dfrac{1.33 \sin 32.0°}{1.66}$

   $\beta = \sin^{-1}\left(\dfrac{1.33 \sin 32.0°}{1.66}\right)$

   = 25.1°

5. The index of refraction for diamond is n_1 = 2.42 and that for air is 1.00. Find the angle of incidence

   α such that the angle of refraction β is 90°.

   $\dfrac{\sin \alpha}{\sin \beta} = \dfrac{n_2}{n_1}$; $\sin \alpha = \dfrac{1.00}{2.42} \sin 90°$; $\alpha = \sin^{-1}\left[\dfrac{1.00}{2.42}(1)\right] = 24.4°$

7. See figure (modified from the figure in the text).

   The eye tends to assume that light travels

   straight. Thus, the ball at B will be interpreted

   as a ball at B′. The ball is actually closer than

   it appears.

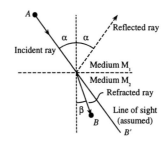

Chapter 2   Trigonometric Functions

9. We use $\sin \dfrac{\theta}{2} = \dfrac{S_w}{S_b}$, where $\theta = 60°$ and $S_w = 20$ km/hr. Then we solve for S_b:

$$\sin \frac{60°}{2} = \frac{20}{S_b}; \quad S_b = \frac{20}{\sin 30°} = 40 \text{ km/hr}$$

11. We use $\sin \dfrac{\theta}{2} = \dfrac{S_w}{S_p}$, where S_p is the speed of the plane. Here, $S_p = 2S_w$. Thus,

$$\sin \frac{\theta}{2} = \frac{S_w}{2S_w}; \quad \sin \frac{\theta}{2} = \frac{1}{2}; \quad \frac{\theta}{2} = 30° = 60°$$

13. We use $\sin \dfrac{\theta}{2} = \dfrac{S_1}{S_p}$, where $\theta = 90°$ and $S_1 = 2 \times 10^{10}$ cm/sec. Then we solve for S_p:

$$\sin \frac{90°}{2} = \frac{2 \times 10^{10}}{S_p}; \quad S_p = \frac{2 \times 10^{10}}{\sin 45°} = 3 \times 10^{10} \text{ cm/sec}$$

15. $d = a + b \sin 4\theta = -2.2 + (-4.5) \sin(4 \cdot 30°) = -2.2 - 4.5 \sin 120° = -6°$

EXERCISE 2.5 Exact Value for Special Angles and Real Numbers

Note for Problems 1–11: The reference angle α is the angle (always taken positive) between the terminal side of θ and the horizontal axis.

1. $\alpha = \theta = 60°$

3. $\alpha = |-60°| = 60°$

5. $\alpha = |-\pi/3| = \pi/3$

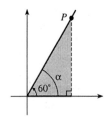
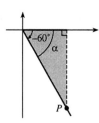
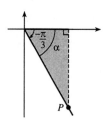

7. $\alpha = \pi - \dfrac{3\pi}{4} = \dfrac{\pi}{4}$

9. $\alpha = 210° - 180° = 30°$

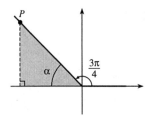
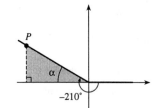

11. $\alpha = \dfrac{5\pi}{4} - \pi = \dfrac{\pi}{4}$

13. $(a, b) = (1, 0), \ R = 1$

$$\cos 0° = \frac{a}{R} = \frac{1}{1} = 1$$

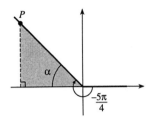

15. Use the special 30° – 60° triangle as the reference triangle. Use the sides of the reference triangle to determine P(a, b) and R. Then use Definition 1.

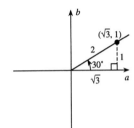

$$(a, b) = \left(\sqrt{3},\ 1\right),\ R = 2$$
$$\sin 30° = \frac{b}{R} = \frac{1}{2}$$

17. (a, b) = (0, 1), R = 1
$$\sin \frac{\pi}{2} = \frac{b}{R} = \frac{1}{1} = 1$$

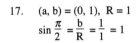

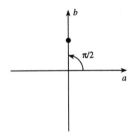

19. Use the special 45° triangle as the reference triangle. Use the sides of the reference triangle to determine P(a, b) and R. Then use Definition 1.

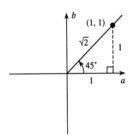

$$(a, b) = (1, 1),\ R = \sqrt{2}$$
$$\tan 45° = \frac{b}{a} = \frac{1}{1} = 1$$

21. Use the special 30° – 60 triangle as the reference triangle. Use the sides of the reference triangle to determine P(a, b). Then use Definition 1.

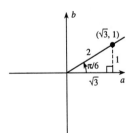

$$(a, b) = \left(\sqrt{3},\ 1\right),\ R = 2$$
$$\tan \frac{\pi}{6} = \frac{b}{a} = \frac{1}{\sqrt{3}}\ \text{or}\ \frac{\sqrt{3}}{3}$$

23.

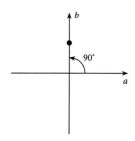

$(a, b) = (0, 1), \ R = 1$

$$\cos 90° = \frac{a}{R} = \frac{0}{1} = 0$$

25. Locate the 30° – 60° reference triangle, determine (a, b) and R, then evaluate.

27.

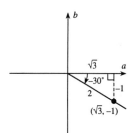

$$\sin(-30°) = \frac{-1}{2} = -\frac{1}{2}$$

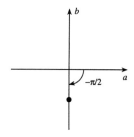

$(a, b) = (0, -1), R = 1$

$$\cos \frac{-\pi}{2} = \frac{0}{1} = 0$$

29. Locate the 30° – 60° reference triangle, determine (a, b) and R, then evaluate.

31. Locate the 30° – 60° reference triangle, determine (a, b) and R, then evaluate.

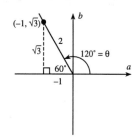

$$\tan 120° = \frac{\sqrt{3}}{-1} = -\sqrt{3}$$

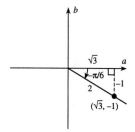

$$\cos \frac{-\pi}{6} = \frac{\sqrt{3}}{2}$$

Chapter 2   Trigonometric Functions

33.

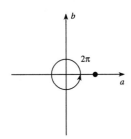

$(a, b) = (1, 0), \ R = 1$

$\cot 2\pi = \dfrac{1}{0}$ Not defined.

35. Locate the 30° – 60° reference triangle, determine (a, b) and R, then evaluate.

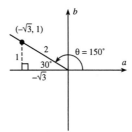

$\cot 150° = \dfrac{-\sqrt{3}}{1} = -\sqrt{3}$

37. Locate the 30° – 60° reference triangle, determine (a, b) and R, then evaluate.

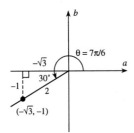

$\sin \dfrac{7\pi}{6} = \dfrac{-1}{2} = -\dfrac{1}{2}$

39.

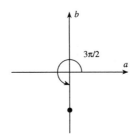

$(a, b) = (0, -1), \ R = 1 \quad \sin \dfrac{3\pi}{2} = \dfrac{-1}{1} = -1$

41. Locate the 45° reference triangle, determine (a, b) and R, then evaluate.

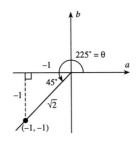

$\sin 225° = \dfrac{-1}{\sqrt{2}} = -\dfrac{1}{\sqrt{2}}$ or $-\dfrac{\sqrt{2}}{2}$

43. Locate the 45° reference triangle, determine (a, b) and R, then evaluate.

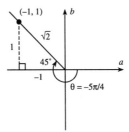

$\cot \dfrac{-5\pi}{4} = \dfrac{-1}{1} = -1$

45. Locate the 30° – 60° reference triangle, determine (a, b) and R, then evaluate.

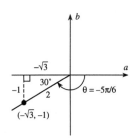

$$\cos\frac{-5\pi}{6} = \frac{-\sqrt{3}}{2} = -\frac{\sqrt{3}}{2}$$

47. Locate the 30° – 60° reference triangle, determine (a, b) and R, then evaluate.

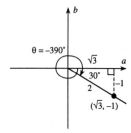

$$\cot(-390°) = \frac{\sqrt{3}}{-1} = -\sqrt{3}$$

49. Locate the 45° reference triangle, determine (a, b) and R, then evaluate.

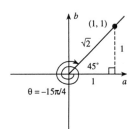

$$\sin\frac{-15\pi}{4} = \frac{1}{\sqrt{2}} \text{ or } \frac{\sqrt{2}}{2}$$

51. Locate the 30° – 60° reference triangle, determine (a, b) and R, then evaluate.

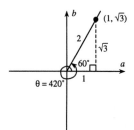

$$\csc 420° = \frac{2}{\sqrt{3}}$$

53.

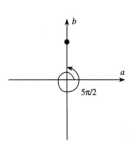

$$(a, b) = (0, 1), \ R = 1$$
$$\cos\frac{5\pi}{2} = \frac{0}{1} = 0$$

55.

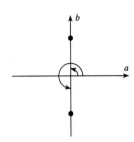

Tangent is undefined for all angles for
which a = 0.

(A) The angles θ, $0° \le \theta < 360°$,

   for which a = 0 are 90° and 270°.

(B) The angles θ, $0 \le \theta < 2\pi$,

for which a = 0 are $\dfrac{\pi}{2}$ and $\dfrac{3\pi}{2}$.

57.

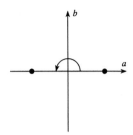

Cosecant is undefined for all angles for
which b = 0.

(A) The angles θ, $0° \le \theta < 360°$,

   for which b = 0 are 0° and 180°.

(B) The angles θ, $0 \le \theta < 2\pi$,

   for which b = 0 are 0 and π.

59. Draw a reference triangle in the first
   quadrant with side opposite reference angle 1
   and hypotenuse 2. Observe that this is a
   special 30° – 60° triangle.

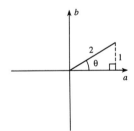

(A) $\theta = 30°$ (B) $\theta = \dfrac{\pi}{6}$

61. Draw a reference triangle in the second
   quadrant with side adjacent reference angle –1
   and hypotenuse 2. Observe that this is a
   special 30° – 60° triangle.

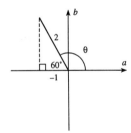

(A) $\theta = 120°$ (B) $\theta = \dfrac{2\pi}{3}$

63. Draw a reference triangle in the second quadrant with side opposite reference angle $\sqrt{3}$ and side adjacent -1. Observe that this is a special $30° - 60°$ triangle.

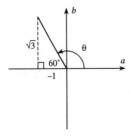

    (A) $\theta = 120°$

    (B) $\theta = \dfrac{2\pi}{3}$

65. We can draw reference triangles in both quadrants III and IV with sides opposite reference angle $-\sqrt{3}$ and hypotenuse 2. Each triangle is a special $30° - 60°$ triangle.

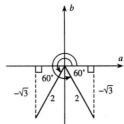

    $\theta = 240°$ or $\theta = 300°$

67. We can draw a reference triangle in the second quadrant with side opposite reference angle 1 and side adjacent $-\sqrt{3}$. We can also draw a reference triangle in the fourth quadrant with side opposite reference angle -1 and side adjacent $\sqrt{3}$. Each triangle is a special $30° - 60°$ triangle.

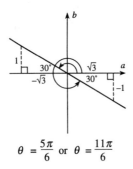

$$\theta = \frac{5\pi}{6} \text{ or } \theta = \frac{11\pi}{6}$$

69. (A) Since $\dfrac{7}{x} = \sin 30°$ and $\sin 30° = \dfrac{1}{2}$; $\dfrac{7}{x} = \dfrac{1}{2}$; $x = 14$

    Since $\dfrac{7}{y} = \tan 30°$ and $\tan 30° = \dfrac{1}{\sqrt{3}}$; $\dfrac{7}{y} = \dfrac{1}{\sqrt{3}}$; $y = 7\sqrt{3}$

    (B) Since $\dfrac{x}{4} = \sin 45°$ and $\sin 45° = \dfrac{1}{\sqrt{2}}$; $\dfrac{x}{4} = \dfrac{1}{\sqrt{2}}$; $x = \dfrac{4}{\sqrt{2}}$

    Since $\dfrac{y}{4} = \cos 45°$ and $\cos 45° = \dfrac{1}{\sqrt{2}}$; $\dfrac{y}{4} = \dfrac{1}{\sqrt{2}}$; $y = \dfrac{4}{\sqrt{2}}$

    (C) Since $\dfrac{5}{x} = \sin 60°$ and $\sin 60° = \dfrac{\sqrt{3}}{2}$; $\dfrac{5}{x} = \dfrac{\sqrt{3}}{2}$; $x = \dfrac{10}{\sqrt{3}}$

    Since $\dfrac{5}{y} = \tan 60°$ and $\tan 60° = \sqrt{3}$; $\dfrac{5}{y} = \sqrt{3}$; $y = \dfrac{5}{\sqrt{3}}$

71. $A = \dfrac{nr^2}{2} \sin \dfrac{2\pi}{n}$.

In this problem, n = 3, r = 2 cm, thus

$$A = \dfrac{3(2)^2}{2} \sin \dfrac{2\pi}{3} = 6 \sin \dfrac{2\pi}{3}$$

To find $\sin \dfrac{2\pi}{3}$, draw a 30° – 60° reference

triangle, determine (a, b) and R, then evaluate.

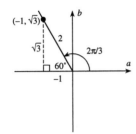

$$\sin \dfrac{2\pi}{3} = \dfrac{\sqrt{3}}{2}$$

$$A = 6 \sin \dfrac{2\pi}{3} = 6\left(\dfrac{\sqrt{3}}{2}\right) = 3\sqrt{3} \ \text{cm}^2$$

73. $A = \dfrac{nr^2}{2} \sin \dfrac{2\pi}{n}$.

In this problem, n = 6, r = 10 in., thus

$$A = \dfrac{6(10)^2}{2} \sin \dfrac{2\pi}{6} = 300 \sin \dfrac{\pi}{3}$$

To find $\sin \dfrac{\pi}{3}$, draw a 30° – 60° reference

triangle, determine (a, b) and R, then evaluate.

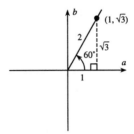

$$\sin \dfrac{\pi}{3} = \dfrac{\sqrt{3}}{2}$$

$$A = 300 \sin \dfrac{\pi}{3} = 300\left(\dfrac{\sqrt{3}}{2}\right) = 150\sqrt{3} \ \text{in}^2$$

EXERCISE 2.6 Circular Functions

1. (A) Since the circumference of a unit circle is 2π, one-half the circumference is $\dfrac{1}{2} \cdot 2\pi$ or π.

(B) Since the circumference of a unit circle is 2π, three-quarters the circumference is $\dfrac{3}{4} \cdot 2\pi$ or $\dfrac{3\pi}{2}$.

3. (A)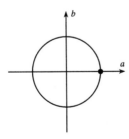

(B)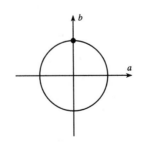

(1, 0)   (0,1)

(C)

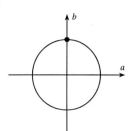

(0, 1)

(D)

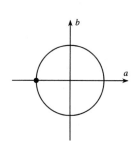

(-1, 0)

(E)

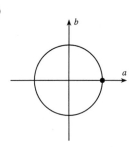

(1, 0)

(F)

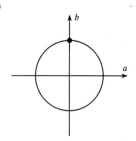

(0, 1)

5. (A) As x varies from 0 to $\frac{\pi}{2}$, y = sin x varies from 0 to 1.

　　(B) As x varies from $\frac{\pi}{2}$ to π, y = sin x varies from 1 to 0.

　　(C) As x varies from π to $\frac{3\pi}{2}$, y = sin x varies from 0 to –1.

　　(D) As x varies from $\frac{3\pi}{2}$ to 2π, y = sin x varies from –1 to 0.

　　(E) As x varies from 2π to $\frac{5\pi}{2}$, y = sin x varies from 0 to 1.

7. (A) As x varies from 0 to $-\frac{\pi}{2}$, y = cos x varies from 1 to 0.

　　(B) As x varies from $-\frac{\pi}{2}$ to $-\pi$, y = cos x varies from 0 to –1.

　　(C) As x varies from $-\pi$ to $-\frac{3\pi}{2}$, y = cos x varies from –1 to 0.

　　(D) As x varies from $-\frac{3\pi}{2}$ to -2π, y = cos x varies from 0 to 1.

　　(E) As x varies from -2π to $-\frac{5\pi}{2}$, y = cos x varies from 1 to 0.

9. sin x = 1 requires b = 1, thus P(a, b) = (0, 1).
　　This occurs when $\theta = \frac{\pi}{2}$ and when $\theta = \frac{\pi}{2} + 2\pi$ or $\frac{5\pi}{2}$.

11. sin x = 0 requires b = 0, thus P(a, b) = (1, 0) or (–1, 0)

　　P(a, b) = (1, 0) when θ = 0, 2π, or 4π　　　　　　　P(a, b) = (–1, 0) when $\theta = \pi$ or 3π

13. $\tan x = 0$ requires $\dfrac{b}{a} = 0$, thus $b = 0$, thus $P(a, b) = (1, 0)$ or $(-1, 0)$

    $P(a, b) = (1, 0)$ when $\theta = 0$, 2π, or 4π $P(a, b) = (-1, 0)$ when $\theta = \pi$ or 3π

15. $\sin x = -1$ requires $b = -1$, thus $P(a, b) = (0, -1)$. This occurs when $\theta = \dfrac{3\pi}{2}$ and when

    $\theta = \dfrac{3\pi}{2} + 2\pi$ or $\dfrac{7\pi}{2}$.

17. $\cos x = 1$ requires $a = 1$, thus $P(a, b) = (1, 0)$. This occurs when $\pi = -2\pi$, 0, or 2π.

19. $\cos x = 0$ requires $a = 0$, thus $P(a, b) = (0, 1)$ or $(0, -1)$

    $P(a, b) = (0, 1)$ when $\theta = \dfrac{\pi}{2}$ and also when $\theta = \dfrac{\pi}{2} - 2\pi = -\dfrac{3\pi}{2}$

    $P(a, b) = (0, -1)$ when $\theta = \dfrac{3\pi}{2}$ and also when $\theta = \dfrac{3\pi}{2} - 2\pi = -\dfrac{\pi}{2}$

21. $\tan x$ is not defined when $\dfrac{b}{a}$ is not defined. This occurs when $a = 0$, thus, $P(a, b) = (0, 1)$ or $(0, -1)$.

    $P(a, b) = (0, 1)$ when $\theta = \dfrac{\pi}{2}$ and also when $\theta = \dfrac{\pi}{2} + 2\pi$ or $\dfrac{5\pi}{2}$

    $P(a, b) = (0, -1)$ when $\theta = \dfrac{3\pi}{2}$ and also when $\theta = \dfrac{3\pi}{2} + 2\pi$ or $\dfrac{7\pi}{2}$

23. $\csc x$ is not defined when $\dfrac{1}{b}$ is not defined. This occurs when $b = 0$, thus $P(a, b) = (1, 0)$ or $(-1, 0)$.

    $P(a, b) = (1, 0)$ when $\theta = 0$, 2π, or 4π $P(a, b) = (-1, 0)$ when $\theta = \pi$ or 3π

25. Start at $(1, 0)$ and proceed counterclockwise (0.8 is positive) until an arc length of 0.8 has

    been covered. The point at the terminal end of the arc has coordinates $(0.7, 0.7) = (a, b)$.

    Thus, $\sin 0.8 = b = 0.7$.

27. Start at $(1, 0)$ and proceed counterclockwise (2.3 is positive) until an arc length of 2.3 has

    been covered. The point at the terminal end of the arc has coordinates $(-0.7, 0.7) = (a, b)$.

    Thus, $\cos 2.3 = a = -0.7$.

29. Start at $(1, 0)$ and proceed clockwise (-0.9 is negative) until an arc length of 0.9 has

    been covered. The point at the terminal end of the arc has coordinates $(0.6, -0.8) = (a, b)$.

    Thus, $\sin(-0.9) = -0.8$.

31. Start at $(1, 0)$ and proceed counterclockwise (2.2 is positive) until an arc length of 2.2 has

    been covered. The point at the terminal end of the arc has coordinates $(-0.6, 0.8) = (a, b)$.

    Thus, $\sec 2.2 = \dfrac{1}{a} = \dfrac{1}{-0.6} = -2$ (to one significant digit).

33. Start at $(1, 0)$ and proceed counterclockwise (0.8 is positive) until an arc length of 0.8 has

    been covered. The point at the terminal end of the arc has coordinates $(0.7, 0.7) = (a, b)$.

    Thus, $\tan 0.8 = \dfrac{b}{a} = \dfrac{0.7}{0.7} = 1$

35. Start at $(1, 0)$ and proceed clockwise (-0.4 is negative) until an arc length of 0.4 has been covered. The point at the terminal end of the arc has coordinates $(0.9, -0.4) = (a, b)$. Thus, $\cot(-0.4) = \dfrac{a}{b} = \dfrac{0.9}{-0.4} = -2$ (to one significant digit).

37. $\sin(-0.2103) = \sin(-0.2103 \text{ rad}) = -0.2088$

39. $\sec 1.432 = \dfrac{1}{\cos 1.432} = \dfrac{1}{\cos(1.432 \text{ rad})} = 7.228$

41. $\tan 4.704 = \tan(4.704 \text{ rad}) = 119.2$ 43. $\cos 105.2 = \cos(105.2 \text{ rad}) = -0.04334$.

45. $\cot(-0.03333) = \dfrac{1}{\tan(-0.03333)} = \dfrac{1}{\tan(-0.03333 \text{ rad})} = -29.99$

47. $\csc 6.2 = \dfrac{1}{\sin 6.2} = \dfrac{1}{\sin(6.2 \text{ rad})} = -12.04$

49. $\cos \dfrac{3\pi}{4} = \cos\left(\dfrac{3\pi}{4} \text{ rad}\right)$

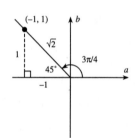

Locate the 45° reference triangle, determine (a, b) and R, then evaluate.

$$\cos \dfrac{3\pi}{4} = \dfrac{-1}{\sqrt{2}} = -\dfrac{1}{\sqrt{2}} \text{ or } \dfrac{-\sqrt{2}}{2}$$

51. $\csc\left(-\dfrac{\pi}{4}\right) = \csc\left(-\dfrac{\pi}{4} \text{ rad}\right)$

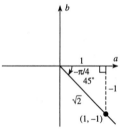

Locate the 45° reference triangle, determine (a, b) and R, then evaluate.

$$\csc\left(-\dfrac{\pi}{4}\right) = \dfrac{\sqrt{2}}{-1} = -\sqrt{2}$$

53. $\csc\dfrac{7\pi}{2} = \csc\left(\dfrac{7\pi}{2}\,\text{rad}\right) = \dfrac{1}{-1} = -1$

$(a, b) = (0, -1),\ \ R = 1$

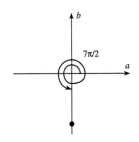

55. $\cot\left(-\dfrac{\pi}{3}\right) = \cot\left(-\dfrac{\pi}{3}\,\text{rad}\right)$

Locate $30° - 60°$ reference triangle,

determine (a, b) and R, then evaluate.

$\cot\left(-\dfrac{\pi}{3}\right) = \dfrac{1}{-\sqrt{3}} = -\dfrac{1}{\sqrt{3}}$ or $-\dfrac{\sqrt{3}}{3}$

57. $\sec\left(\dfrac{4\pi}{3}\right) = \sec\left(\dfrac{4\pi}{3}\,\text{rad}\right)$

Locate the $30° - 60°$ reference triangle,

determine (a, b) and R, then evaluate.

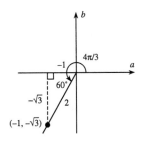

$\sec\left(\dfrac{4\pi}{3}\right) = \dfrac{2}{-1} = -2$

59. $\tan\left(-\dfrac{5\pi}{2}\right) = \tan\left(-\dfrac{5\pi}{2}\,\text{rad}\right) = \dfrac{-1}{0}$

Not defined.

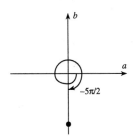

61. (A) – (D) All should equal 0.9525, because the sine function is periodic with period 2π.

63. Calculator in radian mode:

(A) $\tan 1 = 1.6$
$\dfrac{\sin 1}{\cos 1} = 1.6$

(B) $\tan 5.3 = -1.5$
$\dfrac{\sin 5.3}{\cos 5.3} = -1.5$

(C) $\tan(-2.376) = 0.96$
$\dfrac{\sin(-2.376)}{\cos(-2.376)} = 0.96$

65. Calculator in radian mode:

(A) $\sin(-3) = -0.14$

$-\sin 3 = -0.14$

(B) $\sin[-(-12.8)] = 0.23$

$-\sin(-12.8) = 0.23$

(C) $\sin(-407) = 0.99$

$-\sin(407) = 0.99$

Note: Some older calculators cannot evaluate sin(–407) and, instead, signal an error. If this occurs, use the periodicity of the sine function and evaluate sin(–407 + 2π · k), where k is an appropriate integer.

67. Calculator in radian mode:

(A) $\sin^2 1 + \cos^2 1 = (0.841 \ldots)^2 + (0.540 \ldots)^2 = 1.0$

(B) $\sin^2(-8.6) + \cos^2(-8.6) = (-0.734 \ldots)^2 + (-0.678 \ldots)^2 = 1.0$

(C) $\sin^2(263) + \cos^2(263) = (-0.779 \ldots)^2 + (0.626 \ldots)^2 = 1.0$

69. $\sin x \csc x = \sin x \dfrac{1}{\sin x}$ Use Identity (1)

$= 1$

71. $\cot x \sec x = \dfrac{\cos x}{\sin x} \cdot \dfrac{1}{\cos x}$ Use Identities (5) and (2)

$= \dfrac{1}{\sin x}$ Use Identity (1)

$= \csc x$

73. $\dfrac{\sin x}{1 - \cos^2 x} = \dfrac{\sin x}{\sin^2 x + \cos^2 x - \cos^2 x}$ Use Identity (9)

$= \dfrac{\sin x}{\sin^2 x}$ Use Identity (1)

$= \dfrac{1}{\sin x} = \csc x$

75. $\cot (-x) \sin(-x) = \dfrac{\cos(-x)}{\sin(-x)} \sin(-x)$ Use Identity (5)

$= \cos (-x)$

$= \cos x$ Use Identity (7)

77. (A) Identity (4)   (B) Identity (9)   (C) Identity (2)

79. 2π

81. $S_1 = 1$
$S_2 = S_1 + \cos S_1 = 1 + \cos 1 = 1.540302$
$S_3 = S_2 + \cos S_2 = 1.540302 + \cos 1.540302 = 1.570792$
$S_4 = S_3 + \cos S_3 = 1.570792 + \cos 1.570792 = 1.570796$
$S_5 = S_4 + \cos S_4 = 1.570796 + \cos 1.570796 = 1.570796$
$\dfrac{\pi}{2} = 1.570796$

CHAPTER 2   REVIEW EXERCISE

1. (A) $\dfrac{\theta_{deg}}{180°} = \dfrac{\theta_{rad}}{\pi \text{ rad}}$

$\theta_{rad} = \dfrac{\pi \text{ rad}}{180°} \theta_{deg}$

$= \dfrac{\pi}{180} 60 = \dfrac{\pi}{3}$

(B) $\dfrac{\theta_{deg}}{180°} = \dfrac{\theta_{rad}}{\pi \text{ rad}}$

$\theta_{rad} = \dfrac{\pi \text{ rad}}{180°} \theta_{deg}$

$= \dfrac{\pi}{180} 45 = \dfrac{\pi}{4}$

(C) $\dfrac{\theta_{deg}}{180°} = \dfrac{\theta_{rad}}{\pi \text{ rad}}$

$\theta_{rad} = \dfrac{\pi \text{ rad}}{180°} \theta_{deg}$

$= \dfrac{\pi}{180} 90 = \dfrac{\pi}{2}$

2. (A) $\dfrac{\theta_{\text{deg}}}{180°} = \dfrac{\theta_{\text{rad}}}{\pi\ \text{rad}}$

    $\theta_{\text{deg}} = \dfrac{180°}{\pi\ \text{rad}}\,\theta_{\text{rad}}$

    $= \dfrac{180}{\pi} \cdot \dfrac{\pi}{6}$

    $= 30°$

(B) $\dfrac{\theta_{\text{deg}}}{180°} = \dfrac{\theta_{\text{rad}}}{\pi\ \text{rad}}$

    $\theta_{\text{deg}} = \dfrac{180°}{\pi\ \text{rad}}\,\theta_{\text{rad}}$

    $= \dfrac{180}{\pi} \cdot \dfrac{\pi}{2}$

    $= 90°$

(C) $\dfrac{\theta_{\text{deg}}}{180°} = \dfrac{\theta_{\text{rad}}}{\pi\ \text{rad}}$

    $\theta_{\text{deg}} = \dfrac{180°}{\pi\ \text{rad}}\,\theta_{\text{rad}}$

    $= \dfrac{180}{\pi} \cdot \dfrac{\pi}{4}$

    $= 45°$

3. (A) $\dfrac{\theta_{\text{deg}}}{180°} = \dfrac{\theta_{\text{rad}}}{\pi\ \text{rad}}$

    $\theta_{\text{deg}} = \dfrac{180°}{\pi\ \text{rad}}\,\theta_{\text{rad}}$

    $= \dfrac{180}{\pi} \cdot 15.26 = 874.3°$

(B) $\dfrac{\theta_{\text{deg}}}{180°} = \dfrac{\theta_{\text{rad}}}{\pi\ \text{rad}}$

    $\theta_{\text{rad}} = \dfrac{\pi\ \text{rad}}{180°}\,\theta_{\text{deg}}$

    $= \dfrac{\pi}{180}(-389.2) = -6.793\ \text{rad}$

4. $V = R\omega = 25(7.4) = 185\ \text{ft/min}$

5. $\omega = \dfrac{V}{R} = \dfrac{415}{5.2} = 80\ \text{rad/hr}$

6. $P(a, b) = (-4, 3)$

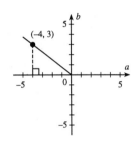

    $R = \sqrt{a^2 + b^2} = \sqrt{(-4)^2 + 3^2} = \sqrt{25} = 5$

    $\sin\theta = \dfrac{b}{R} = \dfrac{3}{5}$

    $\tan\theta = \dfrac{b}{a} = \dfrac{3}{-4} = -\dfrac{3}{4}$

7. (A) Use the reciprocal relationship $\cot\theta = \dfrac{1}{\tan\theta}$.

    Degree mode: $\cot 53°40' = \cot(53.666...°)$ Convert to decimal degrees.

        $= \dfrac{1}{\tan(53.666...°)} = 0.7355$

(B) Use the reciprocal relationship $\csc\theta = \dfrac{1}{\sin\theta}$.

    Degree mode: $\csc 67°10' = \csc(67.166...°)$ Convert to decimal degrees.

        $= \dfrac{1}{\sin(67.1666...°)} = 1.085$

8. (A) Degree mode: $\cos 23.5° = 0.9171$ (B) Degree mode: $\tan 42.3° = 0.9099$

9. (A) Radian mode: $\cos 0.35 = \cos(0.35\ \text{rad})$ (B) Radian mode: $\tan 1.38 = \tan(1.38\ \text{rad})$

        $= 0.9394$ $= 5.177$

Chapter 2   Trigonometric Functions

10. The reference angle α is the angle (always taken positive) between the terminal side of θ and the horizontal axis.

(A)

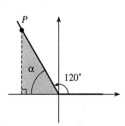

(B)

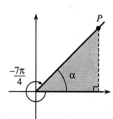

$$\alpha = 180° - 120° = 60°$$

$$\alpha = 2\pi - \left|-\frac{7\pi}{4}\right| = \frac{\pi}{4}$$

11. (A) Use the special $30° - 60°$ triangle as the reference triangle. Use the sides of the reference triangle to determine P(a, b) and R. Then use Definition 1.

(B) Use the special $45°$ triangle as the reference triangle. Use the sides of the reference triangle to determine P(a, b) and R. The use Definition 1.

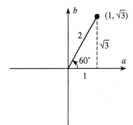

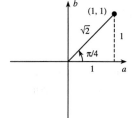

$$(a, b) = \left(1, \sqrt{3}\right), \ \ R = 2$$

$$\sin 60° = \frac{b}{R} = \frac{\sqrt{3}}{2}$$

$$(a, b) = (1, 1), \ \ R = \sqrt{2}$$

$$\cos\frac{\pi}{4} = \frac{a}{R} = \frac{1}{\sqrt{2}} \ \text{ or } \ \frac{\sqrt{2}}{2}$$

(C)

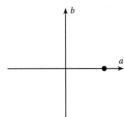

$$(a, b) = (1, 0), \ \ R = 1$$

$$\tan 0° = \frac{0}{1} = 0$$

12. (A)

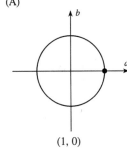

(1, 0)

(B)

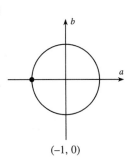

(–1, 0)

(C)

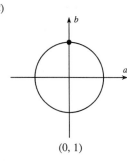

(0, 1)

(D)

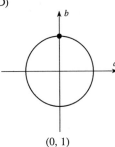

(0, 1)

(E)

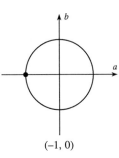

(–1, 0)

(F)

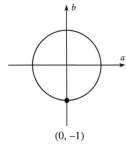

(0, –1)

13. (A) As x varies from 0 to $\frac{\pi}{2}$, y = sin x varies from 0 to 1.

(B) As x varies from $\frac{\pi}{2}$ to π, y = sin x varies from 1 to 0.

(C) As x varies from π to $\frac{3\pi}{2}$, y = sin x varies from 0 to –1.

(D) As x varies from $\frac{3\pi}{2}$ to 2π, y = sin x varies from –1 to 0.

(E) As x varies from 2π to $\frac{5\pi}{2}$, y = sin x varies from 0 to 1.

(F) As x varies from $\frac{5\pi}{2}$ to 3π, y = sin x varies from 1 to 0.

14. (A) As x varies from 0 to $-\frac{\pi}{2}$, y = cos x varies from 1 to 0.

(B) As x varies from $-\frac{\pi}{2}$ to $-\pi$, y = cos x varies from 0 to –1.

(C) As x varies from $-\pi$ to $-\frac{3\pi}{2}$, y = cos x varies from –1 to 0.

(D) As x varies from $-\frac{3\pi}{2}$ to -2π, y = cos x varies from 0 to 1.

(E) As x varies from -2π to $-\frac{5\pi}{2}$, y = cos x varies from 1 to 0.

(F) As x varies from $-\frac{5\pi}{2}$ to -3π, y = cos x varies from 0 to –1.

15. Since the central angle subtended by a circumference has degree measure 360°, the central angle subtended by an arc $\frac{7}{60}$ of a circumference has degree measure $\frac{7}{60}$ (360°) = 42°.

Chapter 2   Trigonometric Functions

16. $s = R\theta$, $\theta = \dfrac{s}{R} = \dfrac{24}{8} = 3$ rad

17. $s = R\theta = 4(1.5) = 6$ cm

18. $\dfrac{\theta_{\text{deg}}}{180°} = \dfrac{\theta_{\text{rad}}}{\pi \text{ rad}}$

$\theta_{\text{rad}} = \dfrac{\pi \text{ rad}}{180°}\,\theta_{\text{deg}} = \dfrac{\pi}{180}(212) = \dfrac{53\pi}{45}$

19. $\dfrac{\theta_{\text{deg}}}{180°} = \dfrac{\theta_{\text{rad}}}{\pi \text{ rad}}$

$\theta_{\text{deg}} = \dfrac{180°}{\pi \text{ rad}}\,\theta_{\text{rad}} = \dfrac{180}{\pi} \cdot \dfrac{\pi}{12} = 15°$

20.   (A)   (B)   (C)

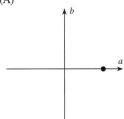

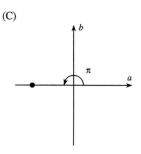

(a, b) = (1, 0)
R = 1
$\tan 0 = \dfrac{b}{a} = \dfrac{0}{1} = 0$
Not defined.

(a, b) = (0, 1)
R = 1
$\tan \dfrac{\pi}{2} = \dfrac{b}{a} = \dfrac{1}{0}$

(a, b) = (−1, 0)
R = 1
$\tan \pi = \dfrac{b}{a} = \dfrac{0}{-1} = 0$

(D)

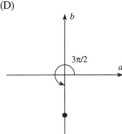

(a, b) = (0, −1)
R = 1
$\tan \dfrac{3\pi}{2} = \dfrac{b}{a} = \dfrac{-1}{0}$
Not defined.

21.   (A) 732° is coterminal with (732 − 360°) = 372°, and thus with (372 − 360)° = 12°. Since 12° is between 0° and 90°, its terminal side lies in quadrant I.

(B) −7 rad is coterminal with (−7 + 2π) rad ≈ −0.72 rad. Since −0.72 rad is between 0 and −$\dfrac{\pi}{2}$, its terminal side lies in quadrant IV.

22.   The reference angle α is the angle (always taken positive) between the terminal side of θ and the horizontal axis.

(A)

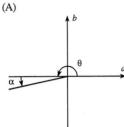

(B)

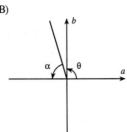

(C)

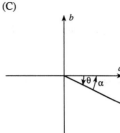

$\alpha = 187.4° - 180°$

$= 7.4°$

$\alpha = 180° - 103°20'$

$= 76°40'$

$\alpha = |-37°40'|$

$= 37°40'$

23. The reference angle α is the angle (always taken positive) between the terminal side of θ and the horizontal axis.

(A)

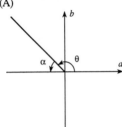

(B)

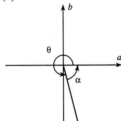

(C)

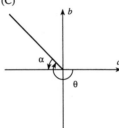

$\alpha = \pi - 2.39$ rad

$\approx 3.14 - 2.39$ rad

≈ 0.75 rad

$\alpha = 2\pi - 5.00$ rad

$\approx 6.28 - 5.00$ rad

≈ 1.28 rad

$\alpha = |-4.0 + \pi|$ rad

$\approx |-4.0 + 3.14|$ rad

≈ 0.86 rad

24. Degree mode: $\cos 187.4° = -0.992$ 25. Degree mode: $\tan 187.4° = 0.130$

26. Degree mode: $\sin 103°20' = \sin(103.333...°)$ Convert to decimal degrees.

$$= 0.973$$

27. Use the reciprocal relationship $\sec \theta = \dfrac{1}{\cos \theta}$.

Degree mode: $\sec 103°20' = \sec(103.333...°)$ Convert to decimal degrees.

$$= \frac{1}{\cos(103.333...°)} = -4.34$$

28. Use the reciprocal relationship $\cot \theta = \dfrac{1}{\tan \theta}$.

Degree mode: $\cot(-37°40') = \cot(-37.666...°)$ Convert to decimal degrees.

$$= \frac{1}{\tan(-37.666...°)} = -1.30$$

Chapter 2   Trigonometric Functions

29. Use the reciprocal relationship $\sec \theta = \dfrac{1}{\cos \theta}$.

   Degree mode: $\sec(-37°40') = \sec(-37.666...°)$ Convert to decimal degrees.

   $$= \dfrac{1}{\cos(-37.666...°)} = 1.26$$

30. Radian mode: $\sin 2.39 = 0.683$

31. Use the reciprocal relationship $\cot \theta = \dfrac{1}{\tan \theta}$.   Radian mode: $\cot 2.39 = \dfrac{1}{\tan 2.39} = -1.07$

32. Radian mode: $\cos 5 = 0.284$ 33. Radian mode: $\tan 5 = -3.38$

34. Radian mode: $\sin(-4) = 0.757$

35. Use the reciprocal relationship $\cot \theta = \dfrac{1}{\tan \theta}$.   Radian mode: $\cot(-4) = \dfrac{1}{\tan(-4)} = -0.864$

36. The reference angle α is the angle (always taken positive) between the terminal side of θ and the horizontal axis.

   (A)   (B)   (C)

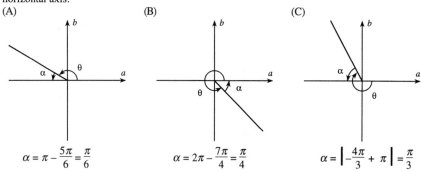

   $\alpha = \pi - \dfrac{5\pi}{6} = \dfrac{\pi}{6}$ $\alpha = 2\pi - \dfrac{7\pi}{4} = \dfrac{\pi}{4}$ $\alpha = \left| -\dfrac{4\pi}{3} + \pi \right| = \dfrac{\pi}{3}$

37. Locate the 30° – 60° reference triangle,   38. See Problem 37.
   determine (a, b) and R, then evaluate. $\tan \dfrac{5\pi}{6} = \dfrac{1}{-\sqrt{3}} = -\dfrac{1}{\sqrt{3}}$

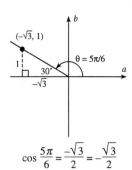

   $\cos \dfrac{5\pi}{6} = \dfrac{-\sqrt{3}}{2} = -\dfrac{\sqrt{3}}{2}$

39. Locate the 45° reference triangle, determine (a, b) and R, then evaluate.

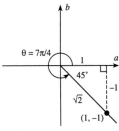

$$\sin \frac{7\pi}{4} = \frac{-1}{\sqrt{2}} = -\frac{1}{\sqrt{2}}$$

40. See Problem 39.

$$\cot \frac{7\pi}{4} = \frac{1}{-1} = -1$$

41.

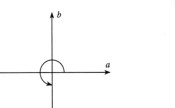

$$(a, b) = (0, -1), \ R = 1$$
$$\sin \frac{3\pi}{2} = \frac{-1}{1} = -1$$

42. See Problem 41.

$$\cos \frac{3\pi}{2} = \frac{0}{1} = 0$$

43. Locate the 30° – 60° reference triangle, determine (a, b) and R, then evaluate.

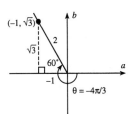

$$\sin \frac{-4\pi}{3} = \frac{\sqrt{3}}{2}$$

44. See Problem 43.

$$\sec \frac{-4\pi}{3} = \frac{2}{-1} = -2$$

45. $(a, b) = (-1, 0), \ R = 1$

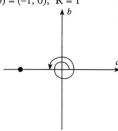

$$\cos 3\pi = \frac{-1}{1} = -1$$

46. See Problem 45.

$$\cot 3\pi = \frac{-1}{0}$$

Not defined.

47. Locate the 30° – 60° reference triangle, determine (a, b) and R, then evaluate.

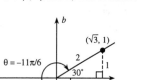

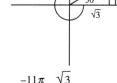

$$\cos \frac{-11\pi}{6} = \frac{\sqrt{3}}{2}$$

48. See Problem 47.

$$\sin \frac{-11\pi}{6} = \frac{1}{2}$$

49. Degree mode: $\sin 384.0314° = 0.40724$

50. Degree mode: $\tan (-198°43'6'') = \tan(-198.71833...°)$ Concert to decimal degrees.

$$= -0.33884$$

51. Radian mode: $\cos 26 = 0.64692$

52. Use the reciprocal relationship $\cot \theta = \dfrac{1}{\tan \theta}$.   Radian mode: $\cot(-68.005) = \dfrac{1}{\tan(-68.005)}$

$$= 0.49639$$

53. If $\sin \theta = -\dfrac{4}{5}$ and the terminal side of θ does not lie in the third quadrant, then it must lie in the fourth quadrant.

$P(a, b) = (3, -4)$

$R = \sqrt{a^2 + b^2} = \sqrt{3^2 + (-4)^2} = \sqrt{25} = 5$

$\cos \theta = \dfrac{a}{R} = \dfrac{3}{5}$

$\tan \theta = \dfrac{b}{a} = \dfrac{-4}{3} = -\dfrac{4}{3}$

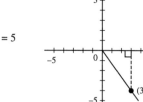

54. Draw a reference triangle in the third quadrant with side opposite reference angle –1 and hypotenuse 2. Observe that this is a special 30° – 60° triangle.

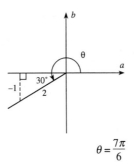

$$\theta = \frac{7\pi}{6}$$

55. Since sin $\theta < 0$ and tan $\theta < 0$, the terminal side of θ lies in quadrant IV. We sketch a reference triangle and label what we know.

    Since sin $\theta = \dfrac{b}{R} = -\dfrac{2}{5} = \dfrac{-2}{5}$, we know that b = –2 and R = 5 (R is never negative).

    Use the Pythagorean theorem to find a:

    $$a^2 + (-2)^2 = 5^2$$
    $$a^2 = 25 - 4 = 21$$
    $$a = \sqrt{21}$$

    a is positive since the terminal side of θ lies

    in quadrant IV. We can now find the other

    five functions using Definition 1:

    $\csc \theta = \dfrac{R}{b} = \dfrac{5}{-2} = -\dfrac{5}{2}$ $\sec \theta = \dfrac{R}{a} = \dfrac{5}{\sqrt{21}}$

    $\cos \theta = \dfrac{a}{R} = \dfrac{\sqrt{21}}{5}$ $\cot \theta = \dfrac{a}{b} = \dfrac{\sqrt{21}}{-2} = -\dfrac{\sqrt{21}}{2}$

    $\tan \theta = \dfrac{b}{a} = \dfrac{-2}{\sqrt{21}} = -\dfrac{2}{\sqrt{21}}$

56. $\tan \theta = \dfrac{b}{a} = -1 = \dfrac{-1}{1}$ or $\dfrac{1}{-1}$

    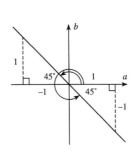

    We can draw a reference triangle in the second quadrant

    with side opposite reference angle 1 and side adjacent –1.

    We can also draw a reference triangle in the fourth quadrant with side opposite reference angle –1 and side adjacent 1.

    $\theta = 135°$ or $315°$

    Each triangle is a special 45° triangle.

Chapter 2   Trigonometric Functions

57. We can draw reference triangles in both quadrants II

    and III with sides adjacent reference angle $-\sqrt{3}$

    and

    hypotenuse 2. Each triangle is a special 30°

    $-$ 60°

    triangle.

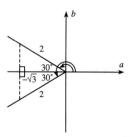

$$\theta = \frac{5\pi}{6} \text{ or } \frac{7\pi}{6}$$

58. (A) $s = R\theta = (12.0)(1.69) \approx 20.3$ cm   (B) $s = \dfrac{\pi}{180}R\theta = \dfrac{\pi}{180}(12.0)(22.5) \approx 4.71$ cm

59. (A) $A = \dfrac{1}{2}R^2\theta$ (B) $A = \dfrac{\pi}{360}R^2\theta = \dfrac{\pi}{360}(40)^2(135) \approx 1{,}880$ ft^2

    $$\left[R = \frac{1}{2}D = \frac{1}{2}(80) = 40\,\text{ft}\right]$$

    $$A = \frac{1}{2}(40)^2(0.773) \approx 618 \text{ ft}^2$$

60. Since the cities have the same longitude, θ is

    given by their difference in latitude.

    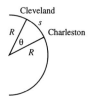

    $$\theta = 41°28' - 38°21' = 3°7' = \left(3 + \frac{7}{60}\right)^{\circ}$$

    $$s = \frac{\pi}{180}R\theta = \frac{\pi}{180}(3{,}964 \text{ miles})\left(3 + \frac{7}{60}\right)$$

    $$= 215.6 \text{ miles}$$

61. $\omega = \dfrac{\theta}{t} = \dfrac{6.43}{15.24} = 0.422$ rad/sec

62. (A) Calculator in radian mode: $\cos 7 = 0.754$

    (B) $-$ (E)   All should equal 0.754, because the cosine function is periodic with period 2π.

63. Calculator in radian mode:

    (A) $\tan(-7) = -0.871$ (B) $\tan[-(-17.9)] = -1.40$ (C) $\tan[-(-2{,}135)] = -3.38$

    $-\tan 7 = -0.871$ $-\tan(-17.9) = -1.40$ $-\tan(-2{,}135) = -3.38$

    Note: Some older calculators cannot evaluate $\tan(-2{,}135)$ and, instead, signal an error. If this
    occurs, use the periodicity of the tangent function and evaluate $\tan(2{,}135 - 2\pi \cdot k)$, where k
    is an appropriate integer.

64. (D) is not an identity, since it is not true for all values of the variable x.

65. $(\csc x)(\cot x)(1 - \cos^2 x) = \dfrac{1}{\sin x} \dfrac{\cos x}{\sin x}(1 - \cos^2 x)$ Use Identities (1) and (5)

$= \dfrac{\cos x}{\sin^2 x}(\sin^2 x + \cos^2 x - \cos^2 x)$ Use Identity (9)

$= \dfrac{\cos x}{\sin^2 x} \cdot \sin^2 x = \cos x$

66. $\cot(-x)\sin(-x) = \dfrac{\cos(-x)}{\sin(-x)}\sin(-x)$ Use Identity (5)

$= \cos(-x) = \cos x$ Use Identity (7)

67. Using the definition of the circular functions, we see that the coordinates of the point in its terminal position are $(a, b) = (\cos 1.4, \sin 1.4) \approx (0.170, 0.985)$

68. Since $A = \dfrac{1}{2}R^2\theta$ and $s = R\theta$, we can eliminate θ between these two equations as follows:

$\theta = \dfrac{s}{R}$ $A = \dfrac{1}{2}R^2\left(\dfrac{s}{R}\right) = \dfrac{1}{2}Rs.$ Then, $s = \dfrac{2A}{R} = \dfrac{2(342.5)}{12} \approx 57$ m

69. From the figure, we note $s = R\theta$.

$R = \sqrt{a^2 + b^2} = \sqrt{4^2 + 5^2} = \sqrt{41}$

$\tan\theta = \dfrac{b}{a} = \dfrac{5}{4}$

$\theta = \tan^{-1}\dfrac{5}{4}$

Thus, $s = \sqrt{41}\ \tan^{-1}\dfrac{5}{4} \approx 5.74$ units

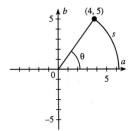

(Calculator in radian mode.)

70. We use $s = R\theta$ with $R = \dfrac{1}{2}d = 5$ cm and $s =$

10 m = 1000 cm. Then, $\theta = \dfrac{s}{R} = \dfrac{1000}{5} =$

200 rad.

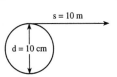

Since 1 revolution corresponds to

1 circumference = 2π radians,

200 radians corresponds to

$\dfrac{200}{2\pi} = \dfrac{100}{\pi} \approx 31.8$ revolutions

71. Since the three gear wheels are coupled together, each must turn through the same distance (arc length). Thus,

$s = R_1\theta_1$ $s = R_2\theta_2$ $s = R_3\theta_3$ $R_1 = 30$ cm $R_2 = 20$ cm $R_3 = 10$ cm

$R_1\theta_1 = R_2\theta_2$

$\theta_2 = \dfrac{R_1}{R_2}\theta_1 = \dfrac{30}{20}(5\text{ revolutions}) = 7.5$ revolutions

Chapter 2   Trigonometric Functions

$$R_3\theta_3 = R_1\theta_1$$

$$\theta_3 = \frac{R_1}{R_3}\theta_1 = \frac{30}{10}(5 \text{ revolutions}) = 15 \text{ revolutions}$$

72. We use $\omega = \dfrac{V}{R}$ with V = 70 ft/sec and R $= \dfrac{27 \text{ in.}}{2} = \dfrac{27 \text{ in.}}{2} \cdot \dfrac{1 \text{ in.}}{12 \text{ ft}} = \dfrac{27}{24}$ ft.

Then $\omega = \dfrac{V}{R} = \dfrac{70}{27/24} = 62$ rad/sec

73. We use V = Rω.

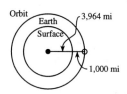

R = radius of orbit = 3,964 + 1,000 = 4,964 mi

$$\omega = \frac{\theta}{t} = \frac{2\pi}{114/60 \text{ hr}} = \frac{120\pi}{114} \text{ rad/hr}$$

$$V = (4,964 \text{ mi})\left(\frac{120\pi}{114} \text{ rad/hr}\right) = 16,400 \text{ mi/hr}$$

74. I = 30 sin(120πt − 60π) = 30 sin[120π(0.015) − 60π] = −17.6 amp

75. (A) We note:

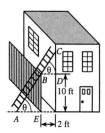

The length of the ladder AC = AB + BC

In triangle ABE, csc $\theta = \dfrac{AB}{BE} = \dfrac{AB}{10 \text{ ft}}$

In triangle BCD, sec $\theta = \dfrac{BC}{BD} = \dfrac{BC}{2 \text{ ft}}$

Then AB = 10 csc θ, BC = 2 sec θ, hence length of ladder

AC = 10 csc θ + 2 sec θ

(B) Radian mode:

$\theta = 0.9$ AC = 10 csc 0.9 + 2 sec 0.9 = 16.0

$\theta = 1.0$ AC = 10 csc 1.0 + 2 sec 1.0 = 15.6

$\theta = 1.1$ AC = 10 csc 1.1 + 2 sec 1.1 = 15.6

$\theta = 1.2$ AC = 10 csc 1.2 + 2 sec 1.2 = 16.2

$\theta = 1.3$ AC = 10 csc 1.3 + 2 sec 1.3 = 17.9

76. Use $\dfrac{n_2}{n_1} = \dfrac{\sin \alpha}{\sin \beta}$, where n_2 = 1.33, n_1 = 1.00, and α = 31.7°

Solve for β: $\dfrac{1.33}{1.00} = \dfrac{\sin 31.7°}{\sin \beta}$

$$\sin \beta = \frac{\sin 31.7°}{1.33}$$

$$\beta = \sin^{-1}\left(\frac{\sin 31.7°}{1.33}\right) = 23.3°$$

77. Find the angle of incidence α such that the angle of refraction is 90°.

$$\frac{\sin \alpha}{\sin \beta} = \frac{n_2}{n_1} \qquad \sin \alpha = \frac{n_2}{n_1}\sin \beta = \frac{1.00}{1.52}\sin 90° \qquad \alpha = \sin^{-1}\left[\frac{1.00}{1.52}(1)\right] = 41.1°$$

78.   We use $\sin\dfrac{\theta}{2}=\dfrac{S_w}{S_b}$, where $\theta=51°$ and $S_b=25$ mph. Then we solve for S_w:

$$\sin\frac{51°}{2}=\frac{S_w}{25} \qquad S_w = 25\sin 25.5° = 11 \text{ mph}$$

Chapter 3   Graphing Trigonometric Functions

EXERCISE 3.1   Basic Graphs

1. $2\pi, 2\pi, \pi$ 3. (A) 1 unit   (B) Indefinitely far   (C) Indefinitely far

5. $-\dfrac{3\pi}{2}, -\dfrac{\pi}{2}, \dfrac{\pi}{2}, \dfrac{3\pi}{2}$ 7. $-2\pi, -\pi, 0, \pi, 2\pi$ 9. The graph has no x intercepts; sec x is never 0.

11. (A) None; sin x is always defined.   (B) $-2\pi, -\pi, 0, \pi, 2\pi$ (C) $-\dfrac{3\pi}{2}, -\dfrac{\pi}{2}, \dfrac{\pi}{2}, \dfrac{3\pi}{2}$

13.

x	0	0.1	0.2	0.3	0.4	0.5	0.6	0.7	0.8
cos x	1	1.0	0.98	0.96	0.92	0.88	0.83	0.76	0.70

x	0.9	1.0	1.1	1.2	1.3	1.4	1.5	1.6
cos x	0.62	0.54	0.45	0.36	0.27	0.17	0.07	−0.03

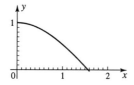

15.

x	−1.4	−1.2	−1.0	−0.8	−0.6	−0.4	−0.2	0
tan x	−5.8	−2.6	−1.6	−1.0	−0.68	−0.42	−0.20	0

x	0.2	0.4	0.6	0.8	1.0	1.2	1.4
tan x	0.20	0.42	0.68	1.0	1.6	2.6	5.8

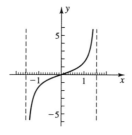

17.

x	0.2	0.4	0.6	0.8	1.0	1.2	1.4	1.6
csc x	5.0	2.6	1.8	1.4	1.2	1.1	1.0	1.0

x	1.8	2.0	2.2	2.4	2.6	2.8	3.0
csc x	1.0	1.1	1.2	1.5	1.9	3.0	7.1

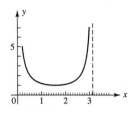

19.

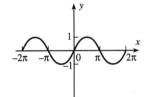

21.

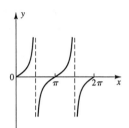

23.   The dashed line shows y = sin x
in this interval. The solid line is
y = csc x.

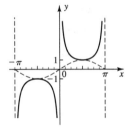

25. Depending on the particular calculator used, either an error message will occur or some very large

number will occur because of round-off error, because:

(A) 0 is not in the domain of the cotangent function.

(B) $\frac{\pi}{2}$ is not in the domain of the tangent function.

(C) π is not in the domain of the cosecant function.

27. (A)

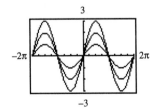

63

Chapter 3   Graphing Trigonometric Functions

(B) The highest point on the graph of y = sin x has y coordinate 1.The lowest point has y coordinate
     –1.The highest point on the graph of y = 2 sin x has y coordinate 2.The lowest point has y
     coordinate –2. The highest point on the graph of y = 3 sin x has y coordinate 3. The lowest
     point has y coordinate –3.

(C) The highest point on the graph of y = p sin x, p > 0, has y coordinate p.
     The lowest point has y coordinate –p.

29. (A)

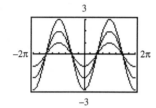

(B) The highest point on the graph of y = –cos x has y coordinate 1.

     The lowest point has y coordinate –1.

     The highest point on the graph of

     y = –2 cos x has y coordinate 2.

     The lowest point has y coordinate –2.

     The highest point on the graph of y = –3 cos x has y coordinate 3.

     The lowest point has y coordinate –3.

(C) The highest point on the graph of y = p cos x, p < 0, has y coordinate –p.
     The lowest point has y coordinate p.

31. (A)

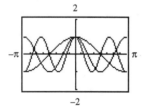

(B) One period of y = cos x appears.

     (–π to π)

     Two periods of y = cos 2x appear.

     (–π to 0, 0 to π)
     Three periods of y = cos 3x appear.
     $\left(-\pi \text{ to } -\dfrac{\pi}{3}, \; -\dfrac{\pi}{3} \text{ to } \dfrac{\pi}{3}, \; \dfrac{\pi}{3} \text{ to } \pi\right)$

(C) n periods of y = cos nx would appear.

33. (A)

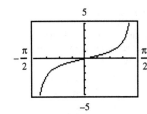

$y = \tan x$

$y = \tan 2x$

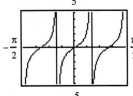

$y = \tan 3x$

(B)   One period of $y = \tan x$ appears. $\left(-\dfrac{\pi}{2} \text{ to } \dfrac{\pi}{2}\right)$

Two periods of $y = \tan 2x$ appear.

$$\left(-\frac{\pi}{2} \text{ to } 0, \ 0 \text{ to } \frac{\pi}{2}\right)$$

Three periods of $y = \tan 3x$ appear.

$$\left(-\frac{\pi}{2} \text{ to } -\frac{\pi}{6}, \ -\frac{\pi}{6} \text{ to } \frac{\pi}{6}, \ \frac{\pi}{6} \text{ to } \frac{\pi}{2}\right)$$

(C)   n periods of $y = \tan nx$ would appear.

35.

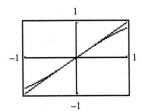

EXERCISE 3.2 Graphing $y = k + A \sin Bx$ and $y = k + A \cos Bx$

1.

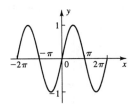

Chapter 3   Graphing Trigonometric Functions

3. $y = 3 \cos x$. Amplitude $= |3| = 3$. Period $= \dfrac{2\pi}{B} = \dfrac{2\pi}{1} = 2\pi$.

One full cycle of the graph is completed as x goes from 0 to 2π. Block out this interval, divide it into four equal parts, locate high and low points, and locate x intercepts. Then complete the graph.

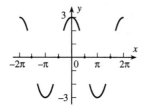 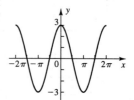

5. $y = -2 \sin x$. Amplitude $= |-2| = 2$. Period $= \dfrac{2\pi}{B} = \dfrac{2\pi}{1} = 2\pi$.

Since A = –2 is negative, the basic curve for y = sin x is turned upside down. One full cycle of the graph is completed as x goes from 0 to 2π. Block out this interval, divide it into four equal parts, locate high and low points, and locate x intercepts. Then complete the graph.

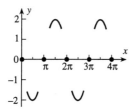 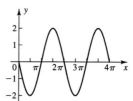

7. $y = \dfrac{1}{2} \sin x$. Amplitude $= \left|\dfrac{1}{2}\right| = \dfrac{1}{2}$. Period $= \dfrac{2\pi}{B} = \dfrac{2\pi}{1} = 2\pi$.

One full cycle of the graph is completed as x goes from 0 to 2π. Block out this interval, divide it into four equal parts, locate high and low points, and locate x intercepts. Then complete the graph.

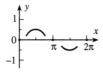 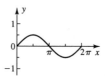

9. $y = \cos 2x$. Amplitude $= |A| = |1| = 1$. Period $= \dfrac{2\pi}{2} = \pi$.

One full cycle of the graph is completed as x goes from 0 to π. Block out this interval, divide it into four equal parts, locate high and low points, and locate x intercepts. Then complete the graph.

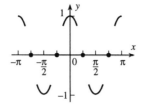 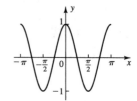

11. $y = \sin 2\pi x$. Amplitude $= |A| = |1| = 1$. Period $= \dfrac{2\pi}{2\pi} = 1$.

One full cycle of this graph is completed as x goes from 0 to 1. Block out this interval, divide it

into four equal parts, locate high and low points, and locate x intercepts. Then complete the graph.

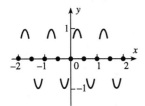   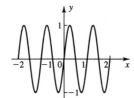

13. $y = \cos \dfrac{x}{4}$. Amplitude $= |A| = |1| = 1$. Period $= \dfrac{2\pi}{1/4} = 8\pi$.

One full cycle of this graph is completed as x goes from 0 to 8π. Block out this interval, divide it
into four equal parts, locate high and low points, and locate x intercepts. Then complete the graph.

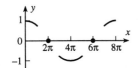   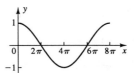

15. $y = 2 \sin 4x$. Amplitude $= |2| = 2$. Period $= \dfrac{2\pi}{4} = \dfrac{\pi}{2}$.

One full cycle of this graph is completed as x goes from 0 to $\dfrac{\pi}{2}$. Block out this interval, divide it

into four equal parts, locate high and low points, and locate x intercepts. Then complete the graph.

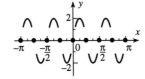   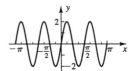

17. $y = \dfrac{1}{3} \cos 2\pi x$. Amplitude $= \left|\dfrac{1}{3}\right| = \dfrac{1}{3}$. Period $= \dfrac{2\pi}{2\pi} = 1$.

One full cycle of this graph is completed as x goes from 0 to 1. Block out this interval, divide it
into four equal parts, locate high and low points, and locate x intercepts. Then complete the graph.

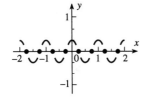   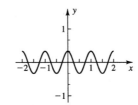

Chapter 3   Graphing Trigonometric Functions

19. $y = -\frac{1}{4} \sin \frac{x}{2}$. Amplitude $= \left|-\frac{1}{4}\right| = \frac{1}{4}$. Period $= \frac{2\pi}{1/2} = 4\pi$.

Since $A = -\frac{1}{4}$ is negative, the basic curve for $y = \sin x$ is turned upside down. One full cycle of the graph is completed as x goes from 0 to 4π. Block out this interval, divide it into four equal parts, locate high and low points, and locate x intercepts. Then complete the graph.

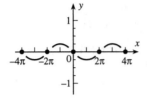   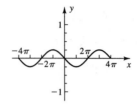

21. $y = -1 + \frac{1}{3} \cos 2\pi x$.

Amplitude $= \left|\frac{1}{3}\right| = \frac{1}{3}$. Period $= \frac{2\pi}{2\pi} = 1$.

$y = \frac{1}{3} \cos 2\pi x$ was graphed in Problem 17.

This graph is the graph of $y = \frac{1}{3} \cos 2\pi x$ moved down $|k| = |-1| = 1$ unit. We start by drawing a

horizontal broken line 1 unit below the x axis, then graph $y = \frac{1}{3} \cos 2\pi x$ relative to the broken line

and the original y axis.

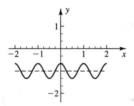

23. $y = 2 - \frac{1}{4} \sin \frac{x}{2}$.

Amplitude $= \left|-\frac{1}{4}\right| = \frac{1}{4}$. Period $= \frac{2\pi}{1/2} = 4\pi$.

$y = -\frac{1}{4} \sin \frac{x}{2}$ was graphed in Problem 19.

This graph is the graph of $y = -\frac{1}{4} \sin \frac{x}{2}$ moved up $k = 2$ units. We start by drawing a horizontal

broken line 2 units above the x axis, then graph $y = -\frac{1}{4} \sin \frac{x}{2}$ relative to the broken line and the

original y axis.

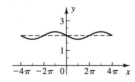

25. Amplitude $= 5 = |A|$. Period $= \dfrac{2\pi}{B} = \pi$. Thus, $B = 2$. The form of the graph is that of the basic sine curve. Thus, $y = |A| \sin Bx = 5 \sin 2x$.

27. Amplitude $= 4 = |A|$. Period $= \dfrac{2\pi}{B} = 4$. Thus, $B = \dfrac{2\pi}{4} = \dfrac{\pi}{2}$. The form of the graph is that of the basic sine curve turned upside down. Thus, $y = -|A| \sin Bx = -4 \sin\left(\dfrac{\pi x}{2}\right)$.

29. Amplitude $= 8 = |A|$. Period $= \dfrac{2\pi}{B} = 8\pi$. Thus, $B = \dfrac{2\pi}{8\pi} = \dfrac{1}{4}$. The form of the graph is that of the basic cosine curve. Thus, $y = |A| \cos Bx = 8 \cos\left(\dfrac{1}{4} x\right)$.

31. Amplitude $= 1 = |A|$. Period $= \dfrac{2\pi}{B} = 6$. Thus, $B = \dfrac{2\pi}{6} = \dfrac{\pi}{3}$. The form of the graph is that of the basic cosine curve turned upside down. Thus, $y = -|A| \cos Bx = -\cos\left(\dfrac{\pi x}{3}\right)$.

33. The graph of $y = \sin x \cos x$ is shown in the figure.
    Amplitude $= 0.5 = |A|$. Period $= \dfrac{2\pi}{B} = \pi$. Thus,
    $B = \dfrac{2\pi}{\pi} = 2$. The form of the graph is that of the
    basic sine curve. Thus,

    $y = |A| \sin Bx = 0.5 \sin 2x$

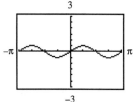

35. The graph of $y = 2 \cos^2 x$ is shown in the figure.
    Amplitude $= \dfrac{1}{2}$ (y coordinate of highest point
    $\qquad$ $-$ y coordinate of lowest point)
    $\qquad = \dfrac{1}{2}(2 - 0) = 1 = |A|$
    Period $= \dfrac{2\pi}{B} = \pi$. Thus, $B = \dfrac{2\pi}{\pi} = 2$.
    The form of the graph is that of the basic cosine
    curve shifted up 1 unit. Thus,

    $y = 1 + |A| \cos Bx = 1 + \cos 2x$

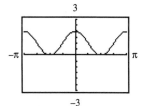

37. The graph of $y = 2 - 4 \sin^2 2x$ is shown in the figure.
    Amplitude $= 2 = |A|$. Period $= \dfrac{2\pi}{B} = \dfrac{\pi}{2}$. Thus,
    $B = 2\pi \div \dfrac{\pi}{2} = 4$. The form of the graph is that of the
    basic cosine curve. Thus,

    $y = |A| \cos Bx = 2 \cos 4x$

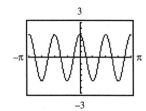

Chapter 3   Graphing Trigonometric Functions

39.(A)

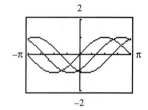

(B) We note that the graph of $y = \sin(x - 1)$ is the same
as the graph of $y = \sin x$ shifted 1 unit to right, and the graph of $y = \sin(x - 2)$ is the same as the
graph of $y = \sin x$ shifted 2 units to the right. Thus, the graph of $y = \sin(x - a)$ is the same as the
graph of $y = \sin x$ shifted a units to the right.

41.(A)

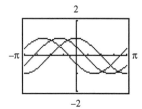

(B) We note that the graph of $y = \cos(x + 1)$ is the same as the graph of $y = \cos x$ shifted 1 unit to the
left, and the graph of $y = \cos(x + 2)$ is the same as the graph of $y = \cos x$ shifted 2 units to the left.
Thus, the graph of $y = \cos(x + a)$ is the same as the graph of $y = \cos x$ shifted a units to the left.

43.(A)

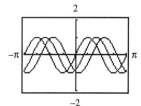

(B) We note that the graph of $y = \sin (2x - 1)$ is the same as the graph of $y = \sin 2x$ shifted $\dfrac{1}{2}$ unit to
the right, and the graph of $y = \sin(2x - 2)$ is the same as the graph of $y = \sin 2x$ shifted 1 unit to the
right. Thus, the graph of $y = \sin(2x - a)$ is the same as the graph of $y = \sin 2x$ shifted $\dfrac{a}{2}$ units to the
right.

45. $E = 110 \sin 120\pi t$. Amplitude $= |110| = 110$.
Period $= \dfrac{2\pi}{120\pi} = \dfrac{1}{60}$ sec.
Frequency $f = \dfrac{1}{p} = \dfrac{1}{1/60} = 60$ Hz

Chapter 3   Graphing Trigonometric Functions

One full cycle of the graph is completed as t goes from 0 to $\frac{1}{60}$. Block out this interval, divide it into four equal parts, locate high and low points, and locate t intercepts. Then complete the graph.

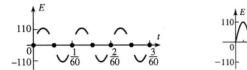 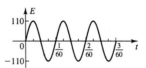

47. *Find A*: The amplitude $|A|$ is given to be 12. Since $E = 12$ when $t = 0$, $A = 12$ (and not -12).

   Find B: We are given that the frequency, f, is 40 Hz. Hence, the period is found using the reciprocal formula: $P = \frac{1}{f} = \frac{1}{40}$ sec. But, $P = \frac{2\pi}{B}$. Thus, $B = \frac{2\pi}{P} = \frac{2\pi}{1/40} = 80\pi$.

   Write the equation: $E = 12 \cos 80\pi t$

49. (A) We use the formula: $\text{Period} = \frac{2\pi}{\sqrt{1,000 \ gA/M}}$ with $g = 9.75$ m/sec^2,

   $A = 3\text{ m} \times 3\text{ m} = 9\text{m}^2$, and Period = 1 sec, and solve for M. Thus, $1 = \frac{2\pi}{\sqrt{(1,000)(9.75)(9)/M}}$;

   $M = \frac{(1,000)(9.75)(9)}{4\pi^2} \approx 2,220$ kg

   (B) $D = 0.2$, $B = \frac{2\pi}{\text{Period}}$, $y = 0.2 \sin 2\pi t$

   (C) One full cycle of this graph is completed as t goes from 0 to 1. Block out this interval, divide it into four equal parts, locate high and low points, and locate t intercepts. The complete this graph.

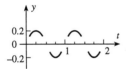 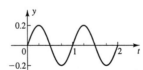

51. $V(t) = 0.45 - 0.35 \cos \frac{\pi t}{2}$. Amplitude = $|-0.35| = 0.35$. Period = $\frac{2\pi}{\pi/2} = 4$.

This graph is the graph of $y = -0.35 \cos \pi t/2$ moved up 0.45 units. We start by drawing a horizontal broken line 0.45 units above the t axis, then graph $-0.35 \cos \pi t/2$ relative to the broken line and the original y axis.

Chapter 3 Graphing Trigonometric Functions

Note: since A = −0.35 is negative, the basic curve for y = sin t is turned upside down. One full cycle of the graph is completed as t goes from 0 to 4. Block out this interval, divide it into four equal parts, locate high and low points, and locate points where the graph crosses the broken line. Then complete the graph.

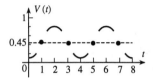

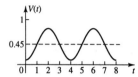

53. Since the rotation is at 4 revolutions per minute, in 1 minute it covers 4 revolutions, or 8π radians. In t minutes, it covers $8\pi t$ radians, thus $\theta = 8\pi t$. To see that $x = 20 \sin 8\pi t$, it might help to look sideways at the wheel. Below, we have indicated a new coordinate system in which θ is in standard position. Then, $\sin \theta = \dfrac{b}{20}$, so $b = 20 \sin \theta = 20 \sin 8\pi t$.

Thus, the coordinate of the shadow = b = x in the author's coordinate system. That is, $x = 20 \sin 8\pi t$.

To graph this, note:

Amplitude = $|A| = |20| = 20$

Period = $\dfrac{2\pi}{8\pi} = \dfrac{1}{4}$.

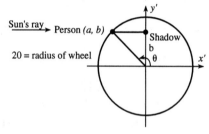

One full cycle of the graph is completed as t goes from 0 to $\dfrac{1}{4}$. Block out this interval, divide it into four equal parts, locate high and low points, and locate t intercepts.

Then complete the graph.

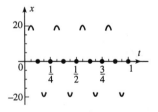

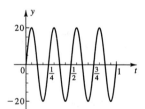

Chapter 3   Graphing Trigonometric Functions

EXERCISE 3.3 Graphing $y = k + A \sin(Bx + C)$ and $y = k + A \cos(Bx + C)$

1. Amplitude $= |A| = |1| = 1$

   Phase Shift and Period: Solve

   $Bx + C = 0$ and $Bx + C = 2\pi$

   $x + \dfrac{\pi}{2} = 0 \qquad x + \dfrac{\pi}{2} = 2\pi$

   $x = -\dfrac{\pi}{2} \qquad\quad x = -\dfrac{\pi}{2} + 2\pi$

   $\uparrow \qquad\qquad\quad \uparrow \quad\;\; \uparrow$

   Phase Shift $\qquad\quad$ Period $= 2\pi$

   Phase Shift $= -\dfrac{\pi}{2}$

   Graph one cycle over the interval from

   $-\dfrac{\pi}{2}$ to $\left(-\dfrac{\pi}{2} + 2\pi\right) = \dfrac{3\pi}{2}$

   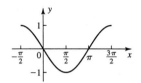

3. Amplitude $= |A| = |1| = 1$

   Phase Shift and Period: Solve

   $Bx + C = 0$ and $Bx + C = 2\pi$

   $x - \dfrac{\pi}{4} = 0 \qquad x - \dfrac{\pi}{4} = 2\pi$

   $x = \dfrac{\pi}{4} \qquad\quad x = \dfrac{\pi}{4} + 2\pi$

   $\uparrow \qquad\qquad\quad \uparrow \quad\;\; \uparrow$

   Phase Shift $\qquad$ Period $= 2\pi$

   Phase Shift $= \dfrac{\pi}{4}$. Graph one cycle over the

   interval from $\dfrac{\pi}{4}$ to $\left(\dfrac{\pi}{4} + 2\pi\right) = \dfrac{9\pi}{4}$.

   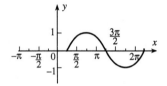

   Extend the graph from $-\pi$ to $\dfrac{\pi}{4}$ and delete the

   portion of the graph from 2π to $\dfrac{9\pi}{4}$, since this

   was not required.

   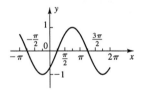

5. Amplitude $= |A| = |3| = 3$

   Phase Shift and Period: Solve

   $Bx + C = 0$ and $Bx + C = 2\pi$

   $x - \dfrac{\pi}{2} = 0 \qquad x - \dfrac{\pi}{2} = 2\pi$

   $x = \dfrac{\pi}{2} \qquad\quad x = \dfrac{\pi}{2} + 2\pi$

   $\uparrow \qquad\qquad\quad \uparrow \quad\;\; \uparrow$

   Phase Shift $\qquad\quad$ Period $= 2\pi$

Chapter 3   Graphing Trigonometric Functions

Phase Shift $= \frac{\pi}{2}$. Graph one cycle over the

interval from $\frac{\pi}{2}$ to $\left(\frac{\pi}{2} + 2\pi \right) = \frac{5\pi}{2}$.

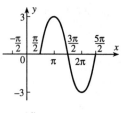

Extend the graph from $-\frac{\pi}{2}$ to $\frac{5\pi}{2}$.

7. Amplitude $= |A| = |1| = 1$
   Phase Shift and Period: Solve
   $Bx + C = 0$ and $Bx + C = 2\pi$

$$2\pi x - \pi = 0 \qquad 2\pi x - \pi = 2\pi$$
$$x = \frac{1}{2} \qquad\qquad x = \frac{1}{2} + 1$$
$$\uparrow \qquad\qquad\qquad \uparrow \quad \uparrow$$
$$\text{Phase Shift} \qquad \text{Period} = 1$$

Phase Shift $= \frac{1}{2}$. Graph one cycle over the

interval from $\frac{1}{2}$ to $\left(\frac{1}{2} + 1 \right) = \frac{3}{2}$.

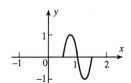

Extend the graph from -1 to 2.

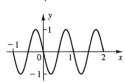

9. Amplitude $= |A| = |4| = 4$
   Phase Shift and Period: Solve
   $Bx + C = 0$ and $Bx + C = 2\pi$

$$\pi x + \frac{\pi}{4} = 0 \qquad \pi x + \frac{\pi}{4} = 2\pi$$
$$x = -\frac{1}{4} \qquad\qquad x = -\frac{1}{4} + 2$$
$$\uparrow \qquad\qquad\qquad \uparrow \quad \uparrow$$
$$\text{Phase Shift} \qquad \text{Period} = 2$$

Phase Shift $= -\frac{1}{4}$. Graph one cycle over the

interval from $-\frac{1}{4}$ to $\left(-\frac{1}{4} + 2\right) = \frac{7}{4}$.

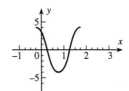

Extend the graph from –1 to 3.

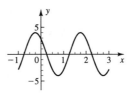

11. Amplitude $= |A| = |-2| = 2$
     Phase Shift and Period: Solve
     $Bx + C = 0$ and $Bx + C = 2\pi$
     $2x + \pi = 0$ $2x + \pi = 2\pi$

     $ x = -\frac{\pi}{2} x = -\frac{\pi}{2} + \pi$
     $ \uparrow \uparrow \uparrow$
      Phase Shift Period $= \pi$

Phase Shift $= -\frac{\pi}{2}$. Graph one cycle (thebasic

cosine curve turned upside down) over the

interval from $-\frac{\pi}{2}$ to $\left(-\frac{\pi}{2} + \pi\right) = \frac{\pi}{2}$

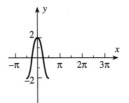

Extend the graph from $-\pi$ to 3π.

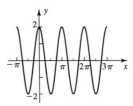

13. $y = -2 + 4\cos\left(\pi x + \frac{\pi}{4}\right)$.

$y = 4\cos\left(\pi x + \frac{\pi}{4}\right)$ was graphed in Problem 9.

This graph is the graph of $y = 4\cos\left(\pi x + \frac{\pi}{4}\right)$

moved down $|k| = |-2| = 2$ units. We start by drawing

a horizontal broken line 2 units below the x axis, then

graph $y = 4\cos\left(\pi x + \frac{\pi}{4}\right)$ relative to the broken line

and the original y axis.

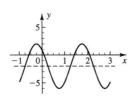

75

15. $y = 3 - 2\cos(2x + \pi)$.

$y = -2\cos(2x + \pi)$ was graphed in Problem 11.

This graph is the graph of $y = -2\cos(2x + \pi)$

moved up $k = 3$ units. We start by drawing a

horizontal broken line 3 units above the x axis,

then graph $y = -2\cos(2x + \pi)$ relative to the
broken line and the original y axis.

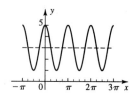

17.   Amplitude $= |A| = |2| = 2$
Phase Shift and Period: Solve
$Bx + C = 0$ and $Bx + C = 2\pi$

$$3x - \frac{\pi}{2} = 0 \qquad 3x - \frac{\pi}{2} = 2\pi$$

$$x = \frac{\pi}{6} \qquad\qquad x = \frac{\pi}{6} + \frac{2\pi}{3}$$

$$\uparrow \qquad\qquad\qquad \uparrow \quad\; \uparrow$$

$$\text{Phase Shift} \qquad \text{Period} = \frac{2\pi}{3}$$

Phase Shift $= \frac{\pi}{6}$. Graph one cycle over the

interval from $\frac{\pi}{6}$ to $\left(\frac{\pi}{6} + \frac{2\pi}{3}\right) = \frac{5\pi}{6}$.

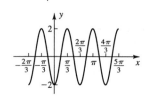

Extend the graph from $-\frac{2\pi}{3}$ to $\frac{5\pi}{3}$.

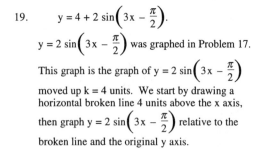

19. $y = 4 + 2\sin\left(3x - \frac{\pi}{2}\right)$.

$y = 2\sin\left(3x - \frac{\pi}{2}\right)$ was graphed in Problem 17.

This graph is the graph of $y = 2\sin\left(3x - \frac{\pi}{2}\right)$

moved up $k = 4$ units. We start by drawing a
horizontal broken line 4 units above the x axis,

then graph $y = 2\sin\left(3x - \frac{\pi}{2}\right)$ relative to the

broken line and the original y axis.

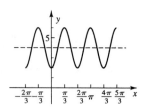

21. We sketch one period of the graph of y = sin x below. It has amplitude 1, period 2π, and phase

    shift 0. To graph $y = \cos\left(x - \dfrac{\pi}{2}\right)$ we compute its amplitude, period, and phase shift:

    Amplitude = |A| = |1| = 1

    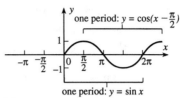

    Phase Shift and Period: Solve

    $$Bx + C = 0 \quad \text{and} \quad Bx + C = 2\pi$$

    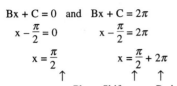

    $$x - \frac{\pi}{2} = 0 \qquad\qquad x - \frac{\pi}{2} = 2\pi$$

    $$x = \frac{\pi}{2} \qquad\qquad x = \frac{\pi}{2} + 2\pi$$

    $\uparrow$ $\uparrow$ $\uparrow$

         Phase Shift   Period

    If we sketch one period of the graph starting at $x = \dfrac{\pi}{2}$ (the phase shift) and ending at $x = \dfrac{\pi}{2} + 2\pi = \dfrac{5\pi}{2}$

    (the phase shift plus one period), we see that it is the same as that of y = sin x over the interval from

    $x = \dfrac{\pi}{2}$ to $x = 2\pi$. Since both curves can be extended indefinitely in both directions, we conclude that

    the graphs are the same, thus $\cos\left(x - \dfrac{\pi}{2}\right) = \sin x$ for all x.

23. Since the maximum deviation from the x axis is 5, we can write:

    Amplitude = |A| = 5. Thus, A = 5 or –5. Since the period is 3 – (–1) = 4, we can write:

    Period $= \dfrac{2\pi}{B} = 4$. Thus, $B = \dfrac{2\pi}{4} = \dfrac{\pi}{2}$. Since we are instructed to choose the phase shift between 0 and

    2, we can regard this graph as containing the basic sine curve with a phase shift of 1. This requires

    us to choose A positive, since the graph shows that as x increases from 1 to 2, y *increases* like the

    basic sine curve (not the upside down sine curve). So, A = 5. Then, $-\dfrac{C}{B} = 1$. Thus,

    $$C = -B = -\frac{\pi}{2} \quad \text{and} \quad y = A\sin(Bx + C) = 5\sin\left(\frac{\pi}{2}x - \frac{\pi}{2}\right).$$

    Check: When x = 0, $y = 5\sin\left(\dfrac{\pi}{2} \cdot 0 - \dfrac{\pi}{2}\right) = 5\sin\left(-\dfrac{\pi}{2}\right) = -5$

            When x = 1, $y = 5\sin\left(\dfrac{\pi}{2} \cdot 1 - \dfrac{\pi}{2}\right) = 5\sin 0 = 0$

25. Since the maximum deviation from the x axis is 2, we can write:

    Amplitude = |A| = 2. Thus, A = 2 or –2. Since the period is $\dfrac{5\pi}{2} - \left(-\dfrac{3\pi}{2}\right) = 4\pi$, we can write:

    Period $= \dfrac{2\pi}{B} = 4\pi$. Thus, $B = \dfrac{2\pi}{4\pi} = \dfrac{1}{2}$. Since we are instructed to choose $-2\pi < \dfrac{C}{B} < 0$, thus,

    $0 < -\dfrac{C}{B} < 2\pi$, that is, the phase shift between 0 and 2π, we regard this graph as containing the basic

    cosine curve, with a phase shift of $\dfrac{3\pi}{2}$. This requires us to choose A positive, since the graph shows

    that as x increases from $3\pi/2$ to 2π, y *decreases* like the basic cosine curve (not the upside down

    cosine curve). So, A = 2. Then, $-\dfrac{C}{B} = \dfrac{3\pi}{2}$. Thus,

    $$C = \frac{3\pi}{2}B = -\frac{3\pi}{2} \cdot \frac{1}{2} = -\frac{3\pi}{4} \quad \text{and} \quad y = A\cos(Bx + C) = 2\cos\left(\frac{1}{2}x - \frac{3\pi}{4}\right).$$

    Check: When x = 0, $y = 2\cos\left(\dfrac{1}{2} \cdot 0 - \dfrac{3\pi}{4}\right) = 2\cos\left(-\dfrac{3\pi}{4}\right) = -\sqrt{2}$

            When $x = \dfrac{\pi}{2}$, $y = 2\cos\left(\dfrac{1}{2} \cdot \dfrac{\pi}{2} - \dfrac{3\pi}{4}\right) = 2\cos\left(-\dfrac{\pi}{2}\right) = 0$

Chapter 3   Graphing Trigonometric Functions

27. This graph of $y = \sin x + \sqrt{3} \cos x$ is shown in the figure.

This graph appears to be a sine wave with amplitude 2 and

period 2π that has been shifted to the left. Thus, we conclude

that $A = 2$ and $B = \dfrac{2\pi}{2\pi} = 1$. To determine C, we use the zoom

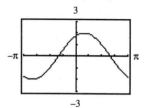

feature or the built-in approximation routine to locate the x

intercept closest to the origin at $x = -1.047$. This is the phase-

shift for the graph.

Substitute $B = 1$ and $x = -1.047$ into the phase-shift equation

$$x = -\frac{C}{B}$$

$$-1.047 = -\frac{C}{1}$$

$$C = 1.047$$

Thus, the equation required is $y = 2 \sin(x + 1.047)$.

29. The graph of $y = \sqrt{2} \sin x - \sqrt{2} \cos x$ is shown in the figure.
This graph appears to be a sine wave with amplitude 2 and
period 2π that has been shifted to the right. Thus, we conclude

that $A = 2$ and $B = \dfrac{2\pi}{2\pi} = 1$.   To determine C, we use the zoom

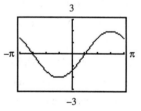

feature or the built-in approximation routine to locate the x
intercept closest to the origin at $x = 0.785$. This is the phase-
shift for the graph.

Substitute $B = 1$ and $x = 0.785$ into the phase-shift equation

$$x = -\frac{C}{B}$$

$$0.785 = -\frac{C}{1}$$

$$C = -0.785$$

Thus, the equation required is $y = 2 \sin(x - 0.785)$.

31. The graph of $y = 1.4 \sin 2x + 4.8 \cos 2x$ is shown in the
figure. This graph appears to be a sine wave with amplitude 5
and period π that has been shifted to the left. Thus, we

conclude that $A = 5$ and $B = \dfrac{2\pi}{\pi} = 2$. To determine C, we use

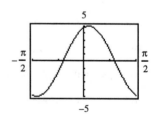

the zoom feature or the built-in approximation routine to
locate the x intercept closest to the origin at $x = -0.644$. This
is the phase-shift for the graph.

Substitute $B = 2$ and $x = -0.644$ into the phase-shift equation

$$x = -\frac{C}{B}$$

$$-0.644 = -\frac{C}{2}$$

$$C = 1.288$$

Thus, the equation required is $y = 5 \sin(x + 1.288)$.

33. The graph of $y = 2 \sin \frac{x}{2} - \sqrt{5} \cos \frac{x}{2}$ is shown in the figure.

This graph appears to be a sine wave with amplitude 3 and period 4π that has been shifted to the right. Thus, we conclude that $A = 3$ and $B = \frac{2\pi}{4\pi} = \frac{1}{2}$. To determine C, we use the zoom feature or the built-in approximation routine to locate the x intercept closest to the origin at $x = 1.682$. This is the phase-shift for the graph.

Substitute $B = \frac{1}{2}$ and $x = 1.682$ into the phase-shift equation

$$x = -\frac{C}{B}$$
$$1.682 = -\frac{C}{1/2}$$
$$C = -\frac{1}{2}(1.682) = -0.841$$

Thus, the equation required is $y = 3 \sin\left(\frac{x}{2} - 0.841\right)$.

35. Amplitude $= |A| = |5| = 5$
Phase Shift and Period: Solve
   $Bx + C = 0$ and $Bx + C = 2\pi$
$\frac{\pi}{6}(t + 3) = 0 \qquad \frac{\pi}{6}(t + 3) = 2\pi$

$\qquad t = -3 \qquad\qquad t = -3 + 12$
$\qquad\quad \uparrow \qquad\qquad\qquad \uparrow \qquad \uparrow$
$\qquad$ Phase Shift $\qquad\qquad$ Period

Phase Shift $= -3$. Period $= 12$ sec

Graph one cycle over the interval from -3 to $(-3 + 12) = 9$.

Extend the graph from 9 to 39, and delete the portion of the graph from -3 to 0, since this was not required.

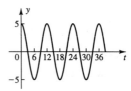

37. Amplitude $= |A| = |30| = 30$
Phase Shift and Period: Solve
   $Bx + C = 0$ and $Bx + C = 2\pi$
$120\pi t - \pi = 0 \qquad 120\pi t - \pi = 2\pi$

$\qquad t = \frac{1}{120} \qquad\qquad t = \frac{1}{120} + \frac{1}{60}$
$\qquad\quad \uparrow \qquad\qquad\qquad\quad \uparrow \qquad \uparrow$
$\qquad$ Phase Shift $\qquad\qquad$ Period

Phase Shift $= \dfrac{1}{120}$. Period $= \dfrac{1}{60}$.

Frequency $= \dfrac{1}{\text{Period}} = \dfrac{1}{1/60} = 60$ Hz

Graph one cycle over the interval

from $\dfrac{1}{120}$ to $\dfrac{1}{120}$ to $\left(\dfrac{1}{120} + \dfrac{1}{60} \right) = \dfrac{3}{120}$.

Extend the graph from 0 to $\dfrac{3}{60}$.

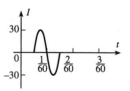

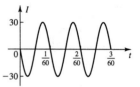

39. From the previous version of this problem (Exercise 3.2, Problem 53), we have A = 20, B = 8π.

Since $\theta = \dfrac{\pi}{2}$ when t = 0, we have Bt + C $= \dfrac{\pi}{2}$ when t = 0, so C $= \dfrac{\pi}{2}$. Therefore, the new equation of

motion is x $= 20\left(8\pi t + \dfrac{\pi}{2} \right)$.

As before, we have: Amplitude = 20; Period $= \dfrac{1}{4}$.

To find the phase shift, we solve Bt + C = 0:

$$8\pi t + \dfrac{\pi}{2} = 0$$

$$t = -\dfrac{1}{16}$$

Phase Shift $= -\dfrac{1}{16}$

Graph one cycle over the interval from $-\dfrac{1}{16}$ to $\left(-\dfrac{1}{16} + \dfrac{1}{4} \right) =$

$\dfrac{3}{16}$

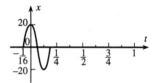

Extend the graph from $\dfrac{3}{16}$ to 1, and delete the

portion of the graph from $-\dfrac{1}{16}$ to 0, since this

was not required.

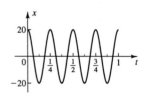

41. (A)

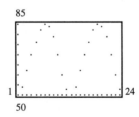

(B) From the table, Max y = 85 and Min y =50. Then,

$$A = \frac{(\text{Max y} - \text{Min y})}{2}$$

$$= \frac{(85 - 50)}{2} = 17.5$$

$$B = \frac{2\pi}{\text{Period}} = \frac{2\pi}{12} = \frac{\pi}{6}$$

$$k = \text{Min y} + A = 50 + 17.5 = 67.5$$

From the plot in (A) or the table in the text, we estimate the smallest positive value of x for which y = k = 67.5 to be approximately 3.9. Then this is the phase-shift for the graph.

Substitute $B = \frac{\pi}{6}$ and x = 3.9 into the phase-shift equation

$$x = -\frac{C}{B}$$

$$3.9 = -\frac{C}{\pi/6}$$

$$C = -\frac{3.9\pi}{6} \approx -2.0.$$

Thus, the equation required is $y = 67.5 + 17.5 \sin\left(\frac{\pi}{6} x - 2.0\right)$

(C)

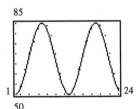

EXERCISE 3.4 Additional Applications

1. Amplitude = |A| = |10| = 10. Phase Shift: Solve Bx + C = 0. $120\pi t - 60\pi = 0$. $t = \frac{1}{2}$.

    Phase Shift = $\frac{1}{2}$. Frequency = $\frac{B}{2\pi} = \frac{120\pi}{2\pi} = 60$ Hz

3. *Find A*: The amplitude |A| is given to be 20. Since A > 0, A = 20.

    Find B: We are given that the frequency f = 30 Hz. But $f = \frac{B}{2\pi}$. Thus, $\frac{B}{2\pi} = 30$, B = 60π

    Write the equation: I = 20 cos 60 πt

5. The height of the wave from trough to crest is the difference in height between the crest (height A)

    and the trough (height –A). In this case, A = 15 ft. A – (–A) = 2A = 2(15 ft) = 30 ft. To find the

    wavelength λ, we note: $\lambda = 5.12T^2$, $T = \frac{2\pi}{B}$, $B = \frac{\pi}{8}$. Thus,

    $T = \frac{2\pi}{\pi/8} = 16$ sec, $\lambda = 5.12(16)^2 \approx 1311$ ft. To find the speed S, we use

    $$S = \sqrt{\frac{g\lambda}{2\pi}} \qquad g = 32 \text{ ft/sec}^2$$

    $$= \sqrt{\frac{32(1311)}{2\pi}} \approx 82 \text{ ft/sec}$$

Chapter 3   Graphing Trigonometric Functions

7. To graph $y = 15 \sin \frac{\pi}{8} t$, we note: Amplitude $= |A| = 15$ ft. Period $= \frac{2\pi}{B} = 16$ sec

One full cycle of the graph is completed as t goes from 0 to 16. Block out this interval, divide it

into four equal parts, locate high and low points, and locate t intercepts. Then complete the graph.

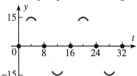   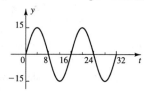

9. (A) *Find A*: The amplitude $|A|$ is given to be 2. Although, on the basis of the given information,

A could be either 2 or –2, it is natural to choose A = 2.

Find B: Since the variable in this problem is r, the distance from the source, the length of one

cycle = wavelength $= \lambda = \frac{2\pi}{B}$. Thus, $B = \frac{2\pi}{\lambda} = \frac{2\pi}{150} = \frac{\pi}{75}$

Write the equation: $y = 2 \sin \frac{\pi}{75} r$

(B) To find the period T, we use: $\lambda = 5.12 T^2$ (λ in feet), $T = \sqrt{\dfrac{\lambda}{5.12}}$

Substituting $\lambda = 150$ miles $= (150)(5280)$ feet, we have $T = \sqrt{\dfrac{(150)(5280)}{5.12}} \approx 393$ sec

11. $y = 25 \sin 2\pi \left(\dfrac{t}{10} - \dfrac{r}{512} \right)$ $r = 1024,\ 0 \le t \le 20$

$= 25 \sin 2\pi \left(\dfrac{t}{10} - \dfrac{1024}{512} \right) = 25 \sin 2\pi \left(\dfrac{\pi t}{5} - 4\pi \right)$

We compute amplitude, period, and phase shift as follows:

Amplitude $= |A| = |25| = 25$

Phase Shift and Period: Solve

	$Bx + C = 0$	and	$Bx + C = 2\pi$
	$\dfrac{\pi t}{5} - 4\pi = 0$		$\dfrac{\pi t}{5} - 4\pi = 2\pi$
	$t = 20$		$t = 20 + 10$
	↑		↑ ↑
			Phase Shift Period

Graph one cycle over the interval from 10 to 20.   Extend the graph from 0 to 10.

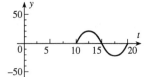   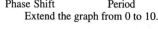 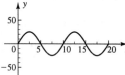

13. Period $= \dfrac{1}{v} = \dfrac{1}{10^8} = 10^{-8}$ sec. To find the wave length λ, we use the formula

$\lambda v = c$ with $c \approx 3 \times 10^8$ m/sec, $\lambda = \dfrac{c}{v} = \dfrac{3 \times 10^8 \text{ m/sec}}{10^8 \text{ Hz}} = 3$ m

Chapter 3   Graphing Trigonometric Functions

15. We first use $\lambda v = c$ to find the frequency and then use $v = \dfrac{B}{2\pi}$ to find B:

$$v = \frac{c}{\lambda} = \frac{3 \times 10^8 \text{ m/sec}}{3 \times 10^{-10}} = 10^{18} \text{ Hz}; \qquad B = 2\pi v = 2\pi \times 10^{18}$$

17. If $y = A \sin 2\pi \times 10^5 t$, then $B = 2\pi \times 10^5$. Since $v = \dfrac{B}{2\pi}$, we have $v = \dfrac{2\pi \times 10^5}{2\pi} = 10^5$ Hz

Since Period $= \dfrac{2\pi}{B}$, we have Period $= \dfrac{2\pi}{2\pi \times 10^5} = 10^{-5}$ sec

Figure 4 (text) shows atmospheric adsorption, for waves of frequency 10^5 Hz, as total. No, such waves cannot pass through the atmosphere.

19. Amplitude = 1, Period = 1. One full cycle of this curve is completed as t goes from 0 to 1. Block out this interval, divide it into four equal parts, locate high and low points and locate t intercepts. Then complete the graph.

    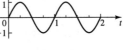

21. Step 1: Graph $y = t$ and its reflection.   Step 2: Sketch the graph of $y = \sin 2\pi t$, but expand high and low points to the height of the envelope.

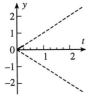   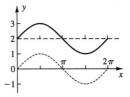

EXERCISE 3.5 Addition of Ordinates

1. We form $y_1 = 2$ and $y_2 = \sin x$. We sketch the graph of each equation in the same coordinate system (dashed lines), then add the ordinates $y_1 + y_2$ (solid curve).

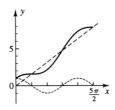

3. We form $y_1 = x$ and $y_2 = \cos x$. We sketch the graph of each equation in the same coordinate system (dashed lines), then add the ordinates $y_1 + y_2$ (solid curve).

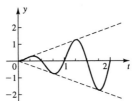

Chapter 3   Graphing Trigonometric Functions

5. We form $y_1 = \dfrac{x}{2}$ and $y_2 = \cos \pi x$. We sketch the graph of each equation in the same coordinate system (dashed lines), then add the ordinates $y_1 + y_2$ (solid curve).

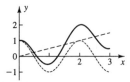

7. We form $y_1 = 3 \cos x$ and $y_2 = \sin 2x$. We sketch the graph of each equation in the same coordinate system (dashed curves), then add the ordinates $y_1 + y_2$ (solid curve).

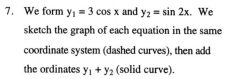
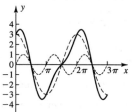

9. We form $y_1 = \sin x$ and $y_2 = 2 \cos 2x$. We sketch the graph of each equation in the same coordinate system (dashed curves), then add the ordinates $y_1 + y_2$ (solid curve).

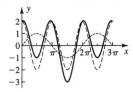

11. We form $y_1 = \sin x$ and $y_2 = \dfrac{1}{3} \sin 3x$. We sketch the graph of each equation in the same coordinate system (dashed curves), then add the ordinates $y_1 + y_2$ (solid curve).

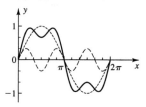

13. We form $y_1 = 0.06 \sin 400\pi t$. $\left(\text{Amplitude} = 0.06, \text{Period} = \dfrac{1}{200} \right)$, and $y_2 = 0.03 \sin 800\pi t$. $\left(\text{Amplitude} = 0.03, \text{Period} = \dfrac{1}{400} \right)$. We sketch the graph of each equation in the same coordinate system (dashed curves), then add the ordinates $y_1 + y_2$ (solid curve).

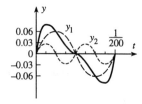

15.

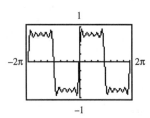

Chapter 3   Graphing Trigonometric Functions

17.

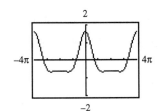

19.

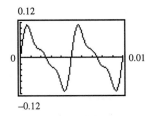

21. *Determine k*: The dashed line indicates that high and low points of the curve are equal distances from the line V = 0.45. Hence, k = 0.45.

Determine A: The maximum deviation from the line V = 0.45 is seen at points such as t = 2, V = 0.8 or t = 4, V = 0.1. Thus, |A| = 0.8 – 0.45 or |A| = 0.45 – 0.1. From either statement, we see, |A| = 0.35. So, A = 0.35 or A = –0.35. Since the portion of the curve as t increases from 0 to 1 has the form of an upside down basic cosine curve (V increasing), A must be negative. A = –0.35.

Determine B: One full cycle of the curve is completed as t varies from 0 to 4 seconds. Hence, the period P = 4 sec. Since $P = \dfrac{2\pi}{B}$, we have $4 = \dfrac{2\pi}{B}$, or $B = \dfrac{2\pi}{4} = \dfrac{\pi}{2}$.

$V = k + A \cos Bt = 0.45 - 0.35 \cos \dfrac{\pi}{2} t.$

23. (A) We form $y_1 = 1200$ and $y_2 = 300 \sin \dfrac{\pi}{4} t$. We sketch the graph of each equation in the same coordinate system (dashed lines), then add the ordinates $y_1 + y_2$ (solid curve).

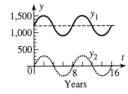

(B) The maximum expected population occurs

at t = 2 and t = 10 (years 1982 and 1990).

1500 deer

(C) The minimum expected population occurs at t = 6 and t = 14 (years 1986 and 1994): 900 deer.

25. (A) We form $S_1 = 5 + \dfrac{t}{52}$ and $S_2 = -4 \cos \dfrac{\pi t}{26}$. We sketch the graph of each equation in the same coordinate system (dashed lines), then add the ordinates $S_1 + S_2$ (solid curve).

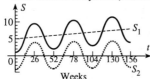

(B) In the 26th week of the 3rd year,

$t = 2 \cdot 52 + 26 = 130$. The sales are given by
$$S = 5 + \frac{130}{52} - 4 \cos \frac{\pi \cdot 130}{26}$$
$$= 5 + 2.5 - 4 \cos 5\pi = 7.5 - 4(-1) = \$11.5 \text{ million}$$

(C) In the 52nd week of the 3rd year, $t = 3 \cdot 52 = 156$. The sales are given by
$$S = 5 + \frac{156}{52} - 4 \cos \frac{\pi \cdot 156}{26} = 5 + 3 - 4 \cos 6\pi = 8 - 4(1) = \$4 \text{ million}$$

27. (A)

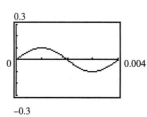

(B)

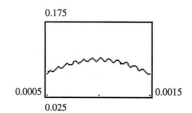

EXERCISE 3.6 Tangent, Cotangent, Secant, and Cosecant Functions Revisited

1.

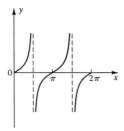

3. The dashed line shows $y = \sin x$ in this interval. The solid line is $y = \csc x$.

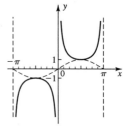

5. Period $= \dfrac{\pi}{B} = \dfrac{\pi}{2}$

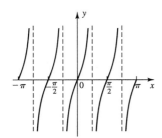

7. Period $= \dfrac{\pi}{B} = \dfrac{\pi}{2\pi} = \dfrac{1}{2}$ The graph of the basic cotangent curve is reflected relative to the x axis.

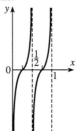

9. Period $= \dfrac{2\pi}{\pi} = 2$

The dashed line shows $y = \cos \pi x$ in this interval. The solid line is $y = \sec \pi x$.

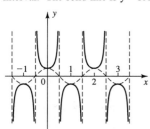

11. Period $= \dfrac{\pi}{1/2} = 2\pi$

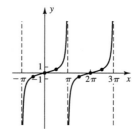

13. Period $= \dfrac{2\pi}{1/2} = 4\pi$

The dashed line shows $y = \dfrac{1}{2}\sin\left(\dfrac{x}{2}\right)$ in this

interval. The solid line is $y = 2\csc\left(\dfrac{x}{2}\right)$.

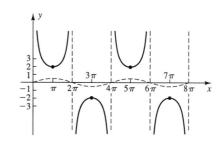

15. We find the period and phase shift by solving $\ x - \dfrac{\pi}{2} = 0\ $ and $\ x - \dfrac{\pi}{2} = \pi$

$$x = \dfrac{\pi}{2} \qquad\qquad x = \dfrac{\pi}{2} + \pi$$

$$\text{Period} = \pi \quad \text{Phase Shift} = \dfrac{\pi}{2}$$

We then sketch one period of the graph starting at $x = \dfrac{\pi}{2}$ (the phase shift) and ending at

$x = \dfrac{\pi}{2} + \pi = \dfrac{3\pi}{2}$ (the phase shift plus one period). Note that a vertical asymptote is at $x = \pi$. We

then extend the graph from $-\pi$ to $\dfrac{\pi}{2}$ and delete the portion of the graph from π to $\dfrac{3\pi}{2}$, since this was

not required.

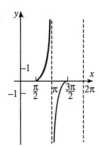

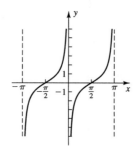

17. We find the period and phase shift by solving

$$2x - \pi = 0 \quad \text{and} \quad 2x - \pi = \pi$$

$$2x = \pi \qquad\qquad 2x = \pi + \pi$$

$$x = \frac{\pi}{2} \qquad\qquad x = \frac{\pi}{2} + \frac{\pi}{2}$$

Period $= \frac{\pi}{2}$, Phase Shift $= \frac{\pi}{2}$

We then sketch one period of the graph

starting at $x = \frac{\pi}{2}$ (the phase shift) and

ending at $x = \frac{\pi}{2} + \frac{\pi}{2} = \pi$ (the phase shift plus one period). Note that vertical asymptotes are at $x = \frac{\pi}{2}$

and $x = \pi$. We then extend the graph from $-\frac{\pi}{2}$ to $\frac{\pi}{2}$ and delete the portion of the graph from $\frac{\pi}{2}$ to π,

since that was not required.

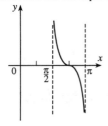

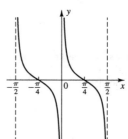

19. We first find the period and phase shift by solving $\pi x - \frac{\pi}{2} = 0$ and $\pi x - \frac{\pi}{2} = 2\pi$

$$x = \frac{1}{2} \qquad\qquad x = \frac{1}{2} + 2$$

$$\text{Period} = 2 \qquad \text{Phase Shift} = \frac{1}{2}$$

Now, since $\csc\left(\pi x - \frac{\pi}{2}\right) = \dfrac{1}{\sin\left(\pi x - \frac{\pi}{2}\right)}$, we graph $y = \sin\left(\pi x - \frac{\pi}{2}\right)$ for one cycle from

$\frac{1}{2}$ to $\frac{1}{2} + 2 = \frac{5}{2}$ with a broken line graph, then take reciprocals. We also place vertical asymptotes

through the x intercepts of the sine graph to guide us when we sketch the cosecant function.

We then extend the one cycle over the required interval from $-\frac{1}{2}$ to $\frac{5}{2}$.

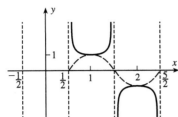
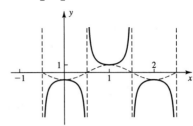

21. We find the period and phase shift by solving $2x + \pi = 0$ and $2x + \pi = \pi$

$$x = -\frac{\pi}{2} \qquad\qquad x = -\frac{\pi}{2} + \frac{\pi}{2}$$

$$\text{Period} = \frac{\pi}{2} \quad \text{Phase Shift} = -\frac{\pi}{2}$$

We then sketch one period of the graph starting at $x = -\frac{\pi}{2}$ (the phase shift) and ending at

$x = -\frac{\pi}{2} + \frac{\pi}{2} = 0$ (the phase shift plus one period). Note that a vertical asymptote is at $x = -\frac{\pi}{4}$.

We then extend the graph from $-\pi$ to π.

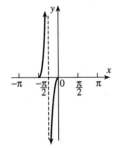
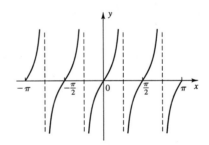

23. We find the period and phase shift by solving $\pi x - \pi = 0$ and $\pi x - \pi = \pi$

$$x = 1 \qquad\qquad x = 1 + 1$$

$$\text{Period} = 1 \quad \text{Phase Shift} = 1$$

We then sketch one period of the graph starting at $x = 1$ (the phase shift) and ending at

$x = 1 + 1 = 2$ (the phase shift plus one period). Note that the graph of the basic cotangent curve is reflected to the x axis. We then extend the graph from –2 to 2.

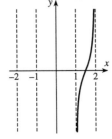
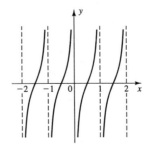

Chapter 3   Graphing Trigonometric Functions

25. We first find the period and phase shift by solving $\pi x - \dfrac{\pi}{2} = 0$ and $\pi x - \dfrac{\pi}{2} = 2\pi$

$$x = \frac{1}{2} \qquad\qquad x = \frac{1}{2} + 2$$

$$\text{Period} = 2 \quad \text{Phase Shift} = \frac{1}{2}$$

Now, since $2\sec\left(\pi x - \dfrac{\pi}{2}\right) = \dfrac{1}{\dfrac{1}{2}\cos\left(\pi x - \dfrac{\pi}{2}\right)}$, we graph $y = \dfrac{1}{2}\cos\left(\pi x - \dfrac{\pi}{2}\right)$ for one cycle from

$\dfrac{1}{2}$ to $\dfrac{1}{2} + 2 = \dfrac{5}{2}$ with a broken line graph, then take reciprocals. We also place vertical asymptotes through the x intercepts of the cosine graph to guide us when we sketch the secant function. We then extend the one cycle over the required interval from −1 to 3.

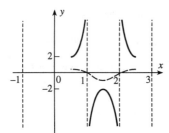 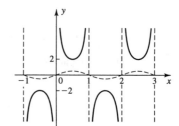

27. The graph of y = csc x − cot x is shown in the figure. This graph appears to have vertical asymptotes at x = −π and x = π, and period 2π. It appears, therefore, to be the same as the graph of y = tan Bx, with $\dfrac{\pi}{B} = 2\pi$, that is $B = \dfrac{1}{2}$.

The required equation is $y = \tan\dfrac{1}{2}x$.

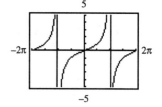

29. This graph appears to have vertical asymptotes at x = −π, $x = -\dfrac{\pi}{2}$, x = 0, $x = \dfrac{\pi}{2}$, and x = π, and period π. Its high and low points appear to have y coordinates of −2 and 2, respectively. It appears, therefore, to be the same as the graph of y = 2 csc Bx, $\dfrac{2\pi}{B} = \pi$, that is, B = 2.

The required equation is y = 2 csc 2x.

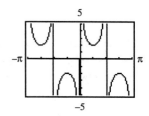

Chapter 3   Graphing Trigonometric Functions

31. The graph of $y = \cot x + \tan x$ is shown in the figure. The graph of $y = \cos 2x + \sin 2x \tan 2x$ is shown in the figure. This graph appears to have vertical asymptotes at $x = -\frac{3\pi}{4}$, $x = -\frac{\pi}{4}$, $x = \frac{\pi}{4}$, and $x = \frac{3\pi}{4}$, and period π. Its high and low points appear to have y coordinates of -1 and 1, respectively. It appears, therefore, to be the same as the graph of $y = \sec Bx$, $\frac{2\pi}{B} = \pi$, that is, $B = 2$. The required equation is $y = \sec 2x$.

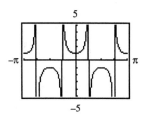

33. The graph of $y = \dfrac{\sin 6x}{1 - \cos 6x}$ is shown in the figure. This graph appears to have vertical asymptotes at $x = -\pi$, $-\frac{2\pi}{3}$, $-\frac{\pi}{3}$, 0, $\frac{\pi}{3}$, $\frac{2\pi}{3}$, and π, and period $\frac{\pi}{3}$. It appears, there-fore, to be the same as the graph of $y = \cot Bx$. $\frac{\pi}{B} = \frac{\pi}{3}$, that is, $B = 3$. The required equation is $y = \cot 3x$.

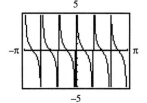

35. (A) In triangle ABC, we can write
$$\tan \theta = \frac{a}{b} = \frac{a}{15}$$
Thus, $a = 15 \tan \theta$, or $a = 15 \tan 2\pi t$.

(B) Period $= \dfrac{\pi}{2\pi} = \dfrac{1}{2}$

One period of the graph would therefore extend from 0 to $\frac{1}{2}$, with a vertical asymptote at $t = \frac{1}{4}$, or 0.25. We sketch half of one period, since the required interval is from 0 to 0.25 only.

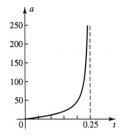

Ordinates can be determined from a calculator, thus:

t	0	0.05	0.10	0.15	0.20	0.24
15 tan 2πt	0	4.9	11	21	46	240

CHAPTER 3   REVIEW EXERCISE

1.

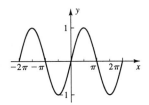

2.

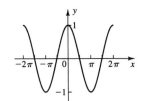

Chapter 3   Graphing Trigonometric Functions

3.

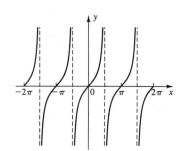

4.

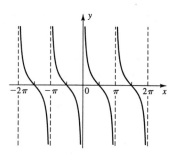

5.

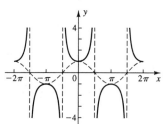

6.

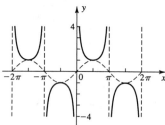

7. Amplitude $= |3| = 3$. Period $= \dfrac{2\pi}{1/2} = 4\pi$.

One full cycle of this graph is completed as x goes from 0 to 4π. Block out this interval, divide it

into four equal parts, locate high and low points, and locate x intercepts. Then complete the graph.

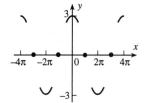

8. Amplitude $= \left|\dfrac{1}{2}\right| = \dfrac{1}{2}$. Period $= \dfrac{2\pi}{2} = \pi$.

One full cycle of this graph is completed as x goes from 0 to π. Block out this interval, divide it

into four equal parts, locate high and low points, and locate x intercepts. then complete the graph.

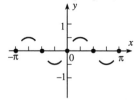

Chapter 3   Graphing Trigonometric Functions

9. We form $y_1 = 4$ and $y_2 = \cos x$.

    We sketch the graph of each equation in the same
    coordinate system (dashed lines), then add the
    ordinates $y_1 + y_2$ (solid curve).

    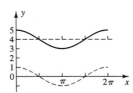

10. Amplitude $= |-2| = 2$. Period $= \dfrac{2\pi}{\pi} = 2$.

    Since $A = -2$ is negative, the basic curve for $y = \sin x$ is turned upside down. One full cycle of
    the graph is completed as x goes from 0 to 2. Block out this interval, divide it into four equal
    parts, locate high and low points, and locate x intercepts. Then complete the graph.

    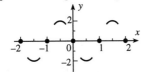   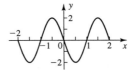

11. Amplitude $= \left|-\dfrac{1}{3}\right| = \dfrac{1}{3}$. Period $= \dfrac{2\pi}{2\pi} = 1$.

    Since $A = -\dfrac{1}{3}$ is negative, the basic curve for $y = \cos x$ is turned upside down. One full cycle of
    the graph is completed as x goes from 0 to 1. Block out this interval, divide it into four equal
    parts, locate high and low points, and locate x intercepts. Then complete the graph.

    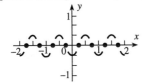   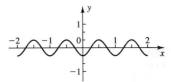

12. $y = -1 + \dfrac{1}{2} \sin 2x$. Amplitude $= \left|\dfrac{1}{2}\right| = \dfrac{1}{2}$. Period $= \dfrac{2\pi}{2} = \pi$,

    $y = \dfrac{1}{2} \sin 2x$ was graphed in Problem 8.

    This graph is the graph of $y = \dfrac{1}{2} \sin 2x$ moved down

    $|k| = |-1| = 1$ unit. We start by drawing a horizontal

    broken line 1 unit below the x axis, then graph

    $y = \dfrac{1}{2} \sin 2x$ relative to the broken line and the original y axis.

    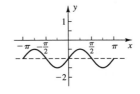

Chapter 3   Graphing Trigonometric Functions

13. $y = 3 - 2\cos\dfrac{\pi}{2}x$. Amplitude $= |-2| = 2$. Period $= \dfrac{2\pi}{\pi/2} = 4$.

This graph is the graph of $y = -2\cos\dfrac{\pi}{2}x$ moved up

3 units. We start by drawing a horizontal broken line

3 units above the x axis, then graph $y = -2\cos\dfrac{\pi}{2}x$

(an upside down cosine curve with amplitude 2 and
period 4) relative to the broken line and the original y axis.

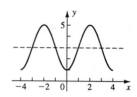

14. Amplitude $= |A| = |1| = 1$. Phase Shift and Period: Solve

$$Bx + C = 0 \quad \text{and} \quad Bx + C = 2\pi$$
$$x - \frac{\pi}{2} = 0 \qquad\qquad x - \frac{\pi}{2} = 2\pi$$
$$x = \frac{\pi}{2} \qquad\qquad x = \frac{\pi}{2} + 2\pi$$

Phase Shift $= \dfrac{\pi}{2}$ Period $= 2\pi$

Graph one cycle over the interval from $\dfrac{\pi}{2}$ to $\left(\dfrac{\pi}{2} + 2\pi\right) = \dfrac{5\pi}{2}$. Then extend the graph from 0 to $\dfrac{\pi}{2}$

and delete the portion of the graph from 2π to $\dfrac{5\pi}{2}$, since this was not required.

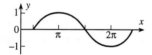   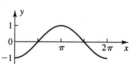

15. Amplitude $= |A| = |1| = 1$. Phase Shift and Period: Solve
$$Bx + C = 0 \quad \text{and} \quad Bx + C = 2\pi$$
$$x + \pi = 0 \qquad\qquad x + \pi = 2\pi$$
$$x = -\pi \qquad\qquad x = -\pi + 2\pi$$
Phase Shift $= -\pi$ Period $= 2\pi$
Graph one cycle over the interval from $-\pi$ to $(-\pi + 2\pi) = \pi$. Then extend the graph from π to 2π and
delete the portion of the graph from $-\pi$ to 0 since this was not required.

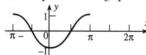   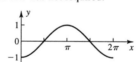

16. Amplitude $= |A| = |-2| = 2$. Phase Shift and Period: Solve

$$Bx + C = 0 \quad \text{and} \quad Bx + C = 2\pi$$
$$\pi x - \pi = 0 \qquad\qquad \pi x - \pi = 2\pi$$
$$x = 1 \qquad\qquad x = 1 + 2$$

Phase Shift $= 1$ Period $= 2$
Graph one cycle (the basic sine curve turned upside down) over the interval from 1 to $(1 + 2) = 3$.
Then extend the graph from 0 to 1 and delete the portion of the graph from 2 to 3 since this was not
required.

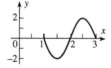   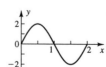

17. Amplitude $= |A| = \left|-\dfrac{1}{4}\right| = \dfrac{1}{4}$.

Phase Shift and Period: Solve

$$Bx + C = 0 \quad \text{and} \quad Bx + C = 2\pi$$
$$2x + \pi = 0 \qquad\qquad 2x + \pi = 2\pi$$
$$x = -\dfrac{\pi}{2} \qquad\qquad x = -\dfrac{\pi}{2} + \pi$$

Phase Shift $= -\dfrac{\pi}{2}$ Period $= \pi$

Graph one cycle (the basic cosine curve turned upside down) over the interval from

$-\dfrac{\pi}{2}$ to $\left(-\dfrac{\pi}{2} + \pi\right) = \dfrac{\pi}{2}$. Then extend the graph from $\dfrac{\pi}{2}$ to 2π and delete the portion of the graph

from $-\dfrac{\pi}{2}$ to 0 since this was not required.

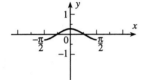
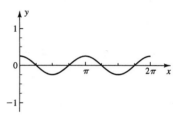

18. $y = -2 \sin (\pi x - \pi)$ was graphed in Problem 16. This graph is the graph of $y = -2 \sin(\pi x - \pi)$

moved up 4 units. We start by drawing a horizontal broken line 4 units above the x axis, then graph

$y = -2 \sin(\pi x - \pi)$ relative to the broken line and the original y axis. $y = 4 - 2 \sin(\pi x - \pi)$

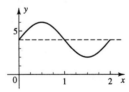

19. Period $= \dfrac{\pi}{2}$

20. Period $= \dfrac{\pi}{\pi} = 1$

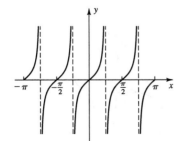
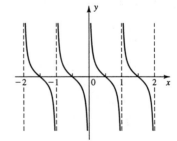

Chapter 3   Graphing Trigonometric Functions

21. Period $= \dfrac{2\pi}{\pi} = 2$. We first sketch a graph of

$y = \dfrac{1}{3} \sin \pi x$ from -1 to 2, which has amplitude

$\dfrac{1}{3}$ and period 2. This curve (dashed curve) can serve

as a guide for $y = 3 \csc \pi x = \dfrac{1}{(1/3) \sin \pi x}$ by

taking reciprocals of ordinates.

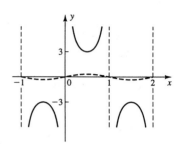

22. Period $= \dfrac{2\pi}{1/2} = 4\pi$. We first sketch a graph of

$y = \dfrac{1}{2} \cos \dfrac{x}{2}$ from $-\pi$ to 3π, which has amplitude

$\dfrac{1}{2}$ and period 4π. This curve (dashed curve) can serve

as a guide for $y = 2 \sec \dfrac{x}{2} = \dfrac{1}{(1/2) \cos (x/2)}$ by

taking reciprocals of ordinates.

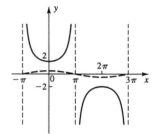

23. We find the period and phase shift by solving

$$x + \dfrac{\pi}{2} = 0 \quad \text{and} \quad x + \dfrac{\pi}{2} = \pi$$

$$x = -\dfrac{\pi}{2} \qquad\qquad x = -\dfrac{\pi}{2} + \pi$$

$$\text{Period} = \pi \quad \text{Phase Shift} = -\dfrac{\pi}{2}$$

We then sketch one period of the graph starting at $x = -\dfrac{\pi}{2}$ (the phase shift) and ending at

$x = -\dfrac{\pi}{2} + \pi = \dfrac{\pi}{2}$ (the phase shift plus one period). Note that a vertical asymptote is at $x = 0$.

We then extend the graph from $-\pi$ to π.

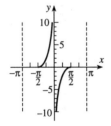
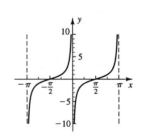

24. We find the period and phase shift by solving $\quad x - \dfrac{\pi}{2} = 0 \quad$ and $\quad x - \dfrac{\pi}{2} = \pi$

$$x = \dfrac{\pi}{2} \qquad\qquad x = \dfrac{\pi}{2} + \pi$$

$$\text{Period} = \pi \qquad \text{Phase Shift} = \dfrac{\pi}{2}$$

We then sketch one period of the graph starting at $x = \dfrac{\pi}{2}$ (the phase shift) and ending at

$x = \dfrac{\pi}{2} + \pi = \dfrac{3\pi}{2}$ (the phase shift plus one period). We then extend the graph from $-\dfrac{\pi}{2}$ to $\dfrac{3\pi}{2}$.

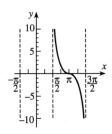

25. We form $y_1 = x$ and $y_2 = \sin x$. We
     sketch the graph of each equation in the
     same coordinate system (dashed lines), then
     add the ordinates $y_1 + y_2$ (solid curve).

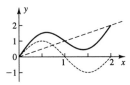

26. We form $y_1 = 2 \sin x$ and $y_2 = \cos 2x$. We
     sketch the graph of each equation in the
     same coordinate system (dashed curves), then
     add the ordinates $y_1 + y_2$ (solid curve).

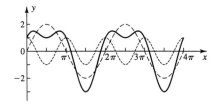

27. See Problem 11.   28. See Problem 20.   29. See Problem 14.

30. Amplitude $= |A| = |-3| = 3$
     To find the period and phase shift, we solve

$$\pi x + \pi = 0 \quad \text{and} \quad \pi x + \pi = 2\pi$$
$$x = -1 \qquad\qquad x = -1 + 2$$
$$\text{Period} = 2 \qquad \text{Phase Shift} = -1$$

31. To find the period and phase shift, we solve

$$\dfrac{\pi}{2} x + \dfrac{\pi}{2} = 0 \quad \text{and} \quad \dfrac{\pi}{2} x + \dfrac{\pi}{2} = \pi$$
$$x = -1 \qquad\qquad x = -1 + 2$$
$$\text{Period} = 2 \qquad \text{Phase Shift} = -1$$

32. To find the period and phase shift, we solve $2x - \pi = 0 \quad$ and $\quad 2x - \pi = 2\pi$

$$x = \dfrac{\pi}{2} \qquad\qquad x = \dfrac{\pi}{2} + \pi$$

$$\text{Period} = \pi \qquad \text{Phase Shift} = \dfrac{\pi}{2}$$

33. Since the maximum deviation from the x axis is 4, we can write: Amplitude $= |A| = 4$.

    Thus, $A = 4$ or -4. Since the period is 2, we can write: Period $= \dfrac{2\pi}{B} = 2$. Thus, $B = \dfrac{2\pi}{2} = \pi$.

    As x increases from 0 to $\dfrac{1}{2}$, y *increases* like the basic sine curve (not the upside down sine curve).

    So, A is positive. $A = 4$; $y = 4 \sin \pi x$.

34. Since the maximum deviation from the x axis is 0.5, we can write: Amplitude $= |A| = 0.5$.

    Thus, $A = 0.5$ or -0.5. Since the period is π, we can write: Period $= \dfrac{2\pi}{B} = \pi$. Thus, $B = \dfrac{2\pi}{\pi} = 2$.

    As x increases from 0 to $\dfrac{\pi}{4}$, y *increases* like the upside down cosine curve. So, A is negative.

    $A = -0.5$; $y = -0.5 \cos 2x$.

35. We first find the period and phase shift by solving $\pi x + \dfrac{\pi}{2} = 0$ and $\pi x + \dfrac{\pi}{2} = \pi$

    $$x = -\dfrac{1}{2} \qquad\qquad x = -\dfrac{1}{2} + 1$$

    $$\text{Period} = 1 \quad \text{Phase Shift} = -\dfrac{1}{2}$$

    We then sketch one period of the graph starting at $x = -\dfrac{1}{2}$ (the phase shift) and ending at

    $x = -\dfrac{1}{2} + 1 = \dfrac{1}{2}$ (the phase shift plus one period). Note that the y axis $(x = 0)$ is a vertical

    asymptote. We then extend the graph from -1 to 1.

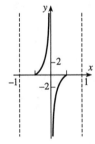
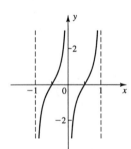

36. We first find the period and phase shift by solving $2x - \pi = 0$ and $2x - \pi = 2\pi$

    $$x = \dfrac{\pi}{2} \qquad\qquad x = \dfrac{\pi}{2} + \pi$$

    $$\text{Period} = \pi \quad \text{Phase Shift} = \dfrac{\pi}{2}$$

    Now, since $2 \sec(2x - \pi) = \dfrac{1}{(1/2) \cos (2x - \pi)}$, we graph $y = \dfrac{1}{2} \cos(2x - \pi)$ for one cycle from $\dfrac{\pi}{2}$ to

    $\dfrac{\pi}{2} + \pi = \dfrac{3\pi}{2}$ with a broken line graph, then take reciprocals. We also place vertical asymptote

    through the x intercepts of the cosine graph to guide us when we sketch the secant function.

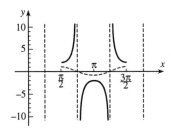

We then extend the one cycle over the required interval from 0 to $\frac{5\pi}{4}$, and delete the portion of the

graph from $\frac{5\pi}{4}$ to $\frac{3\pi}{2}$, since this was not required.

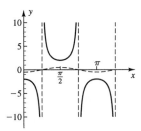

37. Since the maximum deviation from the x axis is 1, we can write: Amplitude $= |A| = 1$. Thus,
      $A = 1$ or -1. Since the period is $\frac{5}{4} - \left(-\frac{3}{4}\right) = 2$, we can write: Period $= \frac{2\pi}{B} = 2$. Thus, $B = \frac{2\pi}{2} = \pi$.
      Since we are instructed to choose the phase shift between 0 and 1, we can regard this graph as con-
      taining the upside down sine curve with a phase shift of $\frac{1}{4}$. This requires us to choose A negative,
      since the graph shows that as x increases from $\frac{1}{4}$ to $\frac{3}{4}$, y *decreases* like the upside down sine curve.
      So, $A = -1$. Then, $-\frac{C}{B} = \frac{1}{4}$. Thus, $C = -\frac{1}{4}B = -\frac{\pi}{4}$. $y = A \sin(Bx + C) = -\sin\left(\pi x - \frac{\pi}{4}\right)$.
      Check: When $x = 0$, $y = -\sin\left(\pi \cdot 0 - \frac{\pi}{4}\right) = -\sin\left(-\frac{\pi}{4}\right) = \frac{\sqrt{2}}{2}$
      When $x = \frac{1}{4}$, $y = -\sin\left(\pi \cdot \frac{1}{4} - \frac{\pi}{4}\right) = -\sin 0 = 0$.

38. The graph of $y = \frac{1}{1 + \tan^2 x}$ is shown in the figure.

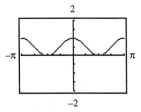

Amplitude $= \frac{1}{2}$ (y coordinate of highest point
          $-$ y coordinate of lowest point)
          $= \frac{1}{2}(1 - 0) = \frac{1}{2} = |A|$
Period $= \frac{2\pi}{2} = \pi$. Thus, $B = \frac{2\pi}{\pi} = 2$.
The form of the graph is that of the basic cosine curve
shifted up $\frac{1}{2}$ unit. Thus, $y = \frac{1}{2} + |A| \cos Bx = \frac{1}{2} + \frac{1}{2} \cos 2x$.

Chapter 3   Graphing Trigonometric Functions

39. The graph of y = 1.2 sin 2x + 1.6 cos 2x is shown in
the figure. This graph appears to be a sine wave with
amplitude 2 and period π that has been shifted to the
left. Thus, we conclude that A = 2 and B = $\frac{2\pi}{\pi}$ = 2.
To determine C, we use the zoom feature or the built-in
approximation routine to locate the x intercept closest to
the origin at x = −0.464. This is the phase-shift for the
graph. Substitute B = 2 and x = −0.464 into the phase-shift
equation x = $-\frac{C}{B}$; −0.464 = $-\frac{C}{2}$; C = 0.928. Thus, the
equation required is y = 2 sin(2x + 0.928).

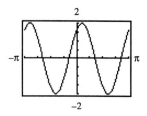

40.

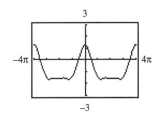

41. (A) The graph of y = $\frac{2 \sin x}{\sin 2x}$ is shown in the figure.
This graph appears to have vertical asymptotes at
x = $-\frac{3\pi}{4}$, x = $-\frac{\pi}{4}$, x = $\frac{\pi}{4}$, and x = $\frac{3\pi}{4}$, and period 2π. Its
high and low points appear to have y coordinates of −1 and
1, respectively. It appears, therefore, to be the same as the
graph of y = sec Bx, $\frac{2\pi}{B}$ = 2π, that is, B = 1. The required
equation is y = sec x.

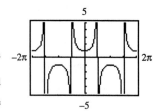

(B) The graph of y = $\frac{2 \cos x}{\sin 2x}$ is shown in the figure. This
graph appears to have vertical asymptotes at x = -2π,
x = $-\pi$, x = 0, x = π, and x = 2π, and period 2π. Its high
and low points appear to have y coordinates of 1 and −1,
respectively. It appears, therefore, to be the same as the
graph of y = csc Bx, $\frac{2\pi}{B}$ = 2π, that is, B = 1. The required
equation is y = csc x.

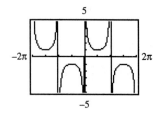

Chapter 3   Graphing Trigonometric Functions

(C) The graph of $y = \dfrac{2\cos^2 x}{\sin 2x}$ is shown in the figure. This graph appears to have vertical asymptotes at $x = -2\pi$, $x = -\pi$, $x = 0$, $x = \pi$, and $x = 2\pi$, and period π. It appears, therefore, to be the same as the graph of $y = \cot Bx$, with $\dfrac{\pi}{B} = \pi$, that is, $B = 1$. The required equation is $y = \cot x$.

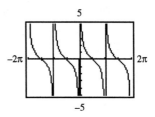

(D) The graph of $y = \dfrac{2\sin^2 x}{\sin 2x}$ is shown in the figure. This graph appears to have vertical asymptotes at $x = -\dfrac{3\pi}{2}$, $x = -\dfrac{\pi}{2}$, $x = \dfrac{\pi}{2}$, and $x = \dfrac{3\pi}{2}$, and period π. It appears, therefore, to be the same as the graph of $y = \tan Bx$, with $\dfrac{\pi}{B} = \pi$, that is, $B = 1$. The required equation is $y = \tan x$.

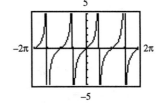

42. *Find A*: The amplitude $|A|$ is given to be 4. Since $y = -4$ (4 cm *below* position at rest) when $t = 0$, $A = -4$ (and not 4).

   Find B: We are given that the frequency, f, is 8 Hz. Hence, the period is found using the reciprocal formula: $P = \dfrac{1}{f} = \dfrac{1}{8}$ sec. But $P = \dfrac{2\pi}{B}$. Thus, $B = \dfrac{2\pi}{P} = \dfrac{2\pi}{1/8} = 16\pi$

   Write the equation: $y = -4 \cos 16\pi t$

43. *Determine K*: The dashed line indicates that high and low points of the curve are equal distances from the line $P = 1$. Hence, $K = 1$.

   Determine A: The maximum deviation from the line $P = 1$ is seen at points such as $n = 0$, $P = 2$ or $n = 26$, $P = 0$. Thus, $|A| = 2 - 1$ or $|A| = 1 - 0$. From either statement, we see $|A| = 1$. So, $A = 1$ or $A = -1$. Since the portion of the curve as n increases from 0 to 26 has the form of the basic cosine curve (P decreasing) — and not the upside down cosine curve — A must be positive. $A = 1$.

   Determine B: One full cycle of the curve is completed as n varies from 0 to 52 weeks. Hence, the Period = 52 weeks. Since Period $= \dfrac{2\pi}{B}$, we have $52 = \dfrac{2\pi}{B}$, or $B = \dfrac{2\pi}{52} = \dfrac{\pi}{26}$.

   $P = K + A \cos Bn = 1 + \cos \dfrac{\pi n}{26}$, $0 \le n \le 104$.

44. (A) We use the formula:

   Period $= \dfrac{2\pi}{\sqrt{1000 \ gA/M}}$ with $g = 9.75$ m/sec^2, $A = \pi\left(\dfrac{1.2}{2} \ m\right)^2$, and

   Period = 0.8 sec, and solve for M. Thus,

   $0.8 = \dfrac{2\pi}{\sqrt{(1000)(9.75) \ \pi(0.6)^2/M}}$ $M = \dfrac{(1000)(9.75) \ \pi(0.6)^2(0.8)^2}{4\pi^2} \approx 179$ kg

   (B) D = amplitude = 0.6 $B = \dfrac{2\pi}{\text{Period}} = \dfrac{2\pi}{0.8} = 2.5\pi$ $y = 0.6 \sin(2.5\pi t)$

Chapter 3   Graphing Trigonometric Functions

(C) One full cycle of this graph is completed as t goes from 0 to 0.8. Block out this interval, divide it into four equal parts, locate high and low points, and locate t intercepts. Then complete the graph.

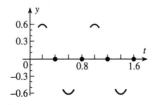 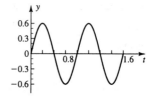

45. (A) Use $\text{Period} = \dfrac{1}{v}$ and $B = \dfrac{2\pi}{\text{Period}}$

   $\text{Period} = \dfrac{1}{280}$ sec $B = \dfrac{2\pi}{1/280} = 560\pi$

   (B) Use $v = \dfrac{1}{\text{Period}}$ and $B = \dfrac{2\pi}{\text{Period}}$

   $v = \dfrac{1}{.0025 \text{ sec}} = 400$ Hz $B = \dfrac{2\pi}{.0025} = 800\pi$

   (C) Use $\text{Period} = \dfrac{2\pi}{B}$ and $v = \dfrac{1}{\text{Period}}$

   $\text{Period} = \dfrac{2\pi}{700\pi} = \dfrac{1}{350}$ sec $v = \dfrac{1}{1/350 \text{ sec}} = 350$ Hz

46. *Find A*: The amplitude $|A|$ is given to be 18. Since $E = 12$ when $t = 0$, $A = 18$ (and not -18).

   Find B: We are given that the frequency, v, is 30 Hz. Hence, the period is found using the reciprocal formula: $P = \dfrac{1}{v} = \dfrac{1}{30}$ sec. But, $P = \dfrac{2\pi}{B}$. Thus, $B = \dfrac{2\pi}{P} = \dfrac{2\pi}{1/30} = 60\pi$

   Write the equation: $y = 18 \cos 60\pi t$

47. $y = 6 \cos \dfrac{\pi}{10}(t - 5)$

   We compute amplitude, period, and phase shift as follows:

   Amplitude $= |A| = |6| = 6$.

   Phase Shift and Period: Solve

   $Bx + C = 0$ and $Bx + C = 2\pi$

   $\dfrac{\pi}{10}(t - 5) = 0$ $\dfrac{\pi}{10}(t - 5) = 2\pi$

   $\qquad t = 5$ $t = 5 + 20$
   $\qquad\quad\uparrow$ $\quad\uparrow\quad\uparrow$

   $\qquad$ Phase Shift $\qquad$ Period

   Graph one cycle over the interval from 5 to 25.

Extend the graph from 0 to 60.

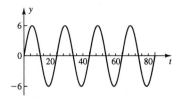

48. *Voltage*: $E = 20 \sin 100\pi t$. Amplitude $= |A| = |20| = 20$. Period $= \dfrac{2\pi}{B} = \dfrac{2\pi}{100\pi} = \dfrac{1}{50}$

Current: $I = 15 \sin\left(100\pi t - \dfrac{\pi}{2}\right)$. Amplitude $= |A| = |15| = 15$

Phase Shift and Period: Solve $Bx + C = 0$ and $Bx + C = 2\pi$

$$100\pi t - \frac{\pi}{2} = 0 \qquad\qquad 100\pi t - \frac{\pi}{2} = 2\pi$$

$$t = \frac{1}{200} \qquad\qquad\qquad t = \frac{1}{200} + \frac{1}{50}$$

$$\qquad\qquad\uparrow \qquad\qquad\qquad\qquad\quad \uparrow \qquad \uparrow$$

$$\qquad\qquad\text{Phase Shift} \qquad\qquad\qquad \text{Period}$$

Graph one cycle of the voltage from 0 to $\dfrac{1}{50}$, then extend the graph from 0 to $\dfrac{3}{50}$. Graph one cycle

of the current from $\dfrac{1}{200}$ to $\dfrac{1}{200} + \dfrac{1}{50} = \dfrac{1}{40}$, then extend the graph from 0 to $\dfrac{3}{50}$.

One cycle of the voltage: One cycle of the current:

                          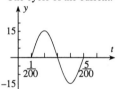

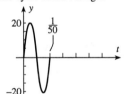

49. The height of the wave from trough to crest is the difference in height between the crest (height A)

and the trough (height $-A$). In this case, $A = 12$ ft. $A - (-A) = 2A = 2(12 \text{ ft}) = 24$ ft.

To find the wavelength λ, we note: $\lambda = 5.12\, T^2$, $T = \dfrac{2\pi}{B}$, $B = \dfrac{\pi}{3}$.

Thus, $T = \dfrac{2\pi}{\pi/3} = 6$ sec, $\lambda = 5.12(6)^2 \approx 184$ ft. To find the speed S, we use

$$S = \sqrt{\frac{g\lambda}{2\pi}} \qquad g = 32 \text{ ft/sec}^2$$

$$\quad = \sqrt{\frac{32(184)}{2\pi}} \approx 31 \text{ ft/sec}$$

Chapter 3   Graphing Trigonometric Functions

50. Period $= \dfrac{1}{v} = \dfrac{1}{10^{15}} = 10^{-15}$ sec

To find the wave length λ, we use the formula $\lambda v = c$ with $c \approx 3 \times 10^8$ m/sec.

$$\lambda = \frac{c}{v} = \frac{3 \times 10^8 \text{ m/sec}}{10^{15} \text{ Hz}} = 3 \times 10^{-7} \text{ m}$$

51. (A) Amplitude = 1 (constant), Period = 1.

One full cycle of this curve is completed as t goes from 0 to 1. Block out this interval, divide it into four equal parts, locate high and low points and locate t intercepts. Then complete the graph.

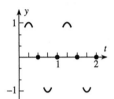   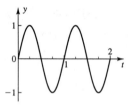

Since the amplitude is constant, this is simple harmonic motion.

(B) Step 1: Graph y = 1 + t and its reflection.
Step 2: Sketch the graph of y = sin 2πt, but expand high and low points to the height of the envelope.

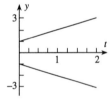   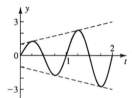

Since the amplitude increases as time increases, this is resonance.

(C) Step 1: Graph $y = \dfrac{1}{1 + t}$ and its reflection.

Step 2: Sketch the graph of y = sin 2πt, but contract high and low points to the height of the envelope.

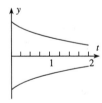   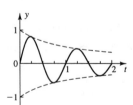

Since the amplitude decreases as time increases, this is damped harmonic motion.

52. We form $S_1 = 5 + t$ and $S_2 = 5 \sin \dfrac{\pi t}{6}$.

    We sketch the graph of each equation in the same coordinate system (dashed lines), then add the ordinates $S_1 + S_2$ (solid curve).

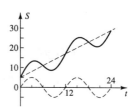

53. (A) The triangle shown in the text figure is a right triangle.

    Hence $\tan \theta = \dfrac{a}{b} = \dfrac{h}{1000}$. Thus, $h = 1000 \tan \theta$

    (B) The period is π. We sketch half of one period, since the required interval is from 0 to $\dfrac{\pi}{2}$ only. There is

    a vertical asymptote at $t = \dfrac{\pi}{2}$. Ordinates can be determined from a calculator, thus:

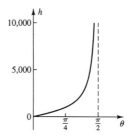

θ	0	$\dfrac{\pi}{8}$	$\dfrac{\pi}{4}$	$\dfrac{3\pi}{8}$	$\dfrac{7\pi}{16}$
$1000 \tan \theta$	0	410	1000	2400	5000

54. (A)

x (months)	1, 13	2, 14	3, 15	4, 16	5, 17	6, 18	7, 19	8, 20	9, 21
y (decimal hours)	17.08	17.63	18.12	18.60	19.07	19.48	19.58	19.25	18.57

x	10, 22	11, 23	12, 24
y	17.78	17.12	16.85

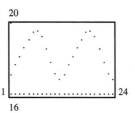

   (B) From the table, Max y = 19.58 and Min y = 16.85. Then,

$$A = \frac{(\text{Max y} - \text{Min y})}{2} = \frac{(19.58 - 16.85)}{2} = 1.37$$

$$B = \frac{2\pi}{\text{Period}} = \frac{2\pi}{12} = \frac{\pi}{6}$$

$$k = \text{Min y} + A = 16.85 + 1.37 = 18.22$$

Chapter 3 Graphing Trigonometric Functions

From the plot in (A) or the table, we estimate the smallest positive value of x for which
y = k = 18.22 to be approximately 3.2. Then this is the phase-shift for the graph. Substitute
$B = \dfrac{\pi}{6}$ and x = 3.2 into the phase-shift equation $x = -\dfrac{C}{B}$; $3.2 = -\dfrac{C}{\pi/6}$; $C = -\dfrac{3.2\pi}{6} \approx -1.7$.

Thus, the equation required is $y = 18.22 + 1.37\sin\left(\dfrac{\pi}{6}x - 1.7\right)$.

(C)

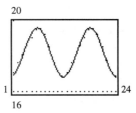

CUMULATIVE REVIEW EXERCISE CHAPTERS 1–3

1. Since one complete revolution has measure 360°, $\dfrac{1}{4}$ revolution has measure $\dfrac{1}{4}(360°) = 90°$.

    Since 90° is one-half of 180°, the corresponding radian measure must be $\dfrac{1}{2}$ of π, or $\dfrac{\pi}{2}$ rad.

2. Since $47' = \dfrac{47°}{60}$, then $21°47' = \left(21 + \dfrac{47}{60}\right)° \approx 21.78°$

3. $\dfrac{\theta_{deg}}{180°} = \dfrac{\theta_{rad}}{\pi\,rad}$; $\theta_{deg} = \dfrac{180°}{\pi\,rad}\theta_{rad}$. If $\theta_{rad} = 1.67$, $\theta_{deg} = \dfrac{180}{\pi}(1.67) = 95.68°$

4. $\dfrac{\theta_{deg}}{180°} = \dfrac{\theta_{rad}}{\pi\,rad}$; $\theta_{rad} = \dfrac{\pi\,rad}{180°}\theta_{deg}$. If $\theta_{deg} = -715.3°$, $\theta_{rad} = \dfrac{\pi}{180}(-715.3) = -12.48$ rad

5. *Solve for the complementary angle:*

    $90° - \theta = 90° - 25° = 65°$

    Solve for b: We will use the sine. Thus,

    $\sin\theta = \dfrac{b}{c}$

    $b = c\sin\theta = (34\ in.)(\sin 25°) = 14$ in.

    Solve for a: We will use the cosine. Thus, $\cos\theta = \dfrac{a}{c}$, $a = c\cos\theta = (34\ in.)(\cos 25°) = 31$ in.

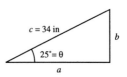

6. $P(a, b) = (8, -15)$

$R = \sqrt{a^2 + b^2} = \sqrt{8^2 + (-15)^2} = 17$

$\sin \theta = \dfrac{b}{R} = \dfrac{-15}{17} = -\dfrac{15}{17}$

$\tan \theta = \dfrac{b}{a} = \dfrac{-15}{8} = -\dfrac{15}{8}$

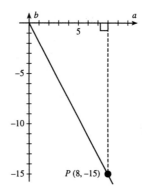

7. (A) Degree mode: $\sin(23°12') = \sin(23.2°)$ Convert to decimal degrees.
$$= 0.3939$$

(B) Use the reciprocal relationship $\sec \theta = \dfrac{1}{\cos \theta}$.

Degree mode: $\sec 145.6° = \dfrac{1}{\cos 145.6°} = -1.212$

(C) Use the reciprocal relationship $\cot \theta = \dfrac{1}{\tan \theta}$

Radian mode: $\cot 0.88 = \dfrac{1}{\tan 0.88} = 0.8267$

8. The reference angle α is the angle (always taken positive) between the terminal side of θ and the horizontal axis.

(A) $\alpha = 2\pi - \dfrac{11\pi}{6} = \dfrac{\pi}{6}$

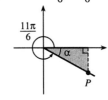

(B) $\alpha = |-225°| - 180° = 45°$

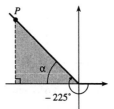

9. (A)

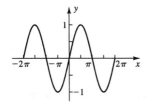

(B)

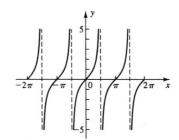

(C)

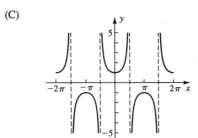

10. If $\tan\theta = 0.9465$, then

$$\theta = \tan^{-1} 0.9465 = 43°30'$$

11. The tip of the second hand travels 1 revolution, or 2π radian, in 60 seconds. In 40 seconds it travels $\dfrac{40}{60}$ revolution, or $\dfrac{40}{60} \cdot 2\pi$ radian, that is, $\dfrac{4\pi}{3}$ radian. Since the distance travelled $s = R\theta$, we have

$$s = R\theta = (5.00 \text{ cm})\left(\frac{4\pi}{3}\text{ rad}\right) = 20.94 \text{ cm}$$

12. The tip of the second hand travels 1 circumference in 1 minute. Thus, its speed is given by

$$V = \frac{d}{t} = \frac{2\pi r}{t} = \frac{2\pi(5.00 \text{ cm})}{1 \text{ min}} = 31.4 \text{ cm/min}$$

13. $\theta = \sin^{-1}(0.4621) = 27.523° = 27°(0.523 \times 60)' = 27°31.362' = 27°31'(0.362 \times 60)'' = 27°31'22''$

14. Label the sides of the triangles as shown:

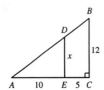

Since triangles ABC and ADE are similar, we have

$$\frac{DE}{AE} = \frac{BC}{AC}$$

$$\frac{x}{10} = \frac{12}{10 + 5}$$

$$x = 10 \cdot \frac{12}{15} = 8$$

15. $\dfrac{\theta_{\text{deg}}}{180°} = \dfrac{\theta_{\text{rad}}}{\pi\text{ rad}}$

$\theta_{\text{rad}} = \dfrac{\pi\text{ rad}}{180°}\theta_{\text{deg}}$

$= \dfrac{\pi}{180}(48) = \dfrac{4\pi}{15}\text{ rad}$

16. Since $s = \dfrac{\pi}{180}R\theta$, we have

$s = \dfrac{\pi}{180}(2.0)(145)$

$= 5.1 \text{ ft}$

17.　The reference angle α is the angle (always taken positive) between the terminal side of θ and the
　　horizontal axis.

(A)

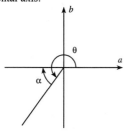

$$\alpha = 237.6° - 180° = 57.6°$$

(B)

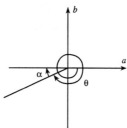

$$\alpha = 360° + 180° - 514°30' = 25°30'$$

(C)

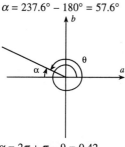

$$\alpha = 2\pi + \pi - 9 = 0.42$$

(D)

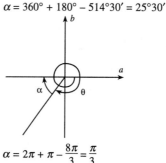

$$\alpha = 2\pi + \pi - \frac{8\pi}{3} = \frac{\pi}{3}$$

18.　(A)　Locate the 45° reference triangle,
　　　　determine (a, b) and R, then evaluate.

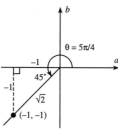

$$\sin \frac{5\pi}{4} = \frac{-1}{\sqrt{2}} = -\frac{1}{\sqrt{2}}$$

(B)　Locate the 30° – 60° reference triangle,
　　determine (a, b) and R, then evaluate.

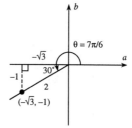

$$\cos \frac{7\pi}{6} = \frac{-\sqrt{3}}{2} = -\frac{\sqrt{3}}{2}$$

(C)　Locate the 30° – 60° reference triangle,
　　determine (a, b) and R, then evaluate.

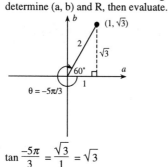

$$\tan \frac{-5\pi}{3} = \frac{\sqrt{3}}{1} = \sqrt{3}$$

(D)　(a, b) = (–1, 0),　R = 1

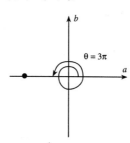

$$\csc 3\pi = \frac{1}{0} \cdot$$

Not defined

Chapter 3   Graphing Trigonometric Functions

19. Since $\cos\theta < 0$ and $\tan\theta < 0$, the terminal side of θ lies in quadrant II. We sketch a reference triangle and label what we know.

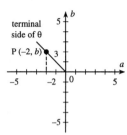

Since $\cos\theta = \dfrac{a}{R} = -\dfrac{2}{3} = \dfrac{-2}{3}$, we know that $a = -2$ and $R = 3$ (R is never negative). Use the Pythagorean theorem to find b:

$$(-2)^2 + b^2 \;=\; 3^2$$
$$b^2 \;=\; 9 - 4 = 5$$
$$b \;=\; \sqrt{5}$$

b is positive since the terminal side of θ lies in quadrant II. We can now find the other five functions using their definitions.

$$\sin\theta = \frac{b}{R} = \frac{\sqrt{5}}{3} \qquad\qquad \sec\theta = \frac{R}{a} = \frac{3}{-2} = -\frac{3}{2}$$

$$\tan\theta = \frac{b}{a} = \frac{\sqrt{5}}{-2} = -\frac{\sqrt{5}}{2} \qquad \csc\theta = \frac{R}{b} = \frac{3}{\sqrt{5}} \qquad \cot\theta = \frac{a}{b} = \frac{-2}{\sqrt{5}} = -\frac{2}{\sqrt{5}}$$

20. Since $C = 2\pi R$, we have $R = \dfrac{C}{2\pi} = \dfrac{24}{2\pi} = \dfrac{12}{\pi}$ cm. Since $\theta = \dfrac{s}{R}$, we have $\theta = \dfrac{7}{12/\pi} = \dfrac{7\pi}{12}$ rad.

Hence, $A = \dfrac{1}{2} R^2 \theta = \dfrac{1}{2}\left(\dfrac{12}{\pi}\right)^2\left(\dfrac{7\pi}{12}\right) = \dfrac{42}{\pi} \approx 13.4$ cm^2

21. $y = 1 - \dfrac{1}{2}\cos 2x$. Amplitude $= \left|-\dfrac{1}{2}\right| = \dfrac{1}{2}$.

Period $= \dfrac{2\pi}{2} = \pi$. This graph is the graph of

$y = -\dfrac{1}{2}\cos 2x$ moved up 1 unit. We start by drawing

a horizontal broken line 1 unit above the x axis, then

graph $y = -\dfrac{1}{2}\cos 2x$ $\left(\text{an upside down cosine curve with amplitude } \dfrac{1}{2} \text{ and period } \pi\right)$ relative to

the broken line and the original y axis.

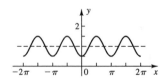

110

22. Amplitude $= |A| = |2| = 2$. Phase Shift and Period: Solve $Bx + C = 0$ and $Bx + C = 2\pi$

$$x - \frac{\pi}{4} = 0 \qquad\qquad x - \frac{\pi}{4} = 2\pi$$

$$x = \frac{\pi}{4} \qquad\qquad x = \frac{\pi}{4} + 2\pi$$

$$\text{Phase Shift} = \frac{\pi}{4} \qquad\qquad \text{Period} = 2\pi$$

Graph one cycle over the interval from $\frac{\pi}{4}$ to $\left(\frac{\pi}{4} + 2\pi\right) = \frac{9\pi}{4}$. Then extend the graph from $-\pi$ to 3π.

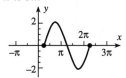   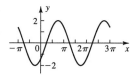

23. Period $= \frac{\pi}{4}$

24. Period $= \frac{2\pi}{1/2} = 4\pi$

    The dashed line shows $y = \sin \frac{x}{2}$ in this interval. The solid line is $y = \csc \frac{x}{2}$.

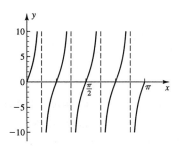   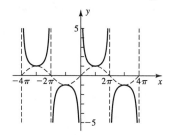

25. Period $= \frac{2\pi}{\pi} = 2$.

    We first sketch a graph of $y = \frac{1}{2} \cos \pi x$ from -2 to 2, which has amplitude $\frac{1}{2}$ and period 2. This curve (dashed curve) can serve as a guide for $y = 2 \sec \pi x = \frac{1}{(1/2) \cos \pi x}$ by taking reciprocals of ordinates.

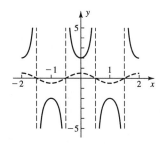

26. We first find the period and phase shift by solving $\pi x + \frac{\pi}{2} = 0$ and $\pi x + \frac{\pi}{2} = \pi$

$$\pi x = -\frac{\pi}{2} \qquad\qquad \pi x = -\frac{\pi}{2} + \pi$$

Chapter 3   Graphing Trigonometric Functions

$$x = -\frac{1}{2} \qquad\qquad x = -\frac{1}{2} + 1$$

$$\text{Period} = 1 \qquad \text{Phase Shift} = -\frac{1}{2}$$

We then sketch one period of the graph starting at $x = -\frac{1}{2}$ (the phase shift) and ending at

$x = -\frac{1}{2} + 1 = \frac{1}{2}$ (the phase shift plus one period). We then extend the graph from −1 to 3.

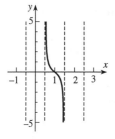 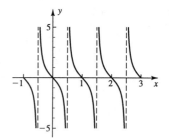

27. We form $y_1 = \sin x$ and $y_2 = \sin 2x$.

    We sketch the graph of each equation in the
    same coordinate system (dashed lines), then
    add the ordinates $y_1 + y_2$ (solid curve).

    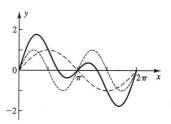

28. Since the maximum deviation from the x axis is 2, we can write: Amplitude $= |A| = 2$. Thus,
    $A = 2$ or -2. Since the period is 1, we can write: Period $= \frac{2\pi}{B} = 1$. Thus, $B = 2\pi$. As x
    increases from 0 to 0.25, y *decreases* like the upside down sine curve. So, A is negative.
    $A = -2$; $y = -2\sin(2\pi x)$.

29. $(\tan x)(\sin x) + \cos x = \dfrac{\sin x}{\cos x} \sin x + \cos x = \dfrac{\sin^2 x}{\cos x} + \dfrac{\cos x}{1} = \dfrac{\sin^2 x}{\cos x} + \dfrac{\cos^2 x}{\cos x}$

    $= \dfrac{\sin^2 x + \cos^2 x}{\cos x} = \dfrac{1}{\cos x} = \sec x$

30. We can draw reference triangles in both
    quadrants III and IV with sides opposite
    reference angle −1 and hypotenuse 2. Each
    triangle is a special $30° - 60°$ triangle.

    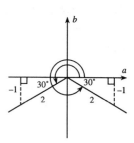

    $\theta = 210°$ or $330°$

Chapter 3   Graphing Trigonometric Functions

31. *Solve for θ*: We will use the tangent. Thus,

$$\tan \theta = \frac{b}{a} = \frac{23.5 \text{ in.}}{37.3 \text{ in.}} = 0.6300$$

$$\theta = \tan^{-1} 0.6300 = 32.2°$$

Solve for the complementary angle:

$$90° - \theta = 90° - 32.2° = 57.8°$$

Solve for c: We use the Pythagorean theorem.

Since $c^2 = a^2 + b^2$.

$$c = \sqrt{a^2 + b^2} = \sqrt{(37.3)^2 + (23.5)^2} = 44.1 \text{ in.}$$

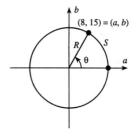

32. $57.8° = 57°(0.8 \times 60)' = 57°50'$ to the nearest 10'.
    $32.2° = 32°(0.2 \times 60)' = 32°10'$ to the nearest 10'.

33. Since $\cos \theta = \frac{a}{R}$ and $\sin \theta = \frac{b}{R}$, we have $a = R \cos \theta = 1 \cos 2.3$ and $b = R \sin \theta = 1 \sin 2.3$.
    Thus, $(a, b) = (1 \cos 2.3, 1 \sin 2.3) = (-0.666, 0.746)$

34. In the figure, we note, $s = R\theta$.

$$R = \sqrt{a^2 + b^2} = \sqrt{8^2 + 15^2} = 17;$$
$$\tan \theta = \frac{b}{a} = \frac{15}{8}; \ \theta = \tan^{-1} \frac{15}{8} \text{ radians}$$

Therefore,

$$s = r\theta = 17 \tan^{-1} \frac{15}{8} \approx 18.37 \text{ units (calculator in radian}$$
$$\text{mode).}$$

35. Using the fundamental identities, we can write $\tan \theta = a$, hence, $\frac{\sin \theta}{\cos \theta} = a$, or $\sin \theta = a \cos \theta$.
    Also, $\sin^2\theta + \cos^2\theta = 1$. Substituting $a \cos \theta$ for $\sin \theta$ and solving for $\cos \theta$, we obtain:

$$(a \cos \theta)^2 + \cos^2 \theta = 1$$
$$(a^2 + 1) \cos^2\theta = 1$$
$$\cos^2\theta = \frac{1}{1 + a^2}$$

Since θ is a first quadrant angle, $\cos \theta$ is positive. Hence, $\cos \theta = \frac{1}{\sqrt{1 + a^2}}$. It follows that

$\sin \theta = a \cos \theta = \frac{a}{\sqrt{1 + a^2}}$. Using the reciprocal identities, we can write

$$\cot \theta = \frac{1}{\tan \theta} = \frac{1}{a}$$

$$\sec \theta = \frac{1}{\cos \theta} = \frac{1}{1/\sqrt{1 + a^2}} = \sqrt{1 + a^2} \qquad \csc \theta = \frac{1}{\sin \theta} = \frac{1}{a/\sqrt{1 + a^2}} = \frac{\sqrt{1 + a^2}}{a}$$

36. Since the maximum deviation from the x axis is 4, we can write:

Amplitude $= |A| = 4$. Thus, $A = 4$ or -4. Since the period is $\frac{4}{3} - \left(-\frac{2}{3}\right) = 2$, we can write:

Period $= \frac{2\pi}{B} = 2$. Thus, $B = \frac{2\pi}{2} = \pi$. Since we are instructed to choose the phase shift between

Chapter 3   Graphing Trigonometric Functions

0 and 1, we can regard this graph as containing the basic sine curve with a phase shift of $\frac{1}{3}$. This requires us to choose A positive, since the graph shows that as x increases from $\frac{1}{3}$ to $\frac{5}{6}$, y *increases* like the basic sine curve (not the upside down sine curve). So, A = 4. Then, $-\frac{C}{B} = \frac{1}{3}$. Thus,

$$C = -\frac{1}{3} B = -\frac{1}{3}\pi .$$
$$y = A \sin(Bx + C) = 4 \sin\left(\pi x - \frac{1}{3}\pi\right).$$

Check: When x = 0, $y = 4 \sin\left(\pi \cdot 0 - \frac{1}{3}\pi\right) = 4 \sin\left(-\frac{1}{3}\pi\right) = -2\sqrt{3}$

When $x = \frac{1}{3}$, $y = 4 \sin\left(\pi \cdot \frac{1}{3} - \frac{1}{3}\pi\right) = 4 \sin 0 = 0$

37.   The graph of $y = \dfrac{\tan^2 x}{1 + \tan^2 x}$ is shown in the figure.

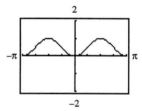

Amplitude $= \dfrac{1}{2}$ (y coordinate of highest point

$= -$ y coordinate of lowest point)

$= \dfrac{1}{2}(1 - 0) = \dfrac{1}{2} = |A|$

Period $= \dfrac{2\pi}{2} = \pi$.

Thus, $B = \dfrac{2\pi}{\pi} = 2$. The form of the graph is that of the upside down cosine curve shifted up $\dfrac{1}{2}$ unit.

Thus, $y = \dfrac{1}{2} - |A| \cos Bx = \dfrac{1}{2} - \dfrac{1}{2}\cos 2x$

38.   The graph of $y = 2.4 \sin \dfrac{x}{2} - 1.8 \cos \dfrac{x}{2}$ is shown in the figure.

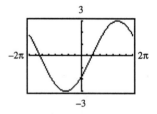

This graph appears to be a sine wave with amplitude 3 and period 4π that has been shifted to the right. Thus, we conclude that A = 3 and $B = \dfrac{2\pi}{4\pi} = \dfrac{1}{2}$. To determine C, we use the zoom feature or the built-in approximation routine to locate the x intercept closest to the origin at x = 1.287. This is the phase-shift for the graph.

Substitute $B = \dfrac{1}{2}$ and x = 1.287 into the phase-shift equation $x = -\dfrac{C}{B}$; $1.287 = -\dfrac{C}{1/2}$; C = −0.6435.

Thus, the equation required is $y = 3 \sin\left(\dfrac{x}{2} - 0.6435\right)$.

39.

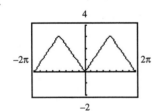

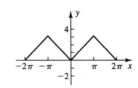

114

40. (A) The graph of $y = \dfrac{\sin 2x}{1 + \cos 2x}$ is shown in the figure.

    The graph appears to have vertical asymptotes and

    $x = -\dfrac{\pi}{2}$ and $x = \dfrac{\pi}{2}$ and period π. It appears, therefore,

    to be the same as the graph of $y = \tan Bx$, with

    $\dfrac{\pi}{B} = \pi$, that is, $B = 1$. The required equation is

    $y = \tan x$.

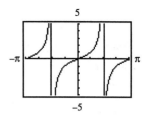

(B) The graph of $y = \dfrac{2 \cos x}{1 + \cos 2x}$ is shown in the figure.

    This graph appears to have vertical asymptotes at

    $x = -\dfrac{\pi}{2}$ and $x = \dfrac{\pi}{2}$ and period 2π. Its high and low

    points appear to have y coordinates of -1 and 1,

    respectively. It appears, therefore, to be the same as

    the graph of $y = \sec Bx$, $\dfrac{2\pi}{B} = 2\pi$, that is, $B = 1$.

        The required equation is $y = \sec x$.

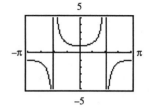

(C) The graph of $y = \dfrac{2 \sin x}{1 - \cos 2x}$ is shown in the figure.

    This graph appears to have vertical asymptotes at

    $x = -\pi$, $x = 0$, and $x = \pi$, and period 2π. Its high and

    low points appear to have y coordinates of -1 and 1,

    respectively. It appears, therefore, to be the same as

    the graph of $y = \csc Bx$, $\dfrac{2\pi}{B} = 2\pi$, that is, $B = 1$.

    The required equation is $y = \csc x$.

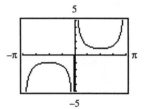

(D) The graph of $y = \dfrac{\sin 2x}{1 - \cos 2x}$ is shown in the figure.

    This graph appears to have vertical asymptotes at

    $x = -\pi$, $x = 0$, and $x = \pi$, and period π. It appears,

    therefore, to be the same as the graph of $y = \cot Bx$,

    with $\dfrac{\pi}{B} = \pi$, that is, $B = 1$. The required equation

    is $y = \cot x$.

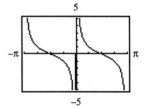

41. We use the diagram and reason as follows: Since the cities

    have the same longitude, θ is given by their distance in

    latitude $\theta = 41°36' - 30°25' = 11°11' = \left(11 + \dfrac{11}{60}\right)^{\circ}$.

    Since $\dfrac{s}{C} = \dfrac{\theta}{360°}$ and $C = 2\pi R$, then $\dfrac{s}{2\pi R} = \dfrac{\theta}{360°}$.

    $s = 2\pi R \cdot \dfrac{\theta}{360°} \approx 2(3.14)(3960 \text{ mi})\dfrac{11 + \dfrac{11}{60}}{360} \approx 773 \text{ mi.}$

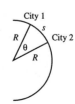

Chapter 3   Graphing Trigonometric Functions

42. Since the two right triangles shown in the figure are similar, we can write $\frac{r}{4} = \frac{6}{9}$, thus, $r = \frac{8}{3}$ cm.

Then, we have $V = \frac{1}{3}\pi r^2 h = \frac{1}{3}\pi\left(\frac{8}{3}\right)^2 6 = \frac{128\pi}{9} \approx 45$ cm³

43. Labelling the diagram as shown, we can write:

$\cos\theta = \frac{AC}{AB}$ θ = angle of elevation

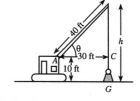

$\theta = \cos^{-1}\frac{AC}{AB} = \cos^{-1}\frac{30}{40} = 41°$

To find h, the altitude of the tip, we note

$h = BC + CG$

$BC^2 = AB^2 - AC^2$ (Pythagorean theorem)

$BC = \sqrt{AB^2 - AC^2} = \sqrt{40^2 - 30^2}$

$= 26$ ft

Then, $h = 26 + 10 = 36$ ft

44. The two right triangles shown in the text figure have corresponding angles equal, hence they are similar. Thus, we can write $\frac{x}{2} = \frac{4-x}{4}$; $2x = 4 - x$; $x = \frac{4}{3}$ ft.

45. From the figure, it is clear that $\tan\theta = \frac{a}{b}$.

Given: percentage of inclination $\frac{a}{b} = 3\% = 0.03$, then

$\tan\theta = 0.03$; $\theta = 1.7°$

Given: angle of inclination $\theta = 3°$, then

$\frac{a}{b} = \tan 3°$; $\frac{a}{b} = 0.05$ or 5%

46. Labelling the text figure as shown, we note:

We are asked for BT = x + 10, the height of the office building, and d, the width of the street.

In right triangle SCT, $\cot 68° = \frac{d}{x}$

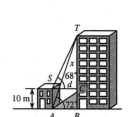

In right triangle ABT, $\cot 72° = \frac{d}{x+10}$

We solve the system of equations

$\cot 68° = \frac{d}{x}$ $\cot 72° = \frac{d}{x+10}$

by clearing of fractions, then eliminating d.

(1) $d = x \cot 68°$ $d = (x+10)\cot 72°$

$x\cot 68° = (x+10)\cot 72°$

$= x\cot 72° + 10\cot 72°$

$x\cot 68° - x\cot 72° = 10\cot 72°$

$x = \frac{10\cot 72°}{\cot 68° - \cot 72°} = 41$ m

Then the height of the office building = x + 10 = 51 m. Substituting in (1),

$d = x\cot 68° = (41$ m$)\cot 68° = 17$ m

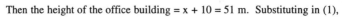

116

Chapter 3 Graphing Trigonometric Functions

47. We redraw and label the figure in the text.

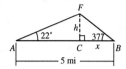

(A) We note: $AC = AB - BC = 5 - x$. We are to find x.

In right triangle BCF, $\tan 37° = \dfrac{h}{x}$

In right triangle ACF, $\tan 22° = \dfrac{h}{5 - x}$

We solve the system of equations

$$\tan 37° = \frac{h}{x} \qquad \tan 22° = \frac{h}{5 - x}$$

by clearing of fractions, then eliminating h.

(1) $h = x \tan 37°$; $h = (5 - x) \tan 22°$;

$x \tan 37° = (5 - x) \tan 22° = 5 \tan 22° - x \tan 22°$;

$x \tan 37° + x \tan 22° = 5 \tan 22°$; $x = \dfrac{5 \tan 22°}{\tan 37° + \tan 22°} = 1.7$ mi

(B) We are to find h. From (1) in part (A), $h = x \tan 37° = 1.7 \tan 37° = 1.3$ mi

48. Since $A = \dfrac{1}{2} R^2 \theta$ and $P = s + 2R = R\theta + 2R$, we can eliminate θ between the two equations and

write $2A = R^2 \theta$; $\theta = \dfrac{2A}{R^2}$; $P = R\left(\dfrac{2A}{R^2}\right) + 2R = \dfrac{2A}{R} + 2R$. Thus,

$$P = \frac{2(15.43)}{4.2} + 2(4.2) \approx 16 \text{ ft}$$

49. (A) For the drive motor, we note: $300 \text{ rpm} = 300 \cdot 2\pi \text{ rad/min} = 600\pi \text{ rad/min}$.

$V = R\omega = (15 \text{ in.})(600\pi \text{ rad/min}) = 9000\pi \text{ in./min}$

This is the linear velocity of the chain, hence the linear velocity of the smaller wheel of the saw.

Then, for the saw, $\omega = \dfrac{V}{R} = \dfrac{9000\pi \text{ in./min}}{30 \text{ in.}} = 300\pi \text{ rad/min} \approx 942 \text{ rad/min}$

(B) For the saw itself, ω is also 300π rad/min

$V = R\omega = (68 \text{ in.})(300\pi \text{ rad/min}) = 20{,}400\pi \text{ in./min} \approx 64{,}088 \text{ in./min}$

50. Use $\dfrac{n_2}{n_1} = \dfrac{\sin \alpha}{\sin \beta}$, where $n_2 = 1.33$, $n_1 = 1.00$, and $\alpha = 38.4°$

Solve for β: $\dfrac{1.33}{1.00} = \dfrac{\sin 38.4°}{\sin \beta}$; $\sin \beta = \dfrac{\sin 38.4°}{1.33}$; $\beta = \sin^{-1}\left(\dfrac{\sin 38.4°}{1.33}\right) = 27.8°$

51. We use $\sin \dfrac{\theta}{2} = \dfrac{S_s}{S_p}$, where S_s is the speed of sound and $S_p = $ speed of the plane $= 1.5 S_s$. Thus,

$\sin \dfrac{\theta}{2} = \dfrac{S_s}{1.5 S_s} = \dfrac{1}{1.5}$; $\dfrac{\theta}{2} \approx 42°$; $\theta \approx 84°$

Chapter 3　Graphing Trigonometric Functions

52. The speed S of a wave traveling with wavelength λ is given approximately by $S = \sqrt{\dfrac{g\lambda}{2\pi}}$.

    Solving for λ in terms of S, we obtain $S^2 = \dfrac{g\lambda}{2\pi}$; $\lambda = \dfrac{2\pi S^2}{g}$. In this case $S = 25$ ft/sec. Thus, the

    wavelength λ is given by $\lambda = \dfrac{2\pi (25)^2}{32} = 123$ ft. The wavelength is related to the period T by

    $\lambda = 5.12T^2$ (approximately). Solving for T in terms of λ, we have

    $$T^2 = \frac{\lambda}{5.12}; \quad T = \sqrt{\frac{\lambda}{5.12}}. \text{ Hence, } T = \sqrt{\frac{123}{5.12}} \approx 4.9 \text{ sec}$$

53. $A1 = \dfrac{1}{2}$ (base)(height) $= \dfrac{1}{2}(1)(\sin x) = \dfrac{1}{2}\sin x$ (the height is the perpendicular distance from P to the

    x axis, thus, sin x).

54. $A2 = \dfrac{1}{2}$ (radius)^2(angle) $= \dfrac{1}{2}(1)^2 x = \dfrac{1}{2}x$

55. $A3 = \dfrac{1}{2}$ (base)(height) $= \dfrac{1}{2}(1) h$. In triangle OAB, $\tan x = \dfrac{h}{1}$, hence, $h = \tan x$. Thus, $A_3 = \dfrac{1}{2}\tan x$.

56. Since $A1 < A2 < A3$, we can write $\dfrac{1}{2}\sin x < \dfrac{1}{2}x < \dfrac{1}{2}\tan x$

    Multiplying by 2, we can write $\sin x < x < \tan x$

    Applying a fundamental identity, we can write $\sin x < x < \dfrac{\sin x}{\cos x}$

    As long as $x > 0$, these quantities are positive. For positive quantities, $a < b < c$ is equivalent to

    $\dfrac{1}{c} < \dfrac{1}{b} < \dfrac{1}{a}$. Hence, $\dfrac{\cos x}{\sin x} < \dfrac{1}{x} < \dfrac{1}{\sin x}$. If $x > 0$, $\sin x > 0$, and we can multiply all parts of this

    double inequality by sin x without altering the sense of the inequalities. Thus,

    $$\sin x \cdot \frac{\cos x}{\sin x} \; < \; \sin x \cdot \frac{1}{x} \; < \; \sin x \cdot \frac{1}{\sin x}$$

    $$\cos x \; < \; \frac{\sin x}{x} \; < 1, \qquad x > 0$$

57.

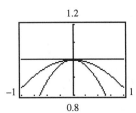

58. Amplitude $= |-5| = 5$. Period $= \dfrac{2\pi}{10} = \dfrac{\pi}{5}$ sec. Frequency $= \dfrac{1}{\text{Period}} = \dfrac{1}{\pi/5} = \dfrac{5}{\pi}$ Hz.

    Since $A = -5$ is negative, the basic curve for $y = \cos t$ is turned upside down. One full cycle of the

    graph is completed as t goes from 0 to $\dfrac{\pi}{5}$. Block out this interval, divide it into four equal parts,

    locate high and low points, and locate t intercepts. Then complete the graph.

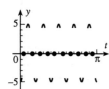
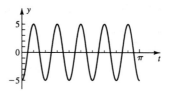

59. $I = 12 \sin(60\pi t - \pi)$
We compute amplitude, period, frequency, and phase shift as
follows: Amplitude $= |A| = |12| = 12$.

Phase Shift and Period: Solve

$$Bx + C \; = \; 0 \text{ and } Bx + C \; = \; 2\pi$$

$$60\pi t - \pi \; = \; 0 \qquad 60\pi t - \pi \; = \; 2\pi$$

$$60\pi t \; = \; \pi \qquad 60\pi t \; = \; \pi + 2\pi$$

$$t \; = \; \frac{1}{60} \qquad t \; = \; \frac{1}{60} + \frac{1}{30}$$

$$\qquad \underset{\text{Phase Shift}}{\uparrow} \qquad \underset{\text{Period}}{\uparrow \quad \uparrow}$$

$$\text{Frequency} = \frac{1}{\text{Period}} = \frac{1}{1/30} = 30 \text{ Hz}$$

Graph one cycle over the interval from   Extend the graph from 0 to 0.1.

$$\frac{1}{60} \text{ to } \frac{1}{60} + \frac{1}{30} \; \left(= \frac{1}{20} \right)$$

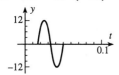
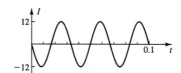

60. To find the period T, we use the formula $\lambda v = c$, with $v = \dfrac{1}{\text{Period}} = \dfrac{1}{T}$ and solve for T.

$$\lambda \frac{1}{T} = c; \; \lambda = Tc; \; T = \frac{\lambda}{c}. \; \text{ Hence, } T = \frac{6 \times 10^{-5} \text{ m}}{3 \times 10^{8} \text{ m/sec}} = 2 \times 10^{-13} \text{ sec.}$$

61. (A) In the figure, note that ABC and

CDE are right triangles, and that

AE = AC + CE = a + b. Then,

$\csc \theta = \dfrac{a}{100}$ from triangle ABC,

so a = 100 csc θ, sec $\theta = \dfrac{b}{70}$ from

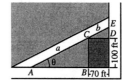

triangle CDE, so b = 70 sec θ.
Thus, AE = 100 csc θ + 70 sec θ.

(B) We form $y_1 = 100 \csc \theta$ and $y_2 = 70 \sec \theta$. We sketch the graph of each equation in the same coordinate system (dashed curves), then add the ordinates $y_1 + y_2$ (solid curve).

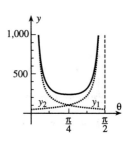

62. (A) In the right triangle in the figure, $\tan \theta = \dfrac{d}{50}$.

Hence, $d = 50 \tan \theta$.

(B) 20 rpm $= 20(2\pi)$ rad/min $= 40\pi$ rad/min. Since $\theta = \omega t$, and $\omega = 40\pi$ rad/min, $\theta = 40\pi t$.

(C) Substituting the expression for θ from part (B) into $d = 50 \tan \theta$, we obtain $d = 50 \tan 40\pi t$.

(D) Period $= \dfrac{\pi}{40\pi} = \dfrac{1}{40} = 0.025$.

One period of the graph would therefore extend from 0 to 0.025, with a vertical asymptote at $t = \dfrac{1}{80}$, or 0.0125. We sketch half of one period, since the required interval is from 0 to 0.0125 only. Ordinates can be determined from calculator.

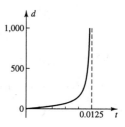

Thus,

t	0	0.0025	0.005	0.0075	0.01	0.011
$50 \tan 40\pi t$	0	16.2	36.3	68.8	154	262

63. (A)

x (months)	1, 13	2, 14	3, 15	4, 16	5, 17	6, 18	7, 19	8, 20	9, 21
y (temperatures)	19	23	32	45	55	65	71	69	62

x	10, 22	11, 23	12, 24
y	51	37	25

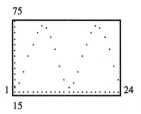

(B) From the table, Max y = 71 and Min y = 19. Then,

$$A = \frac{(Max\ y - Min\ y)}{2} = \frac{(71 - 19)}{2} = 26$$

$$B = \frac{2\pi}{Period} = \frac{2\pi}{12} = \frac{\pi}{6}$$

$$k = Min\ y + A = 19 + 26 = 45$$

From the plot in (A) or the table, we estimate the smallest positive value of x for which y = k = 45 to be approximately 4.2. Then, this is the phase-shift for the graph.

Substitute $B = \frac{\pi}{6}$ and x = 4.2 into the phase-shift equation

$$x = -\frac{C}{B}\ ,\ 4.2 = \frac{-C}{\pi/6}\ ,\ C = -\frac{4.2\pi}{6} \approx -2.2$$

Thus, the equation required is $y = 45 + 26\ \sin\left(\frac{\pi x}{6} - 2.2\right)$.

(C)

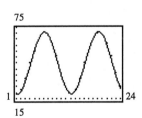

Chapter 4   Identities

EXERCISE 4.1 Fundamental Identities and Their Use

1. *Find tan x* : $\tan x = \dfrac{\sin x}{\cos x} = \dfrac{-2/3}{\sqrt{5/3}} = -\dfrac{2}{\sqrt{5}}$ *Find cot x* : $\cot x = \dfrac{1}{\tan x} = \dfrac{1}{-2/\sqrt{5}} = -\dfrac{\sqrt{5}}{2}$

   Find csc x : $\csc x = \dfrac{1}{\sin x} = \dfrac{1}{-2/3} = -\dfrac{3}{2}$ *Find sec x* : $\sec x = \dfrac{1}{\cos x} = \dfrac{1}{\sqrt{5/3}} = \dfrac{3}{\sqrt{5}}$

3. *Find cos x* : Since $\tan x = \dfrac{\sin x}{\cos x}$, we can write $\cos x = \dfrac{\sin x}{\tan x}$; $\cos x = \dfrac{\sin x}{\tan x} = \dfrac{-2/\sqrt{5}}{2} = -\dfrac{1}{\sqrt{5}}$

   Find csc x : $\csc x = \dfrac{1}{\sin x} = \dfrac{1}{-2/\sqrt{5}} = -\dfrac{\sqrt{5}}{2}$

   Find sec x : $\sec x = \dfrac{1}{\cos x} = \dfrac{1}{-1/\sqrt{5}} = -\sqrt{5}$ *Find cot x* : $\cot x = \dfrac{1}{\tan x} = \dfrac{1}{2}$

5. $\tan u \cot u = \tan u \dfrac{1}{\tan u}$ Reciprocal identity

            $= 1$ Algebra

7. $\tan x \csc x = \dfrac{\sin x}{\cos x} \dfrac{1}{\sin x}$ Quotient and reciprocal identities

            $= \dfrac{1}{\cos x}$ Algebra

            $= \sec x$ Reciprocal identity

9. $\dfrac{\sec^2 x - 1}{\tan x} = \dfrac{\tan^2 x + 1 - 1}{\tan x}$ Pythagorean identity

            $= \dfrac{\tan^2 x}{\tan x}$ Algebra

            $= \tan x$ Algebra

11. $\dfrac{\sin^2\theta}{\cos\theta} + \cos\theta = \dfrac{\sin^2\theta}{\cos\theta} + \dfrac{\cos\theta}{1}$ Algebra

            $= \dfrac{\sin^2\theta}{\cos\theta} + \dfrac{\cos^2\theta}{\cos\theta}$ Algebra

            $= \dfrac{\sin^2\theta + \cos^2\theta}{\cos\theta}$ Algebra

            $= \dfrac{1}{\cos\theta}$ Pythagorean identity

            $= \sec\theta$ Reciprocal identity

   Key algebraic steps: $\dfrac{a^2}{b} + b = \dfrac{a^2}{b} + \dfrac{b}{1} = \dfrac{a^2}{b} + \dfrac{b^2}{b} = \dfrac{a^2 + b^2}{b}$

13. $\dfrac{1}{\sin^2\beta} - 1 = \left(\dfrac{1}{\sin\beta}\right)^2 - 1$ Algebra

       $= \csc^2\beta - 1$ Reciprocal identity

       $= 1 + \cot^2\beta - 1$ Pythagorean identity

       $= \cot^2\beta$ Algebra

15. $\dfrac{(1-\cos x)^2 + \sin^2 x}{1-\cos x} = \dfrac{1 - 2\cos x + \cos^2 x + \sin^2 x}{1-\cos x}$ Algebra

       $= \dfrac{1 - 2\cos x + 1}{1-\cos x}$ Pythagorean identity

       $= \dfrac{2 - 2\cos x}{1-\cos x}$ Algebra

       $= \dfrac{2(1-\cos x)}{1-\cos x}$ Algebra

       $= 2$ Algebra

Key algebraic step: $(1-a)^2 = 1 - 2a + a^2$

17. *Find cos x :* We start with the Pythagorean identity $\sin^2 x + \cos^2 x = 1$ and solve for cos x.
$$\cos x = \pm\sqrt{1 - \sin^2 x}$$

Since sin x is positive and tan x is negative, x is associated with the second quadrant, where cos x is negative; hence, $\cos x = -\sqrt{1 - \sin^2 x} = -\sqrt{1 - \left(\tfrac{1}{4}\right)^2} = -\sqrt{\tfrac{15}{16}} = -\dfrac{\sqrt{15}}{4}$

Find sec x : $\sec x = \dfrac{1}{\cos x} = \dfrac{1}{-\sqrt{15}/4} = -\dfrac{4}{\sqrt{15}}$ *Find csc x :* $\csc x = \dfrac{1}{\sin x} = \dfrac{1}{1/4} = 4$

Find tan x : $\tan x = \dfrac{\sin x}{\cos x} = \dfrac{1/4}{-\sqrt{15}/4} = -\dfrac{1}{\sqrt{15}}$ *Find cot x :* $\cot x = \dfrac{1}{\tan x} = \dfrac{1}{-1/\sqrt{15}} = -\sqrt{15}$

19. *Find sec x :* We start with the Pythagorean identity $\tan^2 x + 1 = \sec^2 x$ and solve for sec x.
$$\sec x = \pm\sqrt{\tan^2 x + 1}$$

Since sin x and tan x are both negative, x is associated with the fourth quadrant, where sec x is positive; hence, $\sec x = \sqrt{\tan^2 x + 1} = \sqrt{(-2)^2 + 1} = \sqrt{5}$

Find cos x : $\cos x = \dfrac{1}{\sec x} = \dfrac{1}{\sqrt{5}}$

Find sin x : Since $\tan x = \dfrac{\sin x}{\cos x}$, we can write $\sin x = \cos x \tan x = \dfrac{1}{\sqrt{5}}(-2) = -\dfrac{2}{\sqrt{5}}$

Find csc x : $\csc x = \dfrac{1}{\sin x} = \dfrac{1}{-2/\sqrt{5}} = -\dfrac{\sqrt{5}}{2}$ *Find cot x :* $\cot x = \dfrac{1}{\tan x} = \dfrac{1}{-2} = -\dfrac{1}{2}$

Chapter 4   Identities

21. *Find sin x :* Since $\csc x = \dfrac{1}{\sin x}$, we can write $\sin x = \dfrac{1}{\csc x} = \dfrac{1}{3/2} = \dfrac{2}{3}$

Find cos x : We start with the Pythagorean identity $\sin^2 x + \cos^2 x = 1$ and solve for cos x.

$$\cos x = \pm\sqrt{1 - \sin^2 x}$$

Since sin x is positive and tan x is negative, x is associated with the second quadrant, where cos x is negative; hence, $\cos x = -\sqrt{1 - \sin^2 x} = -\sqrt{1 - \left(\dfrac{2}{3}\right)^2} = -\sqrt{\dfrac{5}{9}} = -\dfrac{\sqrt{5}}{3}$

Find tan x : $\tan x = \dfrac{\sin x}{\cos x} = \dfrac{2/3}{-\sqrt{5}/3} = -\dfrac{2}{\sqrt{5}}$ *Find sec x :* $\sec x = \dfrac{1}{\cos x} = \dfrac{1}{-\sqrt{5}/3} = -\dfrac{3}{\sqrt{5}}$

Find cot x : $\cot x = \dfrac{1}{\tan x} = \dfrac{1}{-2/\sqrt{5}} = -\dfrac{\sqrt{5}}{2}$

23. $\csc(-y)\cos(-y) = \dfrac{1}{\sin(-y)}\cos(-y)$ Reciprocal identity

    $\qquad\qquad = -\dfrac{1}{\sin y}\cos y$ Identities for negatives

    $\qquad\qquad = -\dfrac{\cos y}{\sin y}$ Algebra

    $\qquad\qquad = -\cot y$ Quotient identity

25. $\cot x \cos x + \sin x = \dfrac{\cos x}{\sin x}\cos x + \sin x$ Quotient identity

    $\qquad\qquad = \dfrac{\cos^2 x}{\sin x} + \dfrac{\sin x}{1}$ Algebra

    $\qquad\qquad = \dfrac{\cos^2 x}{\sin x} + \dfrac{\sin^2 x}{\sin x}$ Algebra

    $\qquad\qquad = \dfrac{\cos^2 x + \sin^2 x}{\sin x}$ Algebra

    $\qquad\qquad = \dfrac{1}{\sin x}$ Pythagorean identity

    $\qquad\qquad = \csc x$ Reciprocal identity

Key algebraic steps: $\dfrac{a}{b} a + b = \dfrac{a}{b} \cdot \dfrac{a}{1} + \dfrac{b}{1} = \dfrac{a^2}{b} + \dfrac{b}{1} = \dfrac{a^2}{b} + \dfrac{b^2}{b} = \dfrac{a^2 + b^2}{b}$

27. $\dfrac{\cot(-\theta)}{\csc\theta} + \cos\theta = \dfrac{\dfrac{\cos(-\theta)}{\sin(-\theta)}}{\dfrac{1}{\sin\theta}} + \cos\theta$ Quotient and reciprocal identities

$\qquad = \dfrac{\dfrac{\cos\theta}{-\sin\theta}}{\dfrac{1}{\sin\theta}} + \cos\theta$ Identities for negatives

$\qquad = \dfrac{\cos\theta}{-\sin\theta}\cdot\dfrac{\sin\theta}{1} + \cos\theta$ Algebra

$\qquad = -\cos\theta + \cos\theta$ Algebra

$\qquad = 0$ Algebra

29. $\dfrac{\cot x}{\tan x} + 1 = \cot x \div \tan x + 1$ Algebra

$\qquad = \cot x + \dfrac{1}{\cot x} + 1$ Reciprocal identity

$\qquad = \cot x \cdot \dfrac{\cot x}{1} + 1$ Algebra

$\qquad = \cot^2 x + 1$ Algebra

$\qquad = \csc^2 x$ Pythagorean identity

31. $\sec w \csc w - \sec w \sin w = \dfrac{1}{\cos w}\cdot\dfrac{1}{\sin w} - \dfrac{1}{\cos w}\cdot\sin w$ Reciprocal identities

$\qquad = \dfrac{1}{\cos w \sin w} - \dfrac{\sin w}{\cos w}$ Algebra

$\qquad = \dfrac{1}{\cos w \sin w} - \dfrac{\sin^2 w}{\cos w \sin w}$ Algebra

$\qquad = \dfrac{1 - \sin^2 w}{\cos w \sin w}$ Algebra

$\qquad = \dfrac{\cos^2 w}{\cos w \sin w}$ Pythagorean identity (solved for $1 - \sin^2 x = \cos^2 x$)

$\qquad = \dfrac{\cos w}{\sin w}$ Algebra

$\qquad = \cot w$ Quotient identity

33. (A) By the Pythagorean identity, $\sin^2\alpha + \cos^2\alpha = 1$ for any α. Therefore,

$$\sin^2\frac{x}{2} + \cos^2\frac{x}{2} = 1 \text{ (this is independent of x)}.$$

(B) By the Pythagorean identity, $\sec^2\alpha = 1 + \tan^2\alpha$ for any α. Therefore,

$$\sec^2\frac{x}{2} - \tan^2\frac{x}{2} = 1 + \tan^2\frac{x}{2} - \tan^2\frac{x}{2} = 1 \text{ (this is independent of x)}.$$

35. $\sqrt{1 - \cos^2 x} = \sqrt{\sin^2 x}$ by the Pythagorean identity, Therefore, $\sqrt{1 - \cos^2 x} = \sin x$ when, and only when, $\sqrt{\sin^2 x} = \sin x$. The latter statement is true whenever sin x is positive, that is, when x is in quadrant I or II.

37. $\sqrt{1 - \sin^2 x} = \sqrt{\cos^2 x}$ by the Pythagorean identity. Therefore, $\sqrt{1 - \sin^2 x} = -\cos x$ when, and only when, $\sqrt{\cos^2 x} = -\cos x$. The latter statement is true whenever cos x is negative, that is, when x is in quadrant II or III.

39. $\sqrt{1 - \sin^2 x} = \sqrt{\cos^2 x}$ by the Pythagorean identity. $\sqrt{\cos^2 x} = |\cos x|$ is always true. Hence, $\sqrt{1 - \sin^2 x} = |\cos x|$ in all quadrants.

41. $\dfrac{\sin x}{\sqrt{1 - \sin^2 x}} = \dfrac{\sin x}{\sqrt{\cos^2 x}}$ by the Pythagorean identity. $\tan x = \dfrac{\sin x}{\cos x}$ by the Quotient identity.

    Therefore, $\dfrac{\sin x}{\sqrt{1 - \sin^2 x}} = \dfrac{\sin x}{\sqrt{\cos^2 x}} = \dfrac{\sin x}{\cos x} = \tan x$ will be true whenever the middle two quantities are equal, that is, when, and only when, $\sqrt{\cos^2 x} = \cos x$. This statement is true whenever cos x is positive, that is, when x is in quadrant I or IV.

43. $\begin{aligned} \sqrt{a^2 - u^2} &= \sqrt{a^2 - (a \sin x)^2} && \text{using the given substitution} \\ &= \sqrt{a^2 - a^2 \sin^2 x} && \text{Algebra} \\ &= \sqrt{a^2(1 - \sin^2 x)} && \text{Algebra} \\ &= \sqrt{a^2 \cos^2 x} && \text{Pythagorean identity} \\ &= |a|\,|\cos x| && \text{Algebra} \\ &= a \cos x && \text{since } a > 0 \text{ and x is in quadrant I or IV} \end{aligned}$

    $\left(\text{given } -\dfrac{\pi}{2} < x < \dfrac{\pi}{2} \right)$, thus, cos x > 0.

45. $\begin{aligned} \sqrt{a^2 + u^2} &= \sqrt{a^2 + (a \tan x)^2} && \text{using the given substitution} \\ &= \sqrt{a^2 + a^2 \tan^2 x} && \text{Algebra} \\ &= \sqrt{a^2(1 + \tan^2 x)} && \text{Algebra} \\ &= \sqrt{a^2 \sec^2 x} && \text{Pythagorean identity} \\ &= |a|\,|\sec x| && \text{Algebra} \\ &= a \sec x && \text{since } a > 0 \text{ and x is in quadrant I} \end{aligned}$

    $\left(\text{given } 0 < x < \dfrac{\pi}{2} \right)$, thus, sec x > 0.

47. Following the hint, we write x = 5 cos t and y = 2 sin t in the form $\dfrac{x}{5} = \cos t$ and $\dfrac{y}{2} = \sin t$. Then, $\left(\dfrac{x}{5}\right)^2 + \left(\dfrac{y}{2}\right)^2 = (\cos t)^2 + (\sin t)^2 = 1$ by the Pythagorean identity. Thus, $\dfrac{x^2}{25} + \dfrac{y^2}{4} = 1$.

EXERCISE 4.2 Verifying Trigonometric Identities

1. Since $\dfrac{1}{\sin x}$ = csc x by the reciprocal identity, $\sin x = \dfrac{1}{\csc x}$. (A)

3. Since $\dfrac{1}{\cos x}$ = sec x by the reciprocal identity, $\cos x = \dfrac{1}{\sec x}$. (E)

5. Since $\dfrac{\cos x}{\sin x}$ = cot x by the quotient identity, this matches (H).

7. Since $\cos^2 x = 1 - \sin^2 x$ by the Pythagorean identity, this matches (J).

9. Since $\tan^2 x + 1 = \sec^2 x$ by the Pythagorean identity, this matches (L).

11. Since $\sin^2 x = 1 - \cos^2 x$ by the Pythagorean identity, this matches (K).

13. $\begin{aligned} \cos x \sec x &= \cos x \, \dfrac{1}{\cos x} \\ &= 1 \end{aligned}$

        Reciprocal identity

        Algebra

15. $\begin{aligned} \tan x \cos x &= \dfrac{\sin x}{\cos x} \cos x \\ &= \sin x \end{aligned}$

        Quotient identity

        Algebra

17. $\begin{aligned} \tan x &= \dfrac{\sin x}{\cos x} \\ &= \sin x \cdot \dfrac{1}{\cos x} \\ &= \sin x \sec x \end{aligned}$

        Quotient identity

        Algebra

        Reciprocal identity

19. $\begin{aligned} \csc(-x) &= \dfrac{1}{\sin(-x)} \\ &= \dfrac{1}{-\sin x} \\ &= -\dfrac{1}{\sin x} \\ &= -\csc x \end{aligned}$

        Reciprocal identity

        Identities for negatives

        Algebra

        Reciprocal identity

21. $\begin{aligned} \dfrac{\sin \alpha}{\cos \alpha \tan \alpha} &= \dfrac{\sin \alpha}{\cos \alpha \, \dfrac{\sin \alpha}{\cos \alpha}} \\ &= \dfrac{\sin \alpha}{\sin \alpha} \\ &= 1 \end{aligned}$

        Quotient identity

        Algebra

        Algebra

23. $\begin{aligned} \dfrac{\cos \beta \sec \beta}{\tan \beta} &= \dfrac{\cos \beta \, \dfrac{1}{\cos \beta}}{\tan \beta} \\ &= \dfrac{1}{\tan \beta} \\ &= \cot \beta \end{aligned}$

        Reciprocal identity

        Algebra

        Reciprocal identity

Chapter 4   Identities

25. $\sec \theta(\sin \theta + \cos \theta)$ $=$ $\sec \theta \sin \theta + \sec \theta \cos \theta$ Algebra

$= \dfrac{1}{\cos \theta} \sin \theta + \dfrac{1}{\cos \theta} \cos \theta$ Reciprocal identity

$= \dfrac{\sin \theta}{\cos \theta} + 1$ Algebra

$= \tan \theta + 1$ Quotient identity

27. $\dfrac{\cos^2 t - \sin^2 t}{\sin t \cos t}$ $=$ $\dfrac{\cos^2 t}{\sin t \cos t} - \dfrac{\sin^2 t}{\sin t \cos t}$ Algebra

$= \dfrac{\cos t}{\sin t} - \dfrac{\sin t}{\cos t}$ Algebra

$= \cot t - \tan t$ Quotient identity

Key algebraic steps: $\dfrac{b^2 - a^2}{ab} = \dfrac{b^2}{ab} - \dfrac{a^2}{ab} = \dfrac{b}{a} - \dfrac{a}{b}$

29. $\dfrac{\cos \beta}{\cot \beta} + \dfrac{\sin \beta}{\tan \beta}$ $=$ $\cos \beta \div \cot \beta + \sin \beta \div \tan \beta$ Algebra

$= \cos \beta \div \dfrac{\cos \beta}{\sin \beta} + \sin \beta \div \dfrac{\sin \beta}{\cos \beta}$ Quotient identity

$= \dfrac{\cos \beta}{1} \cdot \dfrac{\sin \beta}{\cos \beta} + \dfrac{\sin \beta}{1} \cdot \dfrac{\cos \beta}{\sin \beta}$ Algebra

$= \sin \beta + \cos \beta$ Algebra

31. $\sec^2 \theta - \tan^2 \theta$ $=$ $\tan^2 \theta + 1 - \tan^2 \theta$ Pythagorean identity

$= 1$ Algebra

33. $\sin^2 x(1 + \cot^2 x)$ $=$ $\sin^2 x \, \csc^2 x$ Pythagorean identity

$= \sin^2 x \left(\dfrac{1}{\sin x} \right)^2$ Reciprocal identity

$= \dfrac{\sin^2 x}{1} \cdot \dfrac{1}{\sin^2 x}$ Algebra

$= 1$ Algebra

35. $(\csc \alpha + 1)(\csc \alpha - 1)$ $=$ $\csc^2 \alpha - 1$ Algebra

$= 1 + \cot^2 \alpha - 1$ Pythagorean identity

$= \cot^2 \alpha$ Algebra

Key algebraic step: $(x + 1)(x - 1) = x^2 - 1$

37. $\dfrac{\sin t}{\csc t} + \dfrac{\cos t}{\sec t}$ $=$ $\dfrac{\sin t}{1/\sin t} + \dfrac{\cos t}{1/\cos t}$ Reciprocal identities

$= \sin t \div \dfrac{1}{\sin t} + \cos t \div \dfrac{1}{\cos t}$ Algebra

$$= \sin t \cdot \frac{\sin t}{1} + \cos t \cdot \frac{\cos t}{1} \qquad \text{Algebra}$$

$$= \sin^2 t + \cos^2 t \qquad \text{Algebra}$$

$$= 1 \qquad \text{Pythagorean identity}$$

39. $\dfrac{\sin^2 x}{\cos x} + \cos x = \dfrac{\sin^2 x}{\cos x} + \dfrac{\cos x}{1}$ Algebra

$$= \frac{\sin^2 x}{\cos x} + \frac{\cos^2 x}{\cos x} \qquad \text{Algebra}$$

$$= \frac{\sin^2 x + \cos^2 x}{\cos x} \qquad \text{Algebra}$$

$$= \frac{1}{\cos x} \qquad \text{Pythagorean identity}$$

$$= \sec x \qquad \text{Reciprocal identity}$$

Key algebraic steps: $\dfrac{a^2}{b} + b = \dfrac{a^2}{b} + \dfrac{b}{1} = \dfrac{a^2}{b} + \dfrac{b^2}{b} = \dfrac{a^2 + b^2}{b}$

41. $\dfrac{1 - (\cos\theta - \sin\theta)^2}{\cos\theta} = \dfrac{1 - (\cos^2\theta - 2\sin\theta\cos\theta + \sin^2\theta)}{\cos\theta}$ Algebra

$$= \frac{1 - \cos^2\theta + 2\sin\theta\cos\theta - \sin^2\theta}{\cos\theta} \qquad \text{Algebra}$$

$$= \frac{\sin^2\theta + 2\sin\theta\cos\theta - \sin^2\theta}{\cos\theta} \qquad \text{Pythagorean identity}$$

$$= \frac{2\sin\theta\cos\theta}{\cos\theta} \qquad \text{Algebra}$$

$$= 2\sin\theta \qquad \text{Algebra}$$

Key algebraic steps: $1 - (b - a)^2 = 1 - (b^2 - 2ab + a^2) = 1 - b^2 + 2ab - a^2$

$$\frac{a^2 + 2ab - a^2}{b} = \frac{2ab}{b} = 2a$$

43. $\dfrac{\tan w + 1}{\sec w} = \dfrac{\tan w}{\sec w} + \dfrac{1}{\sec w}$ Algebra

$$= \frac{\sin w/\cos w}{1/\cos w} + \cos w \qquad \text{Quotient and reciprocal identities}$$

$$= \frac{\sin w}{\cos w} \cdot \frac{\cos w}{1} + \cos w \qquad \text{Algebra}$$

$$= \frac{\sin w}{\cos w} \cdot \frac{\cos w}{1} + \cos w \qquad \text{Algebra}$$

$$= \sin w + \cos w \qquad \text{Algebra}$$

Key algebraic steps: $\dfrac{a/b}{1/b} + b = \dfrac{a}{b} \div \dfrac{1}{b} + b = \dfrac{a}{b} \cdot \dfrac{b}{1} + b = a + b$

45. $\dfrac{\cos s}{\sin^2 s - 1} = \dfrac{\cos s}{1 - \cos^2 s - 1}$ Pythagorean identity

$= \dfrac{\cos s}{-\cos^2 s}$ Algebra

$= -\dfrac{1}{\cos s}$ Algebra

$= -\sec s$ Reciprocal identity

47. $\dfrac{1}{1 - \cos^2\theta} = \dfrac{1}{\sin^2\theta}$ Pythagorean identity

$= \left(\dfrac{1}{\sin\theta}\right)^2$ Algebra

$= \csc^2\theta$ Reciprocal identity

$= 1 + \cot^2\theta$ Pythagorean identity

49. $\dfrac{\sin^2\beta}{1 - \cos\beta} = \dfrac{1 - \cos^2\beta}{1 - \cos\beta}$ Pythagorean identity

$= \dfrac{(1 - \cos\beta)(1 + \cos\beta)}{1 - \cos\beta}$ Algebra

$= 1 + \cos\beta$ Algebra

Key algebraic steps: $\dfrac{1 - b^2}{1 - b} = \dfrac{(1 - b)(1 + b)}{1 - b} = 1 + b$

51. $\dfrac{2 - \cos^2\theta}{\sin\theta} = \dfrac{2 - (1 - \sin^2\theta)}{\sin\theta}$ Pythagorean identity

$= \dfrac{2 - 1 + \sin^2\theta}{\sin\theta}$ Algebra

$= \dfrac{1 + \sin^2\theta}{\sin\theta}$ Algebra

$= \dfrac{1}{\sin\theta} + \dfrac{\sin^2\theta}{\sin\theta}$ Algebra

$= \dfrac{1}{\sin\theta} + \sin\theta$ Algebra

$= \csc\theta + \sin\theta$ Reciprocal identity

Key algebraic steps: $\dfrac{2 - (1 - a^2)}{a} = \dfrac{2 - 1 + a^2}{a} = \dfrac{1 + a^2}{a} = \dfrac{1}{a} + \dfrac{a^2}{a} = \dfrac{1}{a} + a$

53. $\tan x + \cot x = \dfrac{\sin x}{\cos x} + \dfrac{\cos x}{\sin x}$ Quotient identity

$= \dfrac{\sin^2 x}{\sin x \cos x} + \dfrac{\cos^2 x}{\sin x \cos x}$ Algebra

$= \dfrac{\sin^2 x + \cos^2 x}{\sin x \cos x}$ Algebra

$$= \frac{1}{\sin x \cos x} \qquad \text{Pythagorean identity}$$

$$= \frac{1}{\sin x} \cdot \frac{1}{\cos x} \qquad \text{Algebra}$$

$$= \sec x \csc x \qquad \text{Reciprocal identities}$$

Key algebraic steps: $\dfrac{a}{b} + \dfrac{b}{a} = \dfrac{a^2}{ab} + \dfrac{b^2}{ab} = \dfrac{a^2 + b^2}{ab}$

55. $\dfrac{1 - \csc x}{1 + \csc x} = \dfrac{1 - \dfrac{1}{\sin x}}{1 + \dfrac{1}{\sin x}}$ Reciprocal identity

$$= \dfrac{\sin x \cdot 1 - \sin x \cdot \dfrac{1}{\sin x}}{\sin x \cdot 1 + \sin x \cdot \dfrac{1}{\sin x}} \qquad \text{Algebra}$$

$$= \dfrac{\sin x - 1}{\sin x + 1} \qquad \text{Algebra}$$

57. $\csc^2\alpha - \cos^2\alpha - \sin^2\alpha = \csc^2\alpha - (\cos^2\alpha + \sin^2\alpha) \qquad \text{Algebra}$

$$= \csc^2\alpha - 1 \qquad \text{Pythagorean identity}$$

$$= \cot^2\alpha \qquad \text{Pythagorean identity}$$

59. $(\sin x + \cos x)^2 - 1 = \sin^2 x + 2 \sin x \cos x + \cos^2 x - 1 \qquad \text{Algebra}$

$$= 2 \sin x \cos x + \sin^2 x + \cos^2 x - 1 \qquad \text{Algebra}$$

$$= 2 \sin x \cos x + 1 - 1 \qquad \text{Pythagorean identity}$$

$$= 2 \sin x \cos x \qquad \text{Algebra}$$

61. $(\sin u - \cos u)^2 + (\sin u + \cos u)^2$

$$= \sin^2 u - 2 \sin u \cos u + \cos^2 u + \sin^2 u + 2 \sin u \cos u + \cos^2 u \qquad \text{Algebra}$$

$$= 2 \sin^2 u + 2 \cos^2 u \qquad \text{Algebra}$$

$$= 2(\sin^2 u + \cos^2 u) \qquad \text{Algebra}$$

$$= 2 \cdot 1 \text{ or } 2 \qquad \text{Pythagorean identity}$$

Key algebraic steps: $(a - b)^2 + (a + b)^2 = a^2 - 2ab + b^2 + a^2 + 2ab + b^2$

$$= 2a^2 + 2b^2 = 2(a^2 + b^2)$$

63. $\sin^4 x - \cos^4 x = (\sin^2 x)^2 - (\cos^2 x)^2 \qquad \text{Algebra}$

$$= (\sin^2 x - \cos^2 x)(\sin^2 x + \cos^2 x) \qquad \text{Algebra}$$

$$= (\sin^2 x - \cos^2 x)(1) \qquad \text{Pythagorean identity}$$

$$= \sin^2 x - \cos^2 x \qquad \text{Algebra}$$

$$= (1 - \cos^2 x) - \cos^2 x \qquad \text{Pythagorean identity}$$

$$= 1 - 2 \cos^2 x \qquad \text{Algebra}$$

Key algebraic steps: $a^4 - b^4 = (a^2)^2 - (b^2)^2 = (a^2 - b^2)(a^2 + b^2)$

65. $\dfrac{1 + \cos^2 s}{1 - \cos^4 s} = \dfrac{1 + \cos^2 s}{(1)^2 - (\cos^2 s)^2}$ Algebra

$\qquad\qquad = \dfrac{1 + \cos^2 s}{(1 + \cos^2 s)(1 - \cos^2 s)}$ Algebra

$\qquad\qquad = \dfrac{1}{1 - \cos^2 s}$ Algebra

$\qquad\qquad = \dfrac{1}{\sin^2 s}$ Pythagorean identity

$\qquad\qquad = \left(\dfrac{1}{\sin s}\right)^2$ Algebra

$\qquad\qquad = \csc^2 s$ Reciprocal identity

Key algebraic steps: $\dfrac{1 + b^2}{1 - b^4} = \dfrac{1 + b^2}{(1)^2 - (b^2)^2} = \dfrac{1 + b^2}{(1 + b^2)(1 - b^2)} = \dfrac{1}{1 - b^2}$

67. $\dfrac{\cos x}{1 - \sin x} + \dfrac{\cos x}{1 + \sin x}$

$\qquad = \dfrac{\cos x(1 + \sin x)}{(1 - \sin x)(1 + \sin x)} + \dfrac{\cos x(1 - \sin x)}{(1 + \sin x)(1 - \sin x)}$ Algebra

$\qquad = \dfrac{\cos x(1 + \sin x) + \cos x(1 - \sin x)}{(1 - \sin x)(1 + \sin x)}$ Algebra

$\qquad = \dfrac{\cos x + \sin x \cos x + \cos x - \sin x \cos x}{1 - \sin^2 x}$ Algebra

$\qquad = \dfrac{2 \cos x}{1 - \sin^2 x}$ Algebra

$\qquad = \dfrac{2 \cos x}{\cos^2 x}$ Pythagorean identity

$\qquad = \dfrac{2}{\cos x}$ Algebra

$\qquad = 2 \sec x$ Reciprocal identity

Key algebraic steps: $\dfrac{b}{1 - a} + \dfrac{b}{1 + a} = \dfrac{b(1 + a)}{(1 - a)(1 + a)} + \dfrac{b(1 - a)}{(1 + a)(1 - a)}$

$\qquad\qquad\qquad\qquad = \dfrac{b(1 + a) + b(1 - a)}{(1 - a)(1 + a)} = \dfrac{b + ab + b - ab}{1 - a^2} = \dfrac{2b}{1 - a^2}$

69. $\dfrac{\sin \alpha}{1 - \cos \alpha} - \dfrac{1 + \cos \alpha}{\sin \alpha}$

$\qquad = \dfrac{\sin \alpha \cdot \sin \alpha}{(1 - \cos \alpha)\sin \alpha} - \dfrac{(1 - \cos \alpha)(1 + \cos \alpha)}{(1 - \cos \alpha)\sin \alpha}$ Algebra

$$= \frac{\sin \alpha \sin \alpha - (1 - \cos \alpha)(1 + \cos \alpha)}{(1 - \cos \alpha)\sin \alpha} \qquad \text{Algebra}$$

$$= \frac{\sin^2\alpha - (1 - \cos^2\alpha)}{(1 - \cos \alpha)\sin \alpha} \qquad \text{Algebra}$$

$$= \frac{\sin^2\alpha - \sin^2\alpha}{(1 - \cos \alpha)\sin \alpha} \qquad \text{Pythagorean identity}$$

$$= \frac{0}{(1 - \cos \alpha)\sin \alpha} \qquad \text{Algebra}$$

$$= 0 \qquad \text{Algebra}$$

Key algebraic steps: $\dfrac{a}{1 - b} - \dfrac{1 + b}{a} = \dfrac{a \cdot a}{a(1 - b)} - \dfrac{(1 - b)(1 + b)}{a(1 - b)} = \dfrac{a^2 - (1 - b^2)}{a(1 - b)}$

71. $\dfrac{1}{\csc \theta + \cot \theta} + \dfrac{1}{\csc \theta - \cot \theta}$

$$= \frac{\csc \theta - \cot \theta}{(\csc \theta + \cot \theta)(\csc \theta - \cot \theta)} + \frac{\csc \theta + \cot \theta}{(\csc \theta + \cot \theta)(\csc \theta - \cot \theta)} \qquad \text{Algebra}$$

$$= \frac{\csc \theta - \cot \theta + \csc \theta + \cot \theta}{\csc^2\theta - \cot^2\theta} \qquad \text{Algebra}$$

$$= \frac{2 \csc \theta}{\csc^2\theta - \cot^2\theta} \qquad \text{Algebra}$$

$$= \frac{2 \csc \theta}{1 + \cot^2\theta - \cot^2\theta} \qquad \text{Pythagorean identity}$$

$$= \frac{2 \csc \theta}{1} \qquad \text{Algebra}$$

$$= 2 \csc \theta \qquad \text{Algebra}$$

Key algebraic steps: $\dfrac{1}{a + b} + \dfrac{1}{a - b} = \dfrac{a - b}{(a + b)(a - b)} + \dfrac{a + b}{(a + b)(a - b)}$

$$= \frac{a - b + a + b}{a^2 - b^2} = \frac{2a}{a^2 - b^2}$$

73. $\dfrac{\cos^2 n - 3 \cos n + 2}{\sin^2 n} = \dfrac{(\cos n - 2)(\cos n - 1)}{\sin^2 n} \qquad \text{Algebra}$

$$= \frac{(\cos n - 2)(\cos n - 1)}{1 - \cos^2 n} \qquad \text{Pythagorean identity}$$

$$= \frac{(\cos n - 2)(\cos n - 1)}{(1 - \cos n)(1 + \cos n)} \qquad \text{Algebra}$$

$$= \frac{(\cos n - 2)(-1)}{1 + \cos n} \qquad \text{Algebra}$$

$$= \frac{2 - \cos n}{1 + \cos n} \qquad \text{Algebra}$$

75. $\dfrac{1 - \cot^2 x}{\tan^2 x - 1} = \dfrac{1 - \cot^2 x}{\dfrac{1}{\cot^2 x} - 1}$ Reciprocal identity

$= \dfrac{\cot^2 x(1 - \cot^2 x)}{\cot^2 x \cdot \dfrac{1}{\cot^2 x} - \cot^2 x \cdot 1}$ Algebra

$= \dfrac{\cot^2 x(1 - \cot^2 x)}{1 - \cot^2 x}$ Algebra

$= \cot^2 x$ Algebra

77. $\sec^2 x + \csc^2 x = \dfrac{1}{\cos^2 x} + \dfrac{1}{\sin^2 x}$ Reciprocal identities

$= \dfrac{\sin^2 x}{\cos^2 x \sin^2 x} + \dfrac{\cos^2 x}{\cos^2 x \sin^2 x}$ Algebra

$= \dfrac{\sin^2 x + \cos^2 x}{\cos^2 x \sin^2 x}$ Algebra

$= \dfrac{1}{\cos^2 x \sin^2 x}$ Pythagorean identity

$= \dfrac{1}{\cos^2 x} \cdot \dfrac{1}{\sin^2 x}$ Algebra

$= \sec^2 x \csc^2 x$ Reciprocal identities

79. $(\sec x - \tan x)^2 = \left(\dfrac{1}{\cos x} - \dfrac{\sin x}{\cos x}\right)^2$ Reciprocal and quotient identities

$= \left(\dfrac{1 - \sin x}{\cos x}\right)^2$ Algebra

$= \dfrac{(1 - \sin x)^2}{\cos^2 x}$ Algebra

$= \dfrac{(1 - \sin x)^2}{1 - \sin^2 x}$ Pythagorean identity

$= \dfrac{(1 - \sin x)(1 - \sin x)}{(1 - \sin x)(1 + \sin x)}$ Algebra

$= \dfrac{1 - \sin x}{1 + \sin x}$ Algebra

81. $\dfrac{1 + \sin t}{\cos t} = \dfrac{(1 + \sin t)\cos t}{\cos t \cos t}$ Algebra

$= \dfrac{(1 + \sin t)\cos t}{\cos^2 t}$ Algebra

$= \dfrac{(1 + \sin t)\cos t}{1 - \sin^2 t}$ Pythagorean identity

$$= \frac{(1 + \sin t)\cos t}{(1 + \sin t)(1 - \sin t)} \qquad \text{Algebra}$$

$$= \frac{\cos t}{1 - \sin t} \qquad \text{Algebra}$$

83. $\dfrac{\cos \alpha}{\sec \alpha - 1} = \dfrac{\cos \alpha}{\sec \alpha - 1} \cdot \dfrac{\sec \alpha + 1}{\sec \alpha + 1} \qquad$ Algebra

$$= \frac{\cos \alpha(\sec \alpha + 1)}{\sec^2 \alpha - 1} \qquad \text{Algebra}$$

$$= \frac{\cos \alpha \sec \alpha + \cos \alpha}{\sec^2 \alpha - 1} \qquad \text{Algebra}$$

$$= \frac{\cos \alpha \sec \alpha + \cos \alpha}{1 + \tan^2 \alpha - 1} \qquad \text{Pythagorean identity}$$

$$= \frac{\cos \alpha \sec \alpha + \cos \alpha}{\tan^2 \alpha} \qquad \text{Algebra}$$

$$= \frac{\cos \alpha \cdot \dfrac{1}{\cos \alpha} + \cos \alpha}{\tan^2 \alpha} \qquad \text{Reciprocal identity}$$

$$= \frac{1 + \cos \alpha}{\tan^2 \alpha} \ \text{ or } \ \frac{\cos \alpha + 1}{\tan^2 \alpha} \qquad \text{Algebra}$$

85. There are an infinite number of possibilities.

One choice might be $x = \dfrac{\pi}{6}$: $\ \tan \dfrac{\pi}{6} = \dfrac{1}{\sqrt{3}} \neq \sqrt{3} = \cot \dfrac{\pi}{6}$

87. There are an infinite number of possibilities.

One choice might be $x = 0$: $\ \sin^2 0 - \cos^2 0 = (0)^2 - (1)^2 = -1 \neq 1$

89. There are an infinite number of possibilities.

One choice might be $x = \dfrac{\pi}{2}$: $\ \cot^2 \dfrac{\pi}{2} + \cos \dfrac{\pi}{2} = 0^2 + 0 = 0 \neq 1 = \sin^2 \dfrac{\pi}{2}$

91. $\dfrac{\sin x}{1 - \cos x} - \cot x = \dfrac{\sin x}{1 - \cos x} - \dfrac{\cos x}{\sin x} \qquad$ Quotient identity

$$= \frac{(\sin x)(\sin x)}{\sin x(1 - \cos x)} - \frac{(1 - \cos x)(\cos x)}{(1 - \cos x)(\sin x)} \qquad \text{Algebra}$$

$$= \frac{(\sin x)(\sin x) - (1 - \cos x)(\cos x)}{(1 - \cos x)\sin x} \qquad \text{Algebra}$$

$$= \frac{\sin^2 x - \cos x + \cos^2 x}{(1 - \cos x)\sin x} \qquad \text{Algebra}$$

$$= \frac{\sin^2 x + \cos^2 x - \cos x}{(1 - \cos x)\sin x} \qquad \text{Algebra}$$

$$= \frac{1 - \cos x}{(1 - \cos x)\sin x} \qquad \text{Pythagorean identity}$$

$$= \frac{1}{\sin x} \qquad \text{Algebra}$$

$$= \csc x \qquad \text{Reciprocal identity}$$

Key algebraic steps: $\dfrac{a}{1-b} - \dfrac{b}{a} = \dfrac{aa}{a(1-b)} - \dfrac{b(1-b)}{a(1-b)} = \dfrac{a^2 - b(1-b)}{a(1-b)}$

$$= \frac{a^2 - b + b^2}{a(1-b)} = \frac{a^2 + b^2 - b}{a(1-b)}$$

93. $\dfrac{\cot \beta}{\csc \beta + 1} = \dfrac{\cot \beta(\csc \beta - 1)}{(\csc \beta + 1)(\csc \beta - 1)} \qquad$ Algebra

$$= \frac{\cot \beta(\csc \beta - 1)}{\csc^2\beta - 1} \qquad \text{Algebra}$$

$$= \frac{\cot \beta(\csc \beta - 1)}{1 + \cot^2\beta - 1} \qquad \text{Pythagorean identity}$$

$$= \frac{\cot \beta(\csc \beta - 1)}{\cot^2\beta} \qquad \text{Algebra}$$

$$= \frac{\csc \beta - 1}{\cot \beta} \qquad \text{Algebra}$$

95. $\dfrac{3\cos^2m + 5\sin m - 5}{\cos^2m} = \dfrac{3(1 - \sin^2m) + 5\sin m - 5}{1 - \sin^2m} \qquad$ Pythagorean identity

$$= \frac{3 - 3\sin^2m + 5\sin m - 5}{1 - \sin^2m} \qquad \text{Algebra}$$

$$= \frac{-3\sin^2m + 5\sin m - 2}{1 - \sin^2m} \qquad \text{Algebra}$$

$$= \frac{(-\sin m + 1)(3\sin m - 2)}{(1 - \sin m)(1 + \sin m)} \qquad \text{Algebra}$$

$$= \frac{(1 - \sin m)(3\sin m - 2)}{(1 - \sin m)(1 + \sin m)} \qquad \text{Algebra}$$

$$= \frac{3\sin m - 2}{1 + \sin m} \qquad \text{Algebra}$$

97. In this problem, it is more straightforward to start with the right-hand side of the identity to be verified. The student can confirm that the steps would be valid if reversed.

$$\frac{\tan x + \tan y}{1 - \tan x \tan y} = \frac{\dfrac{\sin x}{\cos x} + \dfrac{\sin y}{\cos y}}{1 - \dfrac{\sin x}{\cos x}\dfrac{\sin y}{\cos y}} \qquad \text{Quotient identity}$$

$$= \frac{\cos x \cos y \left(\dfrac{\sin x}{\cos x} + \dfrac{\sin y}{\cos y} \right)}{\cos x \cos y \left(1 - \dfrac{\sin x \sin y}{\cos x \cos y} \right)} \qquad \text{Algebra}$$

$$= \frac{\cos x \cos y \dfrac{\sin x}{\cos x} + \cos x \cos y \dfrac{\sin y}{\cos y}}{\cos x \cos y - \cos x \cos y \dfrac{\sin x \sin y}{\cos x \cos y}} \qquad \text{Algebra}$$

$$= \frac{\sin x \cos y + \cos x \sin y}{\cos x \cos y - \sin x \sin y} \qquad \text{Algebra}$$

99. The graph of f(x) is shown in the figure.

This graph appears to have vertical asymptotes $x = -\dfrac{\pi}{2}$ and $x = \dfrac{\pi}{2}$, x intercepts $-\pi$, 0, and π, and Period π. It appears that g(x) = tan x would be an appropriate choice.
We verify f(x) = g(x) as follows:

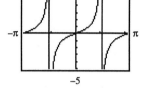

$$f(x) = \frac{1 + \sin x}{2 \cos x} - \frac{\cos x}{2 + 2 \sin x}$$

$$= \frac{(1 + \sin x)(1 + \sin x)}{2 \cos x (1 + \sin x)} - \frac{\cos x \cos x}{2 \cos x (1 + \sin x)} \qquad \text{Algebra}$$

$$= \frac{(1 + \sin x)^2 - \cos^2 x}{2 \cos x (1 + \sin x)} \qquad \text{Algebra}$$

$$= \frac{1 + 2 \sin x + \sin^2 x - \cos^2 x}{2 \cos x (1 + \sin x)} \qquad \text{Algebra}$$

$$= \frac{1 - \cos^2 x + \sin^2 x + 2 \sin x}{2 \cos x (1 + \sin x)} \qquad \text{Algebra}$$

$$= \frac{\sin^2 x + \sin^2 x + 2 \sin x}{2 \cos x (1 + \sin x)} \qquad \text{Pythagorean identity}$$

$$= \frac{2 \sin^2 x + 2 \sin x}{2 \cos x (1 + \sin x)} \qquad \text{Algebra}$$

$$= \frac{2 \sin x (1 + \sin x)}{2 \cos x (1 + \sin x)} \qquad \text{Algebra}$$

$$= \frac{\sin x}{\cos x} \qquad \text{Algebra}$$

$$= \tan x = g(x) \qquad \text{Quotient identity}$$

Chapter 4   Identities

101. The graph of f(x) is shown in the figure.

The graph appears to have vertical asymptotes $x = -\frac{3\pi}{2}$, $-\frac{\pi}{2}$, $\frac{\pi}{2}$, and $\frac{3\pi}{2}$, and Period π. It appears to have high and low points with y coordinates 0 and 2, respectively. It appears that g(x) = 1 + sec x would be an appropriate choice. We verify f(x) = g(x)

as follows:

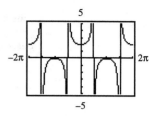

$$f(x) = \frac{\sin x \tan x}{1 - \cos x}$$

$$= \frac{\sin x \cdot \frac{\sin x}{\cos x}}{1 - \cos x} \qquad \text{Quotient identity}$$

$$= \frac{\sin^2 x}{\cos x (1 - \cos x)} \qquad \text{Algebra}$$

$$= \frac{1 - \cos^2 x}{\cos x (1 - \cos x)} \qquad \text{Pythagorean identity}$$

$$= \frac{(1 + \cos x)(1 - \cos x)}{\cos x (1 - \cos x)} \qquad \text{Algebra}$$

$$= \frac{1 + \cos x}{\cos x} \qquad \text{Algebra}$$

$$= \frac{1}{\cos x} + 1 \qquad \text{Algebra}$$

$$= \sec x + 1 = g(x) \qquad \text{Reciprocal identity}$$

103. The graph of f(x) is shown in the figure.

The graph appears to be a basic sine curve with period 2π, phase shift 0, and amplitude 2. It appears that g(x) = 2 sin x would be an appropriate choice. We verify f(x) = g(x) as follows:

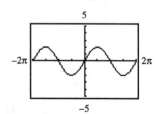

$$f(x) = \frac{3 \sin x - 2 \sin x \cos x}{1 - \cos x} - \frac{1 + \cos x}{\sin x}$$

$$= \frac{(3 \sin x - 2 \sin x \cos x)\sin x - (1 + \cos x)(1 - \cos x)}{(1 - \cos x)\sin x} \qquad \text{Algebra}$$

$$= \frac{3 \sin^2 x - 2 \sin^2 x \cos x - (1 - \cos^2 x)}{(1 - \cos x)\sin x} \qquad \text{Algebra}$$

$$= \frac{3 \sin^2 x - 2 \sin^2 x \cos x - \sin^2 x}{(1 - \cos x)\sin x} \qquad \text{Pythagorean identity}$$

$$= \frac{\sin^2 x(3 - 2\cos x - 1)}{(1 - \cos x)\sin x}$$

Algebra

$$= \frac{\sin^2 x(2 - 2\cos x)}{(1 - \cos x)\sin x}$$

Algebra

$$= 2\sin x = g(x)$$

Algebra

EXERCISE 4.3 Sum, Difference, and Cofunction Identities

1. Use the sum identity for cosine, replacing y with 2π.

$$\cos(x + y) = \cos x \cos y - \sin x \sin y$$

$$\cos(x + 2\pi) = \cos x \cos 2\pi - \sin x \sin 2\pi = \cos x(1) - \sin x(0) = \cos x$$

3. Use the reciprocal identity together with the sum identity for tangent, replacing y with π.

$$\cot(x + \pi) = \frac{1}{\tan(x + \pi)} = \frac{1}{\dfrac{\tan x + \tan \pi}{1 - \tan x \tan \pi}} = \frac{1}{\dfrac{\tan x + 0}{1 - \tan x \cdot 0}} = \frac{1}{\tan x} = \cot x$$

5. Use the sum identity for sine, replacing y with $2k\pi$.

$$\sin(x + y) = \sin x \cos y + \cos x \sin y$$

$$\sin(x + 2k\pi) = \sin x \cos 2k\pi + \cos x \sin 2k\pi = \sin x(1) + \cos x(0) = \sin x$$

7. Use the sum identity for tangent, replacing y with $k\pi$.

$$\tan(x + y) = \frac{\tan x + \tan y}{1 - \tan x \tan y}$$

$$\tan(x + k\pi) = \frac{\tan x + \tan k\pi}{1 - \tan x \tan k\pi} = \frac{\tan x + 0}{1 - \tan x \cdot 0} = \tan x$$

9. $\tan\left(\dfrac{\pi}{2} - 2\right) = \dfrac{\sin\left(\dfrac{\pi}{2} - x\right)}{\cos\left(\dfrac{\pi}{2} - x\right)}$ Quotient identity

$$= \frac{\cos x}{\sin x}$$ Cofunction identities

$$= \cot x$$ Quotient identity

11. $\sec\left(\dfrac{\pi}{2} - x\right) = \dfrac{1}{\cos\left(\dfrac{\pi}{2} - x\right)}$ Reciprocal identity

$$= \frac{1}{\sin x}$$ Cofunction identity

$$= \csc x$$ Reciprocal identity

13. Use the difference identity for sine, replacing y with $45°$.

$$\sin(x - y) = \sin x \cos y - \cos x \sin y$$

$$\sin(x - 45°) = \sin x \cos 45° - \cos x \sin 45° = \sin x \frac{\sqrt{2}}{2} - \cos x \frac{\sqrt{2}}{2} = \frac{\sqrt{2}}{2}(\sin x - \cos x)$$

Chapter 4   Identities

15. Use the sum identity for cosine, replacing y with 180°.

$$\cos(x + y) = \cos x \cos y - \sin x \sin y$$

$$\cos(x + 180°) = \cos x \cos 180° - \sin x \sin 180° = \cos x(-1) - \sin x(0) = -\cos x$$

17. Use the difference identity for tangent, replacing x with $\frac{\pi}{4}$ and y with x.

$$\tan(x - y) = \frac{\tan x - \tan y}{1 + \tan x \tan y}$$

$$\tan\left(\frac{\pi}{4} - x\right) = \frac{\tan \frac{\pi}{4} - \tan x}{1 + \tan \frac{\pi}{4} \tan x} = \frac{1 - \tan x}{1 + 1 \cdot \tan x} = \frac{1 - \tan x}{1 + \tan x}$$

19. Since we can write $75° = 45° + 30°$, the sum of two special angles, we can use the sum identity for sine with $x = 45°$ and $y = 30°$.

$$\sin(x + y) = \sin x \cos y + \cos x \sin y$$

$$\sin(45° + 30°) = \sin 45° \cos 30° + \cos 45° \sin 30°$$

$$= \frac{1}{\sqrt{2}} \cdot \frac{\sqrt{3}}{2} + \frac{1}{\sqrt{2}} \cdot \frac{1}{2} = \frac{\sqrt{3}}{2\sqrt{2}} + \frac{1}{2\sqrt{2}} = \frac{\sqrt{3} + 1}{2\sqrt{2}}$$

21. Using the hint, we write $\frac{\pi}{12} = \frac{\pi}{4} - \frac{\pi}{6}$, the difference of two special angles, and use the difference identity for cosine with $x = \frac{\pi}{4}$ and $y = \frac{\pi}{6}$.

$$\cos(x - y) = \cos x \cos y + \sin x \sin y$$

$$\cos\left(\frac{\pi}{4} - \frac{\pi}{6}\right) = \cos\frac{\pi}{4}\cos\frac{\pi}{6} + \sin\frac{\pi}{4}\sin\frac{\pi}{6} = \frac{1}{\sqrt{2}} \cdot \frac{\sqrt{3}}{2} + \frac{1}{\sqrt{2}} \cdot \frac{1}{2} = \frac{\sqrt{3}}{2\sqrt{2}} + \frac{1}{2\sqrt{2}} = \frac{\sqrt{3} + 1}{2\sqrt{2}}$$

23. Since we can write $60° = 22° + 38°$, we can use the sum identity for sine with $x = 22°$ and $y = 38°$.

$$\sin x \cos y + \cos x \sin y = \sin(x + y)$$

$$\sin 22° \cos 38° + \cos 22° \sin 38° = \sin(22° + 38°) = \sin 60° = \frac{\sqrt{3}}{2}$$

25. Since we can write $60° = 110° - 50°$, we can use the difference identity for tangent with $x = 110°$ and $y = 50°$.

$$\frac{\tan x - \tan y}{1 + \tan x \tan y} = \tan(x - y) \quad \text{and} \quad \frac{\tan 110° - \tan 50°}{1 + \tan 110° \tan 50°} = \tan(110° - 50°) = \tan 60° = \sqrt{3}$$

27. To find $\sin(x - y)$, we start with the difference identity for sine:

$$\sin(x - y) = \sin x \cos y - \cos x \sin y.$$

We know sin x and cos y, but not cos x and sin y. We find the latter two values by using reference triangles and the Pythagorean theorem:

$\sin x = \frac{2}{3}$

$\cos y = -\frac{1}{4}$

140

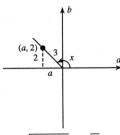

$a = -\sqrt{3^2 - 2^2} = -\sqrt{5}$

$\cos x = -\dfrac{\sqrt{5}}{3}$

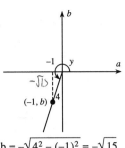

$b = -\sqrt{4^2 - (-1)^2} = -\sqrt{15}$

$\sin y = -\dfrac{\sqrt{15}}{4}$

(handwritten)
$SIN\ x = \cancel{}\ b/R$

$sin y = \cancel{}\ b/R$

$cos\ x = A/R$

$cos y = A/R$

Thus, $\sin(x - y) = \sin x \cos y - \cos x \sin y$

$$= \frac{2}{3}\left(-\frac{1}{4}\right) - \left(-\frac{\sqrt{5}}{3}\right)\left(-\frac{\sqrt{15}}{4}\right) = \frac{-2}{12} - \frac{\sqrt{75}}{12} = \frac{-2 - \sqrt{75}}{12}$$

To find $\tan(x + y)$, we start with the sum identity for tangents: $\tan(x + y) = \dfrac{\tan x + \tan y}{1 - \tan x \tan y}$.

Since $\sin x = \dfrac{2}{3}$ and $\cos x = -\dfrac{\sqrt{5}}{3}$, we know $\tan x = \dfrac{\sin x}{\cos x} = \dfrac{2/3}{-\sqrt{5}/3} = -\dfrac{2}{\sqrt{5}}$. Since

$\sin y = -\dfrac{\sqrt{15}}{4}$ and $\cos y = -\dfrac{1}{4}$, we know $\tan y = \dfrac{\sin y}{\cos y} = \dfrac{-\sqrt{15}/4}{-1/4} = \sqrt{15}$. Thus,

$$\tan(x + y) = \frac{\tan x + \tan y}{1 - \tan x \tan y} = \frac{-\dfrac{2}{\sqrt{5}} + \sqrt{15}}{1 - \left(-\dfrac{2}{\sqrt{5}}\right)\sqrt{15}} = \frac{\sqrt{5}\left(-\dfrac{2}{\sqrt{5}}\right) + \sqrt{5}\sqrt{15}}{\sqrt{5} - \sqrt{5}\left(-\dfrac{2}{\sqrt{5}}\right)\sqrt{15}} = \frac{-2 + \sqrt{75}}{\sqrt{5} + 2\sqrt{15}}$$

29. To find $\sin(x - y)$, we start with the difference identity for sine:

$\sin(x - y) = \sin x \cos y - \cos x \sin y$. We know cos x (and tan y), but not sin x, cos y, or sin y.

We find the latter three values by using reference triangles and the Pythagorean theorem:

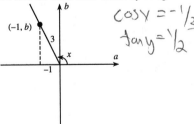

(handwritten)
$cos y = -1/3$
$tan y = 1/2$

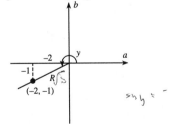

(handwritten) $sin y = -1/s5$

$b = \sqrt{3^2 - (-1)^2} = \sqrt{8}; \ \sin x = \dfrac{\sqrt{8}}{3}$

$R = \sqrt{(-2)^2 + (-1)^2} = \sqrt{5}; \ \sin y = \dfrac{-1}{\sqrt{5}}; \ \cos y = \dfrac{-2}{\sqrt{5}}$

Thus, $\sin(x - y) = \sin x \cos y - \cos x \sin y$

$$= \left(\frac{\sqrt{8}}{3}\right)\left(-\frac{2}{\sqrt{5}}\right) - \left(-\frac{1}{3}\right)\left(-\frac{1}{\sqrt{5}}\right) = \frac{-2\sqrt{8}}{3\sqrt{5}} - \frac{1}{3\sqrt{5}} = \frac{-2\sqrt{8} - 1}{3\sqrt{5}}$$

To find tan(x + y), we start with the sum identity for tangent: $\tan(x+y) = \dfrac{\tan x + \tan y}{1 - \tan x \tan y}$.

Since $\sin x = \dfrac{\sqrt{8}}{3}$, and $\cos x = -\dfrac{1}{3}$, we know $\tan x = \dfrac{\sin x}{\cos x} = \dfrac{\sqrt{8}/3}{-1/3} = -\sqrt{8}$. We also know

that $\tan y = \dfrac{1}{2}$. Thus,

$$\tan(x+y) = \frac{\tan x + \tan y}{1 - \tan x \tan y} = \frac{-\sqrt{8} + \frac{1}{2}}{1 - \left(-\sqrt{8}\right)\left(\frac{1}{2}\right)}$$

$$= \frac{-2\sqrt{8} + 2\left(\frac{1}{2}\right)}{2 - 2\left(-\sqrt{8}\right)\left(\frac{1}{2}\right)} = \frac{-2\sqrt{8} + 1}{2 + \sqrt{8}} = \frac{1 - 4\sqrt{2}}{2 + 2\sqrt{2}}$$

31. $\sin 2x = \sin(x + x)$ Algrebra

    $= \sin x \cos x + \cos x \sin x$ Sum identity for sine

    $= \sin x \cos x + \sin x \cos x$ Algebra

    $= 2 \sin x \cos x$ Algebra

33. $\cot(x-y) = \dfrac{\cos(x-y)}{\sin(x-y)}$ Quotient identity

$$= \frac{\cos x \cos y + \sin x \sin y}{\sin x \cos y - \cos x \sin y}$$ Difference identities for sine and cosine

$$= \frac{\dfrac{\cos x \cos y}{\sin x \sin y} + \dfrac{\sin x \sin y}{\sin x \sin y}}{\dfrac{\sin x \cos y}{\sin x \sin y} - \dfrac{\cos x \sin y}{\sin x \sin y}}$$ Algebra

$$= \frac{\dfrac{\cos x \cos y}{\sin x \sin y} + 1}{\dfrac{\cos y}{\sin y} - \dfrac{\cos x}{\sin x}}$$ Algebra

$$= \frac{\cot x \cot y + 1}{\cot y - \cot x}$$ Quotient identity

35. $\cot 2x = \dfrac{\cos 2x}{\sin 2x}$ Quotient identity

$$= \frac{\cos(x+x)}{\sin(x+x)}$$ Algebra

$$= \frac{\cos x \cos x - \sin x \sin x}{\sin x \cos x + \cos x \sin x}$$ Sum identities for sine and cosine

$$= \frac{\cos x \cos x - \sin x \sin x}{2 \sin x \cos x}$$ Algebra

$$= \frac{\dfrac{\cos x \cos x}{\sin x \sin x} - \dfrac{\sin x \sin x}{\sin x \sin x}}{\dfrac{2 \sin x \cos x}{\sin x \sin x}}$$ Algebra

$$= \frac{\dfrac{\cos x}{\sin x}\dfrac{\cos x}{\sin x} - 1}{2\dfrac{\cos x}{\sin x}}$$ Algebra

$$= \frac{\cot x \cot x - 1}{2 \cot x}$$ Quotient identity

$$= \frac{\cot^2 x - 1}{2 \cot x}$$ Algebra

37. $\dfrac{\tan \alpha + \tan \beta}{\tan \alpha - \tan \beta} = \dfrac{\dfrac{\sin \alpha}{\cos \alpha} + \dfrac{\sin \beta}{\cos \beta}}{\dfrac{\sin \alpha}{\cos \alpha} - \dfrac{\sin \beta}{\cos \beta}}$ Quotient identity

$$= \frac{\dfrac{\cos \alpha \cos \beta}{1} \cdot \dfrac{\sin \alpha}{\cos \alpha} + \dfrac{\cos \alpha \cos \beta}{1} \cdot \dfrac{\sin \beta}{\cos \beta}}{\dfrac{\cos \alpha \cos \beta}{1} \cdot \dfrac{\sin \alpha}{\cos \alpha} - \dfrac{\cos \alpha \cos \beta}{1} \cdot \dfrac{\sin \beta}{\cos \beta}}$$ Algebra

$$= \frac{\sin \alpha \cos \beta + \cos \alpha \sin \beta}{\sin \alpha \cos \beta - \cos \alpha \sin \beta}$$ Algebra

$$= \frac{\sin(\alpha + \beta)}{\sin(\alpha - \beta)}$$ Sum identity for sine

39. $\dfrac{\sin(x - y)}{\cos x \cos y} = \dfrac{\sin x \cos y - \cos x \sin y}{\cos x \cos y}$ Difference identity for sine

$$= \frac{\sin x \cos y}{\cos x \cos y} - \frac{\cos x \sin y}{\cos x \cos y}$$ Algebra

$$= \frac{\sin x}{\cos x} - \frac{\sin y}{\cos y}$$ Algebra

$$= \tan x - \tan y$$ Quotient identities

41. $\tan(x + y) = \dfrac{\sin(x + y)}{\cos(x + y)}$ Quotient identity

$$= \frac{\sin x \cos y + \cos x \sin y}{\cos x \cos y - \sin x \sin y}$$ Sum identities for sine and cosine

$$= \frac{\dfrac{\sin x \cos y}{\sin x \sin y} + \dfrac{\cos x \sin y}{\sin x \sin y}}{\dfrac{\cos x \cos y}{\sin x \sin y} - \dfrac{\sin x \sin y}{\sin x \sin y}}$$ Algebra

$$= \frac{\dfrac{\cos y}{\sin y} + \dfrac{\cos x}{\sin x}}{\dfrac{\cos x \cos y}{\sin x \sin y} - 1}$$

Algebra

$$= \frac{\cot y + \cot x}{\cot y \cot x - 1}$$

Quotient identities

$$= \frac{\cot x + \cot y}{\cot x \cot y - 1}$$

Algebra

43. $\dfrac{\sin(x + h) - \sin x}{h} = \dfrac{\sin x \cos h + \cos x \sin h - \sin x}{h}$

Algebra

$$= \frac{\sin x \cos h - \sin x + \cos x \sin h}{h}$$

Algebra

$$= \frac{\sin x(\cos h - 1) + \cos x \sin h}{h}$$

Algebra

$$= \sin x \frac{\cos h - 1}{h} + \cos x \frac{\sin h}{h}$$

Algebra

45. Following the hint, we note

$\sin 3x \cos x - \cos 3x \sin x = \sin(3x - x)$

by the difference identity for sine. Therefore,

to graph $y = \sin 3x \cos x - \cos 3x \sin x$, we
graph $y = \sin(3x - x)$, that is, $y = \sin 2x$. This is the basic
sine curve with amplitude 1 and period $\dfrac{2\pi}{2} = \pi$.

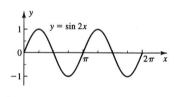

47. Following the hint, we note
$$\sin \frac{\pi x}{4} \cos \frac{3\pi x}{4} + \cos \frac{\pi x}{4} \sin \frac{3\pi x}{4}$$
$$= \sin\left(\frac{\pi x}{4} + \frac{3\pi x}{4}\right) = \sin \pi x$$
by the sum identity for sine. The graph of

$y = \sin \pi x$ is a basic sine curve with

amplitude 1 and period $\dfrac{2\pi}{\pi} = 2$.

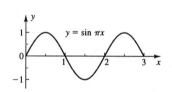

49. $\sin(x + y + z) = \sin[(x + y) + z]$ 　　　　　Algebra

$\qquad\qquad = \sin(x + y)\cos z + \cos(x + y)\sin z$ 　　　Sum identity for sine

$\qquad\qquad = (\sin x \cos y + \cos x \sin y)\cos z$

$\qquad\qquad\quad + (\cos x \cos y - \sin x \sin y)\sin z$ 　　Sum identities for sine and cosine

$\qquad\qquad = \sin x \cos y \cos z + \cos x \sin y \cos z$

$\qquad\qquad\quad + \cos x \cos y \sin z - \sin x \sin y \sin z$ 　　Algebra

51. Use the sum identity for sine, replacing y with $\dfrac{\pi}{3}$.

$\qquad\qquad \sin(x + y) = \sin x \cos y + \cos x \sin y$

$\qquad \sin\left(x + \dfrac{\pi}{3}\right) = \sin x \cos \dfrac{\pi}{3} + \cos x \sin \dfrac{\pi}{3}$

$$= \sin x \cdot \frac{1}{2} + \cos x \cdot \frac{\sqrt{3}}{2}$$

$$= \frac{1}{2} \sin x + \frac{\sqrt{3}}{2} \cos x = y2$$

The graphs are shown in the figure. The graphs of y1 and y2 coincide; the graph of y3 is the horizontal straight line.

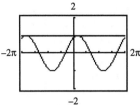

53. Use the difference identity for cosine, replacing y with $\frac{5\pi}{6}$.

$$\cos(x - y) = \cos x \cos y + \sin x \sin y$$

$$\cos\left(x - \frac{5\pi}{6}\right) = \cos x \cos \frac{5\pi}{6} + \sin x \sin \frac{5\pi}{6}$$

$$= \cos x \left(-\frac{\sqrt{3}}{2}\right) + \sin x \frac{1}{2}$$

$$= -\frac{\sqrt{3}}{2} \cos x + \frac{1}{2} \sin x = y2.$$

The graphs are shown in the figure. The graphs of y1 and y2 coincide; the graph of y3 is the horizontal straight line.

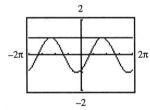

55. Use the sum identity for tangent, replacing y with $\frac{\pi}{4}$.

$$\tan(x + y) = \frac{\tan x + \tan y}{1 - \tan x \tan y}$$

$$\tan\left(x + \frac{\pi}{4}\right) = \frac{\tan x + \tan \frac{\pi}{4}}{1 - \tan x \tan \frac{\pi}{4}}$$

$$= \frac{\tan x + 1}{1 - \tan x \cdot 1}$$

$$= \frac{\tan x + 1}{1 - \tan x} = y2$$

The graphs are shown in the fgure. The graphs of y1 and y2 coincide; the graph of y3 is the horizontal straight line.

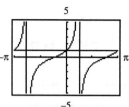

57. $\tan(\theta_2 - \theta_1) = \dfrac{\tan \theta_2 - \tan \theta_1}{1 + \tan \theta_2 \tan \theta_1}$ Difference identity for tangent

$\qquad\qquad = \dfrac{m_2 - m_1}{1 + m_2 m_1}$ Given $m_1 = \tan \theta_1$ and $m_2 = \tan \theta_2$

$\qquad\qquad = \dfrac{m_2 - m_1}{1 + m_1 m_2}$ Algebra

59.   (A) In right triangle ABE, we have (1) $\cot \alpha = \dfrac{AB}{AE} = \dfrac{AB}{h}$

In right triangle BCD, we have (2) $\cot \alpha = \dfrac{BC}{CD} = \dfrac{BC}{H}$

In right triangle EE′D, we have (3) $\tan \beta = \dfrac{E'D}{EE'} = \dfrac{H-h}{AC} = \dfrac{H-h}{AB+BC}$

From (3), $H - h = (AB + BC)\tan \beta$. From (1) and (2), $AB = h \cot \alpha$ and $BC = H \cot \alpha$.

Hence, substituting, we have (4) $H - h = $ $(h \cot \alpha + H \cot \alpha)\tan \beta,$

or $= (h + H)\cot \alpha \tan \beta$

Solving (4) for H, we have $H - h = h \cot \alpha \tan \beta + H \cot \alpha \tan \beta$

$H - H \cot \alpha \tan \beta = h + h \cot \alpha \tan \beta$

$H(1 - \cot \alpha \tan \beta) = h(1 + \cot \alpha \tan \beta)$

$H = h \left(\dfrac{1 + \cot \alpha \tan \beta}{1 - \cot \alpha \tan \beta} \right)$

(B) Start with $H = h \left(\dfrac{1 + \cot \alpha \tan \beta}{1 - \cot \alpha \tan \beta} \right)$

Apply the quotient identities: $H = h \left(\dfrac{1 + \dfrac{\cos \alpha}{\sin \alpha} \dfrac{\sin \beta}{\cos \beta}}{1 - \dfrac{\cos \alpha}{\sin \alpha} \dfrac{\sin \beta}{\cos \beta}} \right)$

Reduce the complex fraction to a simple one: $H = h \left(\dfrac{\sin \alpha \cos \beta + \cos \alpha \sin \beta}{\sin \alpha \cos \beta - \cos \alpha \sin \beta} \right)$

Apply the sum and difference identities for sine: $H = h \left(\dfrac{\sin(\alpha + \beta)}{\sin(\alpha - \beta)} \right)$

(C) Substituting the given values, we have: $H = (5.50 \text{ ft}) \dfrac{\sin(45.00° + 44.92°)}{\sin(45.00° - 44.92°)} = 3{,}940 \text{ ft}$

EXERCISE 4.4 Double-Angle and Half-Angle Identities

1. $\sin 2x = \sin(2 \cdot 60°) = \sin 120° = \dfrac{\sqrt{3}}{2}$; $2 \sin x \cos x = 2 \sin 60° \cos 60° = 2 \cdot \dfrac{\sqrt{3}}{2} \cdot \dfrac{1}{2} = \dfrac{\sqrt{3}}{2}$

3. $\tan x = \tan(2 \cdot 60°) = \tan 120° = -\sqrt{3}$;

$$\frac{2 \tan x}{1 - \tan^2 x} = \frac{2 \tan 60°}{1 - (\tan^2 60°)} = \frac{2 \cdot \sqrt{3}}{1 - (\sqrt{3})^2} = \frac{2 \cdot \sqrt{3}}{1 - 3} = \frac{2\sqrt{3}}{-2} = -\sqrt{3}$$

5. $\sin 105° = \sin \dfrac{210°}{2} = \sqrt{\dfrac{1 - \cos 210°}{2}}$

The positive sign is used since 105° is in the

second quadrant and sine is positive there. We

note that the reference triangle for 210° is a

30° – 60° triangle in the third quadrant. Thus,

$$\sin 105° = \sqrt{\frac{1 - (-\sqrt{3}/2)}{2}} = \sqrt{\frac{2 + \sqrt{3}}{4}} = \frac{\sqrt{2 + \sqrt{3}}}{2}$$

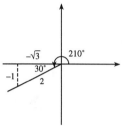

$$\cos 210° = -\cos 30° = -\frac{\sqrt{3}}{2}$$

7. $\tan 15° = \tan \dfrac{30°}{2} = \dfrac{1 - \cos 30°}{\sin 30°} = \dfrac{1 - \sqrt{3}/2}{1/2} = \dfrac{2\left(1 - \dfrac{\sqrt{3}}{2}\right)}{2\left(\dfrac{1}{2}\right)}$

9.

$$\frac{2 \tan x}{\sin 2x} = \frac{2 \tan x}{2 \sin x \cos x} \qquad \text{Double-angle identity}$$

$$= \frac{\tan x}{\sin x \cos x} \qquad \text{Algebra}$$

$$= \frac{\dfrac{\sin x}{\cos x}}{\sin x \cos x} \qquad \text{Quotient identity}$$

$$= \frac{\cos x \dfrac{\sin x}{\cos x}}{\cos x \cdot \sin x \cos x} \qquad \text{Algebra}$$

$$= \frac{\sin x}{\sin x \cos^2 x} \qquad \text{Algebra}$$

$$= \frac{1}{\cos^2 x} \qquad \text{Algebra}$$

$$= \sec^2 x \qquad \text{Reciprocal identity}$$

Chapter 4   Identities

11. $\tan x(1 + \cos 2x)$ = $\tan x(1 + 2\cos^2 x - 1)$ Double-angle identity

         = $\tan x \cdot 2\cos^2 x$ Algebra

         = $\dfrac{\sin x}{\cos x} \cdot 2\cos^2 x$ Quotient identity

         = $\dfrac{\sin x}{\cos x} \cdot \dfrac{2\cos^2 x}{1}$ Algebra

         = $2\sin x \cos x$ Algebra

         = $\sin 2x$ Double-angle identity

13. $2\sin^2 \dfrac{x}{2}$ = $2\left(\pm\sqrt{\dfrac{1 - \cos x}{2}}\right)^2$ Half-angle identity

         = $2\left(\dfrac{1 - \cos x}{2}\right)$ Algebra

         = $\dfrac{1 - \cos x}{1}$ Algebra

         = $\dfrac{1 - \cos x}{1} \cdot \dfrac{1 + \cos x}{1 + \cos x}$ Algebra

         = $\dfrac{1 - \cos^2 x}{1 + \cos x}$ Algebra

         = $\dfrac{\sin^2 x}{1 + \cos x}$ Pythagorean identity

15. $(\sin\theta - \cos\theta)^2$ = $\sin^2\theta - 2\sin\theta\cos\theta + \cos^2\theta$ Algebra

         = $\sin^2\theta + \cos^2\theta - 2\sin\theta\cos\theta$ Algebra

         = $1 - 2\sin\theta\cos\theta$ Pythagorean identity

         = $1 - \sin 2\theta$ Double-angle identity

17. $\cos^2 \dfrac{w}{2}$ = $\left(\pm\sqrt{\dfrac{1 + \cos w}{2}}\right)^2$ Half-angle identity

         = $\dfrac{1 + \cos w}{2}$ Algebra

19. $\cot \dfrac{\alpha}{2}$ = $\dfrac{1}{\tan \dfrac{\alpha}{2}}$ Reciprocal identity

         = $\dfrac{1}{\dfrac{\sin \alpha}{1 + \cos \alpha}}$ Half-angle identity

         = $\dfrac{1 + \cos \alpha}{\sin \alpha}$ Algebra

21. $\tan 2\beta = \dfrac{2 \tan \beta}{1 - \tan^2\beta}$ Double-angle identity

$= \dfrac{\dfrac{2 \tan \beta}{\tan \beta}}{\dfrac{1}{\tan \beta} - \dfrac{\tan^2\beta}{\tan \beta}}$ Algebra

$= \dfrac{2}{\dfrac{1}{\tan \beta} - \tan \beta}$ Algebra

$= \dfrac{2}{\cot \beta - \tan \beta}$ Reciprocal identity

23. $\dfrac{\cos 2t}{1 - \sin 2t} = \dfrac{\cos^2 t - \sin^2 t}{1 - 2 \sin t \cos t}$ Double-angle identity

$= \dfrac{\cos^2 t - \sin^2 t}{\cos^2 t + \sin^2 t - 2 \sin t \cos t}$ Pythagorean identity

$= \dfrac{\cos^2 t - \sin^2 t}{\cos^2 t - 2 \sin t \cos t + \sin^2 t}$ Algebra

$= \dfrac{(\cos t - \sin t)(\cos t + \sin t)}{(\cos t - \sin t)(\cos t - \sin t)}$ Algebra

$= \dfrac{\cos t + \sin t}{\cos t - \sin t}$ Algebra

$= \dfrac{\dfrac{\cos t}{\cos t} + \dfrac{\sin t}{\cos t}}{\dfrac{\cos t}{\cos t} - \dfrac{\sin t}{\cos t}}$ Algebra

$= \dfrac{1 + \tan t}{1 - \tan t}$ Quotient identity

25. $\tan 2x = \tan(x + x)$ Algebra

$= \dfrac{\tan x + \tan x}{1 - \tan x \tan x}$ Sum identity for tangent

$= \dfrac{2 \tan x}{1 - \tan^2 x}$ Algebra

27. $\tan \dfrac{x}{2} = \dfrac{\sin \dfrac{x}{2}}{\cos \dfrac{x}{2}}$ Quotient identity

$= \dfrac{\pm\sqrt{\dfrac{1 - \cos x}{2}}}{\pm\sqrt{\dfrac{1 + \cos x}{2}}}$ Half-angle identities

$$= \pm\sqrt{\frac{1 - \cos x}{1 + \cos x}} \qquad \text{Algebra}$$

$$\left|\tan \frac{x}{2}\right| = \sqrt{\frac{1 - \cos x}{1 + \cos x}} \qquad \text{Algebra}$$

$$= \sqrt{\frac{1 - \cos x}{1 + \cos x} \cdot \frac{1 + \cos x}{1 + \cos x}} \qquad \text{Algebra}$$

$$= \sqrt{\frac{1 - \cos^2 x}{(1 + \cos x)^2}} \qquad \text{Algebra}$$

$$= \sqrt{\frac{\sin^2 x}{(1 + \cos x)^2}} \qquad \text{Pythagorean identity}$$

$$= \left|\frac{\sin x}{1 + \cos x}\right| \qquad \text{Algebra}$$

Since $1 + \cos x \geq 0$ and $\sin x$ has the same sign as $\tan \frac{x}{2}$, we may drop the absolute value signs to obtain

$$\tan \frac{x}{2} = \frac{\sin x}{1 + \cos x} .$$

To show that $\sin x$ has the same sign as $\tan \frac{x}{2}$, we note the following cases:

If $0 < x < \pi$, $\sin x > 0$, then $0 < \frac{x}{2} < \frac{\pi}{2}$, $\tan \frac{x}{2} > 0$.

If $\pi < x < 2\pi$, $\sin x < 0$, then $\frac{\pi}{2} < \frac{x}{2} < \pi$, $\tan \frac{x}{2} < 0$.

The truth of the statement for other values of x follows since $\sin(x + 2k\pi) = \sin x$ and $\tan \frac{x + 2k\pi}{2} = \tan \frac{x}{2}$ by the periodic properties of sine and tangent.

(Note: if $x = 0$ both sides of the proposed identity are 0, if $x = \pi$ both sides are meaningless.)

29. $(\sec 2x)(2 - \sec^2 x) = \dfrac{1}{\cos 2x}\left(2 - \dfrac{1}{\cos^2 x}\right) \qquad \text{Reciprocal identity}$

$$= \frac{1}{\cos 2x}\left(\frac{2}{1} - \frac{1}{\cos^2 x}\right) \qquad \text{Algebra}$$

$$= \frac{1}{\cos 2x}\left(\frac{2 \cos^2 x}{\cos^2 x} - \frac{1}{\cos^2 x}\right) \qquad \text{Algebra}$$

$$= \frac{1}{\cos 2x}\left(\frac{2 \cos^2 x - 1}{\cos^2 x}\right) \qquad \text{Algebra}$$

$$= \frac{1}{\cos 2x} \frac{\cos 2x}{\cos^2 x} \qquad \text{Double-angle identity}$$

$$= \frac{1}{\cos^2 x} \qquad \text{Algebra}$$

$$= \sec^2 x \qquad \text{Reciprocal identity}$$

31. $$\frac{\cot x - \tan x}{\cot x + \tan x} = \frac{\dfrac{\cos x}{\sin x} - \dfrac{\sin x}{\cos x}}{\dfrac{\cos x}{\sin x} + \dfrac{\sin x}{\cos x}}$$ Quotient identity

$$= \frac{\sin x \cos x \dfrac{\cos x}{\sin x} - \sin x \cos x \dfrac{\sin x}{\cos x}}{\sin x \cos x \dfrac{\cos x}{\sin x} + \sin x \cos x \dfrac{\cos x}{\sin x}}$$ Algebra

$$= \frac{\cos^2 x - \sin^2 x}{\cos^2 x + \sin^2 x}$$ Algebra

$$= \frac{\cos 2x}{\cos^2 x + \sin^2 x}$$ Double-angle identity

$$= \frac{\cos 2x}{1}$$ Pythagorean identity

$$= \cos 2x$$ Algebra

33. Following the hint, we note
$$\sin x \cos x = \frac{1}{2} \cdot 2 \sin x \cos x = \frac{1}{2} \sin 2x$$
by the double-angle identity for sine. There-
fore, to graph $y = \sin x \cos x$, we graph
$y = \dfrac{1}{2} \sin 2x$. This is the basic sine curve
with amplitude $\dfrac{1}{2}$ and period $\dfrac{2\pi}{2} = \pi$.

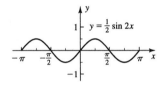

35. Following the hint, we note
$$\frac{\sin x}{1 + \cos x} = \tan \frac{x}{2}$$
by the half-angle identity for tangent. Therefore, to graph $y = \dfrac{\sin x}{1 + \cos x}$, we graph
$y = \tan \dfrac{x}{2}$. This is the basic tangent curve
with period $\dfrac{\pi}{1/2} = 2\pi$.

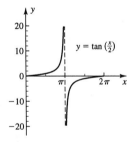

37. Following the hint, we note $1 - 2 \sin^2 x = \cos 2x$
by the double-angle identity for cosine. Hence,
$-2 \sin^2 x = \cos 2x - 1$, $\sin^2 x = -\dfrac{1}{2} \cos 2x + \dfrac{1}{2}$.
Therefore, to graph $y = \sin^2 x$, we graph
$y = \dfrac{1}{2} - \dfrac{1}{2} \cos 2x$. This graph is the graph of
$y = -\dfrac{1}{2} \cos 2x$ moved up $\dfrac{1}{2}$ unit. $y = -\dfrac{1}{2} \cos 2x$ has amplitude $\dfrac{1}{2}$ and period $\dfrac{2\pi}{2} = \pi$, and is the basic
cosine curve turned upside down. We start by drawing a broken line $\dfrac{1}{2}$ unit above the x axis, then

graph $y = -\dfrac{1}{2}\cos 2x$ relative to the broken line and the original y axis.

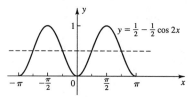

$y = \frac{1}{2} - \frac{1}{2}\cos 2x$

39. First draw a reference triangle in the first quadrant and find cos x and tan x: $a = \sqrt{5^2 - 3^2} = 4$

$\sin x = \dfrac{3}{5}$; $\cos x = \dfrac{4}{5}$; $\tan x = \dfrac{3}{4}$.

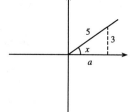

Now use the double-angle identities.

$$\sin 2x = 2\sin x \cos x = 2\left(\frac{3}{5}\right)\left(\frac{4}{5}\right) = \frac{24}{25}$$

$$\cos 2x = 1 - 2\sin^2 x = 1 - 2\left(\frac{3}{5}\right)^2 = 1 - \frac{18}{25} = \frac{7}{25}$$

$$\tan 2x = \frac{2\tan x}{1 - \tan^2 x} = \frac{2\left(\frac{3}{4}\right)}{1 - \left(\frac{3}{4}\right)^2} = \frac{\frac{3}{2}}{1 - \left(\frac{9}{16}\right)} = \frac{24}{7}$$

41. First draw a reference triangle in the second quadrant and find sin x and tan x: $b = \sqrt{5^2 - (-4)^2} = 3$

$\sin x = \dfrac{3}{5}$; $\cos x = -\dfrac{4}{5}$; $\tan x = -\dfrac{3}{4}$. Now use the double-angle identities.

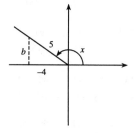

$$\sin 2x = 2\sin x \cos x = 2\left(\frac{3}{5}\right)\left(-\frac{4}{5}\right) = -\frac{24}{25}$$

$$\cos 2x = 2\cos^2 x - 1 = 2\left(-\frac{4}{5}\right)^2 - 1 = \frac{32}{25} - 1 = \frac{7}{25}$$

$$\tan 2x = \frac{2\tan x}{1 - \tan^2 x} = \frac{2\left(-\frac{3}{4}\right)}{1 - \left(-\frac{3}{4}\right)^2} = \frac{-\frac{3}{2}}{1 - \left(\frac{9}{16}\right)} = -\frac{24}{7}$$

43. First draw a reference triangle in the fourth quadrant and
find sin x, cos x, and tan x: $R = \sqrt{5^2 + (-12)^2} = 13$.
$\sin x = -\dfrac{12}{13}$; $\cos x = \dfrac{5}{13}$; $\tan x = -\dfrac{12}{5}$.
Now use the double-angle identities.

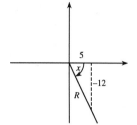

$$\sin 2x = 2 \sin x \cos x = 2\left(-\frac{12}{13}\right)\left(\frac{5}{13}\right) = -\frac{120}{169}$$

$$\cos 2x = 1 - 2 \sin^2 x = 1 - 2\left(-\frac{12}{13}\right)^2 = 1 - 2\left(\frac{144}{169}\right)$$
$$= \frac{169}{169} - \frac{288}{169} = -\frac{119}{169}$$

$$\tan 2x = \frac{2 \tan x}{1 - \tan^2 x} = \frac{2(-12/5)}{1 - (-12/5)^2} = \frac{-24/5}{1 - (144/25)} = \frac{25(-24/5)}{25 - 144} = \frac{120}{119}$$

45. We are given cos x. We can find $\sin \dfrac{x}{2}$ and $\cos \dfrac{x}{2}$ from the half-angle identities, after determining
their sign, as follows: If $0° < x < 90°$, then $0° < \dfrac{x}{2} < 45°$. Thus, $\dfrac{x}{2}$ is in the first quadrant, where
sine and cosine are positive. Using half-angle identities, we obtain:

$$\sin \frac{x}{2} = \sqrt{\frac{1 - \cos x}{2}} = \sqrt{\frac{1 - \frac{1}{3}}{2}} = \sqrt{\frac{1}{3}} \text{ or } \frac{\sqrt{3}}{3}$$

$$\cos \frac{x}{2} = \sqrt{\frac{1 + \cos x}{2}} = \sqrt{\frac{1 + \frac{1}{3}}{2}} = \sqrt{\frac{2}{3}} \text{ or } \frac{\sqrt{6}}{3}$$

47. Draw a reference triangle in the third quadrant and find cos x. a
$= -\sqrt{3^2 - (-1)^2} = -2\sqrt{2}$.
$\cos x = \dfrac{-2\sqrt{2}}{3}$. If $\pi < x < \dfrac{3\pi}{2}$, then

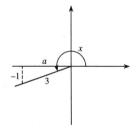

$\dfrac{\pi}{2} < \dfrac{x}{2} < \dfrac{3\pi}{4}$. Thus, $\dfrac{x}{2}$ is in the second quadrant, where sine is
positive and cosine is negative. Using half-angle identities, we
obtain:

$$\sin \frac{x}{2} = \sqrt{\frac{1 - \cos x}{2}} = \sqrt{\frac{1 - \left(-\frac{2\sqrt{2}}{3}\right)}{2}} = \sqrt{\frac{3 + 2\sqrt{2}}{6}}$$

$$\cos \frac{x}{2} = -\sqrt{\frac{1 + \cos x}{2}} = -\sqrt{\frac{1 + \left(-\frac{2\sqrt{2}}{3}\right)}{2}} = -\sqrt{\frac{3 - 2\sqrt{2}}{6}}$$

Chapter 4   Identities

49. Draw a reference triangle in the third quadrant and find cos x.

$R = \sqrt{(-3)^2 + (-4)^2} = 5$. $\cos x = -\frac{3}{5}$. If $-\pi < x < -\frac{\pi}{2}$,

then $-\frac{\pi}{2} < \frac{x}{2} < -\frac{\pi}{4}$. Thus, $\frac{x}{2}$ is in the fourth quadrant, where sine is negative and cosine is positive. Using half-angle identities,

we obtain:

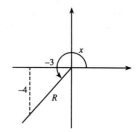

$$\sin \frac{x}{2} = -\sqrt{\frac{1 - \cos x}{2}} = -\sqrt{\frac{1 - \left(-\frac{3}{5}\right)}{2}} = -\sqrt{\frac{4}{5}} \text{ or } -\frac{2\sqrt{5}}{5}$$

$$\cos \frac{x}{2} = \sqrt{\frac{1 + \cos x}{2}} = \sqrt{\frac{1 + \left(-\frac{3}{5}\right)}{2}} = \sqrt{\frac{1}{5}} \text{ or } \frac{\sqrt{5}}{5}$$

51. To obtain sin x and cos x from sin 2x, we use the half-angle identities with x replaced by 2x. Thus,

$$\sin \frac{x}{2} = \pm\sqrt{\frac{1 - \cos x}{2}} \text{ becomes } \sin \frac{2x}{2} = \pm\sqrt{\frac{1 - \cos 2x}{2}} \text{ or } \sin x = \pm\sqrt{\frac{1 - \cos 2x}{2}}$$

$$\cos \frac{x}{2} = \pm\sqrt{\frac{1 + \cos x}{2}} \text{ becomes } \cos \frac{2x}{2} = \pm\sqrt{\frac{1 + \cos 2x}{2}} \text{ or } \cos x = \pm\sqrt{\frac{1 + \cos 2x}{2}}$$

To obtain cos 2x from sin 2x, we draw a reference triangle for 2x in the first quadrant.

$a = \sqrt{5^2 - 3^2} = 4$,

$\cos 2x = \frac{4}{5}$.

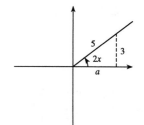

Since $0 < x < \frac{\pi}{4}$, sin x and cos x are positive.

Thus,

$$\sin x = \sqrt{\frac{1 - \cos 2x}{2}} = \sqrt{\frac{1 - \frac{4}{5}}{2}} = \sqrt{\frac{1}{10}} \text{ or } \frac{1}{\sqrt{10}}$$

$$\cos x = \sqrt{\frac{1 + \cos 2x}{2}} = \sqrt{\frac{1 + \frac{4}{5}}{2}} = \sqrt{\frac{9}{10}} \text{ or } \frac{3}{\sqrt{10}}; \quad \tan x = \frac{\sin x}{\cos x} = \frac{\frac{1}{\sqrt{10}}}{\frac{3}{\sqrt{10}}} = \frac{1}{3}$$

53. In Problem 51, we derived the identities:

$$\sin x = \pm\sqrt{\frac{1 - \cos 2x}{2}} \text{ and } \cos x = \pm\sqrt{\frac{1 + \cos 2x}{2}}$$

To obtain cos 2x from sec 2x, we use the reciprocal identity:

$$\sec 2x = \frac{1}{\cos 2x} \text{ and } \cos 2x = \frac{1}{\sec 2x} = \frac{1}{-\frac{5}{3}} = -\frac{3}{5}$$

Since $-\dfrac{\pi}{2} < x < 0$, x is in the fourth quadrant, where sine is negative and cosine is positive. Thus,

$$\sin x = -\sqrt{\dfrac{1 - \cos 2x}{2}} = -\sqrt{\dfrac{1 - \left(-\dfrac{3}{5}\right)}{2}} = -\sqrt{\dfrac{4}{5}} \text{ or } -\dfrac{2}{\sqrt{5}}$$

$$\cos x = \sqrt{\dfrac{1 + \cos 2x}{2}} = \sqrt{\dfrac{1 + \left(-\dfrac{3}{5}\right)}{2}} = \sqrt{\dfrac{1}{5}} \text{ or } \dfrac{1}{\sqrt{5}} ; \quad \tan x = \dfrac{\sin x}{\cos x} = \dfrac{-\dfrac{2}{\sqrt{5}}}{\dfrac{1}{\sqrt{5}}} = -2$$

55. In Problem 51, we derived the identities:

$$\sin x = \pm\sqrt{\dfrac{1 - \cos 2x}{2}} \text{ and } \cos x = \pm\sqrt{\dfrac{1 + \cos 2x}{2}}$$

To obtain cos 2x from tan 2x, we draw a reference
triangle for 2x. Since $-\pi < x < -\dfrac{\pi}{2}$, we have
$-2\pi < 2x < -\pi$. Thus, 2x is in the first or second
quadrant. Since tan 2x is negative, the only possibility
is the second quadrant. $R = \sqrt{(-3)^2 + 4^2} = 5$ and $\cos 2x = -\dfrac{3}{5}$

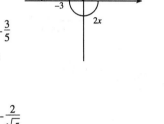

. Since $-\pi < x < -\dfrac{\pi}{2}$, x is in the third quadrant, where sine
and cosine are negative. Thus,

$$\sin x = -\sqrt{\dfrac{1 - \cos 2x}{2}} = -\sqrt{\dfrac{1 - \left(-\dfrac{3}{5}\right)}{2}} = -\sqrt{\dfrac{4}{5}} \text{ or } -\dfrac{2}{\sqrt{5}}$$

$$\cos x = -\sqrt{\dfrac{1 + \cos 2x}{2}} = -\sqrt{\dfrac{1 + \left(-\dfrac{3}{5}\right)}{2}} = -\sqrt{\dfrac{1}{5}} \text{ or } -\dfrac{1}{\sqrt{5}} ; \quad \tan x = \dfrac{\sin x}{\cos x} = \dfrac{-\dfrac{2}{\sqrt{5}}}{-\dfrac{1}{\sqrt{5}}} = 2$$

57. $\begin{aligned}
\sin 3x &= \sin(2x + x) &&\text{Algebra} \\
&= \sin 2x \cos x + \cos 2x \sin x &&\text{Sum identity} \\
&= 2 \sin x \cos x \cos x + (1 - 2 \sin^2 x)\sin x &&\text{Double-angle identities} \\
&= 2 \sin x \cos^2 x + \sin x - 2 \sin^3 x &&\text{Algebra} \\
&= 2 \sin x(1 - \sin^2 x) + \sin x - 2 \sin^3 x &&\text{Pythagorean identity} \\
&= 2 \sin x - 2 \sin^3 x + \sin x - 2 \sin^3 x &&\text{Algebra} \\
&= 3 \sin x - 4 \sin^3 x &&\text{Algebra}
\end{aligned}$

59. $\begin{aligned}
\sin 4x &= \sin 2(2x) &&\text{Algebra} \\
&= 2 \sin 2x \cos 2x &&\text{Double-angle identity} \\
&= 2(2 \sin x \cos x)(1 - 2 \sin^2 x) &&\text{Double-angle identity} \\
&= \cos x(4 \sin x)(1 - 2 \sin^2 x) &&\text{Algebra} \\
&= \cos x(4 \sin x - 8 \sin^3 x) &&\text{Algebra}
\end{aligned}$

Chapter 4　　Identities

61. $\tan 3x = \tan(2x + x)$　　　　　　　　　　　　　　　　　Algebra

$$= \frac{\tan 2x + \tan x}{1 - \tan 2x \tan x}$$　　　　　　　　　　　Sum identity

$$= \frac{\dfrac{2 \tan x}{1 - \tan^2 x} + \tan x}{1 - \dfrac{2 \tan x}{1 - \tan^2 x} \tan x}$$　　　　　　　Double-angle identity

$$= \frac{(1 - \tan^2 x)\dfrac{2 \tan x}{1 - \tan^2 x} + (1 - \tan^2 x)\tan x}{(1 - \tan^2 x) \cdot 1 - (1 - \tan^2 x)\dfrac{2 \tan x}{1 - \tan^2 x} \tan x}$$　　　Algebra

$$= \frac{2 \tan x + (1 - \tan^2 x)\tan x}{1 - \tan^2 x - 2 \tan x \tan x}$$　　　　　　Algebra

$$= \frac{2 \tan x + \tan x - \tan^3 x}{1 - \tan^2 x - 2 \tan^2 x}$$　　　　　　Algebra

$$= \frac{3 \tan x - \tan^3 x}{1 - 3 \tan^2 x}$$　　　　　　　　Algebra

63. $\dfrac{\sin x}{x} = \dfrac{2 \sin \dfrac{x}{2} \cos \dfrac{x}{2}}{x}$　　　　　　　Double-angle identity

$$= \frac{\cos \dfrac{x}{2} \sin \dfrac{x}{2}}{\dfrac{x}{2}}$$　　　　　　　　Algebra

$$= \cos \dfrac{x}{2} \frac{\sin \dfrac{x}{2}}{\dfrac{x}{2}}$$　　　　　　　　Algebra

65. The graphs are shown in the figure. The graphs of y1 and y2 coincide; the graph of y3 is the horizontal straight line.

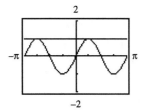

67. The graphs are shown in the figure. The graphs of y1 and y2 coincide; the graph of y3 is the horizontal straight line.

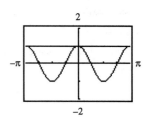

156

69. The graphs are shown in the figure. The graphs of y1 and y2 coincide only on the interval for which the graph shows a horizontal straight line (for y3). This is the interval $0 \leq x \leq 2\pi$.

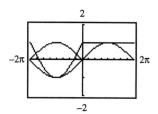

71. The graphs are shown in the figure. The graphs of y1 and y2 coincide only on the intervals for which the graph shows a horizontal straight line (for y3). These are the intervals $-2\pi \leq x \leq -\pi$ and $\pi \leq x \leq 2\pi$.

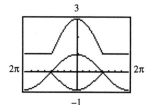

73. The graph of f(x) is shown in the figure. The graph appears to have vertical asymptotes $x = -2\pi$, $x = 0$, and $x = \pi$, x intercepts $-\pi$ and π, and period 2π. It appears that $g(x) = \cot \dfrac{x}{2}$ would be an appropriate choice. We verify $f(x) = g(x)$ as follows:

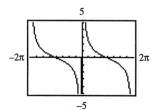

$$
\begin{aligned}
f(x) &= \csc x + \cot x \\[2mm]
&= \frac{1}{\sin x} + \frac{\cos x}{\sin x} && \text{Reciprocal and quotient identities} \\[2mm]
&= \frac{1 + \cos x}{\sin x} && \text{Algebra} \\[2mm]
&= 1 \div \frac{\sin x}{1 + \cos x} && \text{Algebra} \\[2mm]
&= 1 \div \tan \frac{x}{2} && \text{Half-angle identity} \\[2mm]
&= \cot \frac{x}{2} = g(x) && \text{Reciprocal identity}
\end{aligned}
$$

75. The graph of f(x) is shown in the figure. The graph appears to have vertical asymptotes $x = -\pi$, $-\dfrac{\pi}{2}$, 0, $\dfrac{\pi}{2}$, and π, and period π. It appears to have high and low points with y coordinate -1 and 1, respectively. It appears that $g(x) = \csc 2x$ would be an appropriate choice. We verify $f(x) = g(x)$ as follows:

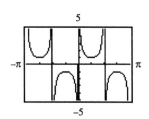

$$f(x) = \frac{\cot x}{1 + \cos 2x}$$

$$= \frac{\dfrac{\cos x}{\sin x}}{1 + \cos 2x} \qquad \text{Quotient identity}$$

$$= \frac{\dfrac{\cos x}{\sin x}}{1 + 2\cos^2 x - 1} \qquad \text{Double-angle identity}$$

$$= \frac{\cos x}{\sin x(2\cos^2 x)} \qquad \text{Algebra}$$

$$= \frac{1}{2\sin x \cos x} \qquad \text{Algebra}$$

$$= \frac{1}{\sin 2x} \qquad \text{Double-angle identity}$$

$$= \csc 2x = g(x) \qquad \text{Reciprocal identity}$$

77. The graph of f(x) is shown in the figure. The graph appears to be a basic cosine curve with
period 2π, amplitude $= \dfrac{1}{2}$ (y max − y min) $= \dfrac{1}{2}[1 - (-3)] = 2$,
displaced downward by $|k| = 1$ unit.
It appears that $g(x) = 2\cos x - 1$ would be an appropriate choice. We verify $f(x) = g(x)$ as follows:

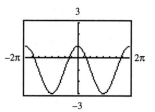

$$f(x) = \frac{1 + 2\cos 2x}{1 + 2\cos x}$$

$$= \frac{1 + 2(2\cos^2 x - 1)}{1 + 2\cos x} \qquad \text{Double-angle identity}$$

$$= \frac{1 + 4\cos^2 x - 2}{1 + 2\cos x} \qquad \text{Algebra}$$

$$= \frac{4\cos^2 x - 1}{2\cos x + 1} \qquad \text{Algebra}$$

$$= \frac{(2\cos x - 1)(2\cos x + 1)}{(2\cos x + 1)} \qquad \text{Algebra}$$

$$= 2\cos x - 1 = g(x) \qquad \text{Algebra}$$

79. Since $2\sin\theta\cos\theta = \sin 2\theta$ by the double-angle identity, we can write

$$d = \frac{2v_0^2 \sin\theta\cos\theta}{32 \text{ ft/sec}^2} = \frac{v_0^2(2\sin\theta\cos\theta)}{32 \text{ ft/sec}^2} = \frac{v_0^2 \sin 2\theta}{32 \text{ ft/sec}^2}$$

81. We note that $\tan\theta = \dfrac{2}{x}$ and $\tan 2\theta = \dfrac{6}{x}$ (see figure).

Using the hint, we have $\tan 2\theta = \dfrac{2\tan\theta}{1-\tan^2\theta}$

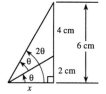

$$\frac{6}{x} = \frac{2\left(\dfrac{2}{x}\right)}{1-\left(\dfrac{2}{x}\right)^2} = \frac{\dfrac{4}{x}}{1-\left(\dfrac{4}{x^2}\right)}$$

$$= \frac{x^2 \cdot \left(\dfrac{4}{x}\right)}{x^2 \cdot 1 - x^2 \cdot \left(\dfrac{4}{x^2}\right)} \qquad x \neq 0$$

$$\frac{6}{x} = \frac{4x}{x^2-4}$$

$$x(x^2-4)\cdot\frac{6}{x} = x(x^2-4)\cdot\frac{4x}{x^2-4} \qquad x \neq 2,\,-2$$

$$6(x^2-4) = x\cdot 4x$$

$$6x^2-24 = 4x^2$$

$$2x^2 = 24$$

$$x = \sqrt{12} \text{ or } 2\sqrt{3} \qquad \text{(We discard the negative solution.)}$$

Then, $\dfrac{2}{x} = \tan\theta$, so $\tan\theta = \dfrac{2}{2\sqrt{3}} = \dfrac{1}{\sqrt{3}}$. To three decimal places $x \approx 3.464$ cm, $\theta = 30.000°$.

83. We label the figure as shown. From the Pythagorean theorem:

Since $AB = s$ and $MB = \dfrac{s}{2}$;

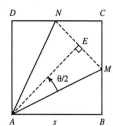

$AM^2 = AB^2 + MB^2 = s^2 + \dfrac{s^2}{4} = \dfrac{5s^2}{4}$; $AM = s\dfrac{\sqrt{5}}{2}$

Since $CN = CM = \dfrac{s}{2}$;

$NM^2 = CN^2 + CM^2 = \left(\dfrac{s}{2}\right)^2 + \left(\dfrac{s}{2}\right)^2 = \dfrac{2s^2}{4}$; $NM = s\dfrac{\sqrt{2}}{2}$

From the fact that $NA = MA$, thus triangle AMN is isosceles:

AE bisects NM, hence $ME = \dfrac{1}{2}MN = \dfrac{1}{2}\cdot s\dfrac{\sqrt{2}}{2} = s\dfrac{\sqrt{2}}{4}$.

From the definition of sine: $\sin\dfrac{\theta}{2} = \dfrac{ME}{MA} = s\dfrac{\sqrt{2}}{4} \div s\dfrac{\sqrt{5}}{2} = \dfrac{1}{2}\dfrac{\sqrt{2}}{\sqrt{5}}$

From the half-angle identity: $\sin\dfrac{\theta}{2} = \sqrt{\dfrac{1-\cos\theta}{2}}$.

Hence, $\sqrt{\dfrac{1-\cos\theta}{2}} = \dfrac{1}{2}\dfrac{\sqrt{2}}{\sqrt{5}}$; $\dfrac{1-\cos\theta}{2} = \dfrac{1}{10}$; $1-\cos\theta = \dfrac{1}{5}$; $\cos\theta = \dfrac{4}{5}$.

(The student may wish to compare with Exercise 1.4, Problem 37.)

Chapter 4   Identities

Product-Sum and Sum-Product Identities

1. $\cos x \cos y = \dfrac{1}{2} [\cos(x + y) + \cos(x - y)]$ Let $x = 7A$ and $y = 5A$

   $\cos 7A \cos 5A = \dfrac{1}{2} [\cos(7A + 5A) + \cos(7A - 5A)] = \dfrac{1}{2} (\cos 12A + \cos 2A) = \dfrac{1}{2} \cos 12A + \dfrac{1}{2} \cos 2A$

3. $\cos x \sin y = \dfrac{1}{2} [\sin(x + y) - \sin(x - y)]$ Let $x = 2\theta$ and $y = 3\theta$

   $\cos 2\theta \sin 3\theta = \dfrac{1}{2} [\sin(2\theta + 3\theta) - \sin(2\theta - 3\theta)] = \dfrac{1}{2} [\sin 5\theta - \sin(-\theta)] = \dfrac{1}{2} (\sin 5\theta + \sin \theta)$

   $\qquad\qquad = \dfrac{1}{2} \sin 5\theta + \dfrac{1}{2} \sin \theta$

5. $\cos x + \cos y = 2 \cos \dfrac{x + y}{2} \cos \dfrac{x - y}{2}$ Let $x = 7\theta$ and $y = 5\theta$

   $\cos 7\theta + \cos 5\theta = 2 \cos \dfrac{7\theta + 5\theta}{2} \cos \dfrac{7\theta - 5\theta}{2} = 2 \cos 6\theta \cos \theta$

7. $\sin x - \sin y = 2 \cos \dfrac{x + y}{2} \sin \dfrac{x - y}{2}$ Let $x = u$ and $y = 5u$

   $\sin u - \sin 5u = 2 \cos \dfrac{u + 5u}{2} \sin \dfrac{u - 5u}{2} = 2 \cos 3u \sin(-2u) = -2 \cos 3u \sin 2u$

9. $\cos x \sin y = \dfrac{1}{2} [\sin(x + y) - \sin(x - y)]$ Let $x = 75°$ and $y = 15°$

   $\cos 75° \sin 15° = \dfrac{1}{2} [\sin(75° + 15°) - \sin(75° - 15°)] = \dfrac{1}{2} [\sin 90° - \sin 60°]$

   $\qquad\qquad = \dfrac{1}{2} \left(1 - \dfrac{\sqrt{3}}{2}\right) = \dfrac{1}{2} \left(\dfrac{2 - \sqrt{3}}{2}\right) = \dfrac{2 - \sqrt{3}}{4}$

11. $\sin x \sin y = \dfrac{1}{2} [\cos(x - y) - \cos(x + y)]$ Let $x = 105°$ and $y = 165°$

   $\sin 105° \sin 165° = \dfrac{1}{2} [\cos(105° - 165°) - \cos(105° + 165°)]$

   $\qquad\qquad = \dfrac{1}{2} [\cos(-60°) - \cos 270°] = \dfrac{1}{2} \left(\dfrac{1}{2} - 0\right) = \dfrac{1}{4}$

13. $\sin x + \sin y = 2 \sin \dfrac{x + y}{2} \cos \dfrac{x - y}{2}$ Let $x = 195°$ and $y = 105°$

   $\sin 195° + \sin 105° = 2 \sin \dfrac{195° + 105°}{2} \cos \dfrac{195° - 105°}{2} = 2 \sin 150° \cos 45°$

   $\qquad\qquad = 2 \left(\dfrac{1}{2}\right) \left(\dfrac{\sqrt{2}}{2}\right) = \dfrac{\sqrt{2}}{2}$

15. $\sin x - \sin y = 2 \cos \dfrac{x + y}{2} \sin \dfrac{x - y}{2}$ Let $x = 75°$ and $y = 165°$

   $\sin 75° - \sin 165° = 2 \cos \dfrac{75° + 165°}{2} \sin \dfrac{75° - 165°}{2} = 2 \cos 120° \sin(-45°)$

   $\qquad\qquad = 2 \left(-\dfrac{1}{2}\right) \left(-\dfrac{\sqrt{2}}{2}\right) = \dfrac{\sqrt{2}}{2}$

17.
$$\cos(x - y) = \cos x \cos y + \sin x \sin y$$
$$\cos(x + y) = \cos x \cos y - \sin x \sin y$$

$$\cos(x - y) - \cos(x + y) = 2 \sin x \sin y \qquad \text{adding the above}$$
$$\sin x \sin y = \frac{1}{2}[\cos(x - y) - \cos(x + y)]$$

19. Let us start with the product-sum identity: $\cos \alpha \sin \beta = \frac{1}{2}[\sin(\alpha + \beta) - \sin(\alpha - \beta)]$

We would like $\alpha + \beta = x$ and $\alpha - \beta = y$, which gives $\alpha = \frac{x + y}{2}$ and $\beta = \frac{x - y}{2}$, as in the text.

Substituting into the product-sum identity, we have $\cos \frac{x + y}{2} \sin \frac{x - y}{2} = \frac{1}{2}(\sin x - \sin y)$, or

$\sin x - \sin y = 2 \cos \frac{x + y}{2} \sin \frac{x - y}{2}$

21. $\dfrac{\cos t - \cos 3t}{\sin t + \sin 3t} = \dfrac{-2 \sin \dfrac{t + 3t}{2} \sin \dfrac{t - 3t}{2}}{2 \sin \dfrac{t + 3t}{2} \cos \dfrac{t - 3t}{2}}$ 　　　Sum-product identities

$= \dfrac{-\sin 2t \sin(-t)}{\sin 2t \cos(-t)}$ 　　　Algebra

$= \dfrac{\sin 2t \sin t}{\sin 2t \cos t}$ 　　　Identities for negatives

$= \dfrac{\sin t}{\cos t}$ 　　　Algebra

$= \tan t$ 　　　Quotient identity

23. $\dfrac{\sin x + \sin y}{\cos x + \cos y} = \dfrac{2 \sin \dfrac{x + y}{2} \cos \dfrac{x - y}{2}}{2 \cos \dfrac{x + y}{2} \cos \dfrac{x - y}{2}}$ 　　　Sum-product identities

$= \dfrac{\sin \dfrac{x + y}{2}}{\cos \dfrac{x + y}{2}}$ 　　　Algebra

$= \tan \dfrac{x + y}{2}$ 　　　Quotient identity

25. $\dfrac{\cos x - \cos y}{\sin x + \sin y} = \dfrac{-2\sin \dfrac{x + y}{2} \sin \dfrac{x - y}{2}}{2 \sin \dfrac{x + y}{2} \cos \dfrac{x - y}{2}}$ 　　　Sum-product identities

$= \dfrac{-\sin \dfrac{x - y}{2}}{\cos \dfrac{x - y}{2}}$ 　　　Algebra

$= -\tan \dfrac{x - y}{2}$ 　　　Quotient identity

27. $\dfrac{\sin x + \sin y}{\sin x - \sin y} = \dfrac{2 \sin \dfrac{x+y}{2} \cos \dfrac{x-y}{2}}{2 \cos \dfrac{x+y}{2} \sin \dfrac{x-y}{2}}$ Sum-product identities

$= \dfrac{\sin \dfrac{x+y}{2} \cos \dfrac{x-y}{2}}{\cos \dfrac{x+y}{2} \sin \dfrac{x-y}{2}}$ Algebra

$= \tan \dfrac{x+y}{2} \cot \dfrac{x-y}{2}$ Quotient identities

$= \tan \dfrac{x+y}{2} \; \dfrac{1}{\tan \dfrac{x-y}{2}}$ Reciprocal identity

$= \dfrac{\tan \dfrac{x+y}{2}}{\tan \dfrac{x-y}{2}}$ Algebra

29. $\sin x \sin y \sin z = \sin x \dfrac{1}{2} [\cos(y-z) - \cos(y+z)]$ Product-sum identity

$= \dfrac{1}{2} \sin x \cos(y-z) - \dfrac{1}{2} \sin x \cos(y+z)$ Algebra

$= \dfrac{1}{2} \left\{ \dfrac{1}{2} [\sin(x+y-z) + \sin(x - \{y-z\})] \right\}$

$\quad - \dfrac{1}{2} \left\{ \dfrac{1}{2} [\sin(x+y+z) + \sin(x - \{y+z\})] \right\}$ Product-sum identities

$= \dfrac{1}{4} \sin(x+y-z) + \dfrac{1}{4} \sin(x-y+z) - \dfrac{1}{4} \sin(x+y+z)$
$\quad - \dfrac{1}{4} \sin(x-y-z)$ Algebra

$= \dfrac{1}{4} [\sin(x+y-z) - \sin(x-y-z) + \sin(z+x-y)$

$\quad - \sin(x+y+z)]$ Algebra

$= \dfrac{1}{4} [\sin(x+y-z) + \sin\{-(x-y-z)\} + \sin(z+x-y)$

$\quad - \sin(x+y+z)]$ Identity for negatives

$= \dfrac{1}{4} [\sin(x+y-z) + \sin(y+z-x) + \sin(z+x-y)$

$\quad - \sin(x+y+z)]$ Algebra

31. (A)   (B) $\cos x \sin y = \frac{1}{2}[\sin(x+y) - \sin(x-y)]$

Let $x = 16\pi X$ and $y = 2\pi X$

$2\cos 16\pi X \sin 2\pi X$

$= 2\left(\frac{1}{2}\right)[\sin(16\pi X + 2\pi X) - \sin(16\pi X - 2\pi X)]$

$= \sin 18\pi X - \sin 14\pi X$

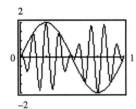
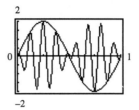

33. (A)   (B) $\sin x \sin y = \frac{1}{2}[\cos(x-y) - \cos(x+y)]$

Let $x = 24\pi X$ and $y = 2\pi X$

$2\sin 24\pi X \sin 2\pi X$

$= 2\left(\frac{1}{2}\right)[\cos(24\pi X - 2\pi X) - \cos(24\pi X + 2\pi X)]$

$= \cos 22\pi X - \cos 26\pi X$

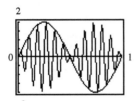
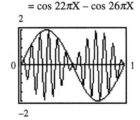

35. The sum of the two tones is

$y = k\sin 522\pi t + k\sin 512\pi t = k(\sin 522\pi t + \sin 512\pi t)$

$= k\left(2\sin\dfrac{522\pi t + 512\pi t}{2}\cos\dfrac{522\pi t - 512\pi t}{2}\right)$ Sum-product identity

This simplifies to $y = 2k\sin 517\pi t \cos 5\pi t$ Algebra

To find the beat frequency, we note

Period of first tone $= \dfrac{2\pi}{B_1} = \dfrac{2\pi}{522\pi} = \dfrac{1}{261}$

Frequency of first tone $= \dfrac{1}{\text{Period}} = \dfrac{1}{1/261} = 261$ Hz

Period of second tone $= \dfrac{2\pi}{B_2} = \dfrac{2\pi}{512\pi} = \dfrac{1}{256}$

Frequency of second tone $= \dfrac{1}{\text{Period}} = \dfrac{1}{1/256} = 256$ Hz

Beat frequency = Frequency of first tone − Frequency of second tone

$f_b = 261$ Hz − 256 Hz = 5 Hz

37. (A)   0.8

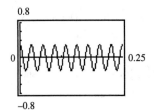

(B)   0.8

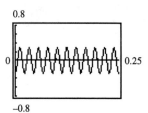

(C)

(D) $\cos x - \cos y = -2 \sin \dfrac{x + y}{2} \sin \dfrac{x - y}{2}$

Let $x = 72\pi t$ and $y = 88\pi t$

$0.3(\cos 72\pi t - \cos 88\pi t)$

$\qquad = (0.3)(-2) \sin \dfrac{72\pi t + 88\pi t}{2} \sin \dfrac{72\pi t - 88\pi t}{2}$

$\qquad = -0.6 \sin 80\pi t \sin(-8\pi t)$

$\qquad = 0.6 \sin 80\pi t \sin 8\pi t$

0.8

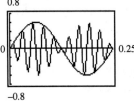

0.8

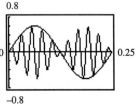

EXERCISE 4.6 From $M \sin Bt + N \cos Bt$ to $A \sin(Bt + C)$

1. $M = 1$ and $N = 1$
    Locate $P(M, N) = P(1, 1)$ to determine
    C: $R = \sqrt{1^2 + 1^2} = \sqrt{2};\ \tan C = 1;\ \sin C = \dfrac{1}{\sqrt{2}};\ C = \dfrac{\pi}{4}$
    (Reference triangle is a special 45° triangle.) Thus, $y = \sin t +$
    $\cos t = \sqrt{2} \sin\left(t + \dfrac{\pi}{4}\right)$
    Amplitude $= |\sqrt{2}| = \sqrt{2}$

    Period and Phase Shift:

    $t + \dfrac{\pi}{4} = 0 \qquad\qquad t + \dfrac{\pi}{4} = 2\pi$

    $\qquad t = -\dfrac{\pi}{4} \qquad\qquad\quad t = -\dfrac{\pi}{4} + 2\pi$

    Period $= 2\pi \qquad$ Phase Shift $= -\dfrac{\pi}{4}$

    Frequency $= \dfrac{1}{\text{Period}} = \dfrac{1}{2\pi}$

3. M = –1 and N = –1
    Locate P(M, N) = P(–1, –1) to determine C:

    $$R = \sqrt{(-1)^2 + (-1)^2} = \sqrt{2}; \ \tan C = 1; \ \sin C = -\frac{1}{\sqrt{2}} \ ;$$

    $C = \frac{5\pi}{4}$. (Reference triangle is a special 45° triangle.) Thus,

    $$y = -\sin t - \cos t = \sqrt{2} \sin\left(t + \frac{5\pi}{4}\right)$$

    Amplitude $= |\sqrt{2}| = \sqrt{2}$

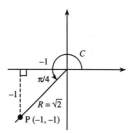

    Period and Phase Shift:

    $$t + \frac{5\pi}{4} = 0 \qquad\qquad t + \frac{5\pi}{4} = 2\pi$$

    $$t = -\frac{5\pi}{4} \qquad\qquad t = -\frac{5\pi}{4} + 2\pi$$

    Period $= 2\pi$ Phase Shift $= -\frac{5\pi}{4}$

    Frequency $= \dfrac{1}{\text{Period}} = \dfrac{1}{2\pi}$

5. M = $-\sqrt{3}$ and N = 1
    Locate P(M, N) = P($-\sqrt{3}$, 1) to determine C:

    $$R = \sqrt{(-\sqrt{3})^2 + 1^2} = 2; \ \sin C = \frac{1}{2} \ ;$$

    $\cos C = -\dfrac{\sqrt{3}}{2}; \ C = \pi - \dfrac{\pi}{6} = \dfrac{5\pi}{6}$. (Reference triangle is a

    special 30° – 60° triangle.) Thus,

    $$y = -\sqrt{3} \sin t + \cos t = 2 \sin\left(t + \frac{5\pi}{6}\right)$$

    Amplitude $= |2| = 2$

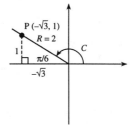

    Period and Phase Shift:

    $$t + \frac{5\pi}{6} = 0 \qquad\qquad t + \frac{5\pi}{6} = 2\pi$$

    $$t = -\frac{5\pi}{6} \qquad\qquad t = -\frac{5\pi}{6} + 2\pi$$

    Period $= 2\pi$ Phase Shift $= -\frac{5\pi}{6}$

    Frequency $= \dfrac{1}{\text{Period}} = \dfrac{1}{2\pi}$

7. M = –1 and N = $-\sqrt{3}$
    Locate P(M, N) = P(–1, $-\sqrt{3}$) to determine

    C: $R = \sqrt{(-1)^2 + (-\sqrt{3})^2} = 2; \ \sin C = -\dfrac{\sqrt{3}}{2} \ ;$

    $\cos C = -\dfrac{1}{2}; \ C = \pi + \dfrac{\pi}{3} = \dfrac{4\pi}{3}$. (Reference triangle is a

    special 30° – 60° triangle.) Thus,

    $$y = -\sin t - \sqrt{3} \cos t = 2 \sin\left(t + \frac{4\pi}{3}\right)$$

Amplitude $= |2| = 2$

Period and Phase Shift:

$$t + \frac{4\pi}{3} = 0 \qquad\qquad t + \frac{4\pi}{3} = 2\pi$$

$$t = -\frac{4\pi}{3} \qquad\qquad t = -\frac{4\pi}{3} + 2\pi$$

$$\text{Period} = 2\pi \quad \text{Phase Shift} = -\frac{4\pi}{3}$$

$$\text{Frequency} = \frac{1}{\text{Period}} = \frac{1}{2\pi}$$

9. $M = 1$ and $N = -1$

Locate $P(M, N) = P(1, -1)$ to determine C:

$$R = \sqrt{1^2 + (-1)^2} = \sqrt{2}; \ \sin C = -\frac{1}{\sqrt{2}}; \ \tan C = -1;$$

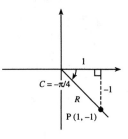

$$C = -\frac{\pi}{4}; \ |C| \text{ is minimum for this choice. Thus,}$$

$$y = \sin t - \cos t = \sqrt{2} \sin\left(t - \frac{\pi}{4} \right)$$

Amplitude $= |\sqrt{2}| = \sqrt{2}$

Period and Phase Shift:

$$t - \frac{\pi}{4} = 0 \qquad\qquad t - \frac{\pi}{4} = 2\pi$$

$$t = \frac{\pi}{4} \qquad\qquad t = \frac{\pi}{4} + 2\pi$$

$$\text{Period} = 2\pi \quad \text{Phase Shift} = \frac{\pi}{4}$$

$$\text{Frequency} = \frac{1}{\text{Period}} = \frac{1}{2\pi}$$

Graph:

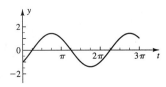

11. $M = 1$ and $N = 1$.

Locate $P(M, N) = P(1, 1)$ to determine C:

$$R = \sqrt{1^2 + 1^2} = \sqrt{2}; \ \sin C = \frac{1}{\sqrt{2}};$$

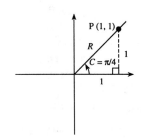

$$\tan C = 1; \ C = \frac{\pi}{4}. \ \text{Thus,}$$

$$y = \sin \pi t + \cos \pi t = \sqrt{2} \sin\left(\pi t + \frac{\pi}{4} \right)$$

Amplitude $= |\sqrt{2}| = \sqrt{2}$

Period and Phase Shift:

$$\pi t + \frac{\pi}{4} = 0 \qquad\qquad \pi t + \frac{\pi}{4} = 2\pi$$

$$t = -\frac{1}{4} \qquad\qquad t = -\frac{1}{4} + 2$$

Period $= 2$ Phase Shift $= -\frac{1}{4}$

Frequency $= \dfrac{1}{\text{Period}} = \dfrac{1}{2}$

Graph:

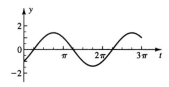

13. $M = \sqrt{3}$ and $N = -1$

Locate $P(M, N) = P(\sqrt{3}, -1)$ to determine C:

$$R = \sqrt{(\sqrt{3})^2 + (-1)^2} = 2; \ \sin C = -\frac{1}{2}; \ \cos C = \frac{\sqrt{3}}{2};$$

$C = -\dfrac{\pi}{6}$; $|C|$ is minimum for this choice. Thus,

$$y = \sqrt{3}\,\sin \pi t - \cos \pi t = 2 \sin\left(\pi t - \frac{\pi}{6}\right)$$

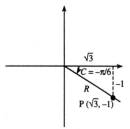

Amplitude $= |2| = 2$

Period and Phase Shift:

$$\pi t - \frac{\pi}{6} = 0 \qquad\qquad \pi t - \frac{\pi}{6} = 2\pi$$

$$t = \frac{1}{6} \qquad\qquad t = \frac{1}{6} + 2$$

Period $= 2$ Phase Shift $= \dfrac{1}{6}$

Frequency $= \dfrac{1}{\text{Period}} = \dfrac{1}{2}$

Graph:

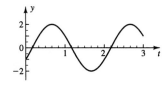

Chapter 4   Identities

15. M = 1 and N = 1

Locate P(M, N) = P(1, 1) to determine C:

$R = \sqrt{1^2 + 1^2} = \sqrt{2}$; $\tan C = 1$; $\sin C = \dfrac{1}{\sqrt{2}}$; $C = \dfrac{\pi}{4}$

Thus, $y = \sin 2\pi t + \cos 2\pi t = \sqrt{2} \sin\left(2\pi t + \dfrac{\pi}{4}\right)$

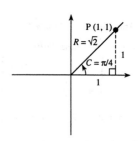

Amplitude $= |\sqrt{2}| = \sqrt{2}$

Period and Phase Shift:

$2\pi t + \dfrac{\pi}{4} = 0 \qquad 2\pi t + \dfrac{\pi}{4} = 2\pi$

$\qquad t = -\dfrac{1}{8} \qquad\qquad t = -\dfrac{1}{8} + 1$

Period = 1 \qquad Phase Shift $= -\dfrac{1}{8}$

Frequency $= \dfrac{1}{\text{Period}} = 1$

Graph:

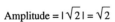

17. M = −1 and N = −√3

Locate P(M, N) = P(−1, −√3) to determine C:

$R = \sqrt{(-1)^2 + (-\sqrt{3})^2} = 2$; $\sin C = -\dfrac{\sqrt{3}}{2}$;

$\cos C = -\dfrac{1}{2}$; $C = -\pi + \dfrac{\pi}{3} = -\dfrac{2\pi}{3}$;

|C| is minimum for this choice. Thus,

$y = -\sin 2\pi t - \sqrt{3}\cos 2\pi t = 2 \sin\left(2\pi t - \dfrac{2\pi}{3}\right)$

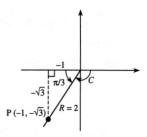

Amplitude $= |2| = 2$

Period and Phase Shift:

$2\pi t - \dfrac{2\pi}{3} = 0 \qquad 2\pi t - \dfrac{2\pi}{3} = 2\pi$

$\qquad t = \dfrac{1}{3} \qquad\qquad t = \dfrac{1}{3} + 1$

Period = 1 \qquad Phase Shift $= \dfrac{1}{3}$

Frequency $= \dfrac{1}{\text{Period}} = 1$

Graph:

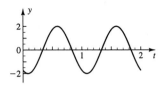

19. M = 4 and N = –3

Locate P(M, N) = P(4, –3) to determine C: $R = \sqrt{4^2 + (-3)^2}$

$= 5$; $\sin C = -\dfrac{3}{5} = -0.6$;

$\cos C = \dfrac{4}{5} = 0.8$. Find the reference angle α. Then $C = -\alpha$.

$\sin \alpha = \dfrac{3}{5} = 0.6$.

$\alpha = \sin^{-1}(0.6) \approx 0.64$. Thus, $C \approx -0.64$. $|C|$ is minimum for this choice of C.

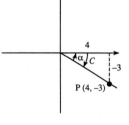

We can now write $y = 4 \sin \pi t - 3 \cos \pi t = 5 \sin(\pi t - 0.64)$

Amplitude $= |5| = 5$

Period and Phase Shift:

$$\pi t - 0.64 = 0 \qquad \pi t - 0.64 = 2\pi$$
$$t = \frac{0.64}{\pi} \qquad\qquad t = \frac{0.64}{\pi} + 2$$

Period $= 2$ Phase Shift $= \dfrac{0.64}{\pi} \approx 0.20$

Frequency $= \dfrac{1}{\text{Period}} = \dfrac{1}{2}$

21. M = –5 and N = 3

Locate P(M, N) = P(–5, 3) to determine C: $R = \sqrt{(-5)^2 + 3^2}$

$= \sqrt{34}$; $\sin C = \dfrac{3}{\sqrt{34}} \approx 0.51$;

$\cos C = -\dfrac{5}{\sqrt{34}} \approx -0.86$. Find the reference angle α. Then C

$= \pi - \alpha$. $\sin \alpha = \dfrac{3}{\sqrt{34}}$;

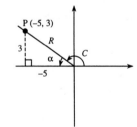

Chapter 4   Identities

$\alpha = \sin^{-1}\left(\dfrac{3}{\sqrt{34}}\right) \approx 0.54.$ Thus, $C = \pi - 0.54 \approx 2.60.$ $|C|$ is minimum for this choice of C.

We can now write $y = -5 \sin 3t + 3 \cos 3t = \sqrt{34} \sin(3t + 2.60).$

Amplitude $= |\sqrt{34}| = \sqrt{34}$

Period and Phase Shift:

$$3t + 2.60 = 0 \qquad\qquad 3t + 2.60 = 2\pi$$
$$t = -\frac{2.60}{3} \qquad\qquad t = -\frac{2.60}{3} + \frac{2\pi}{3}$$

$\text{Period} = \dfrac{2\pi}{3}$ \quad Phase Shift $= -\dfrac{2.60}{3} \approx -0.87$

$\text{Frequency} = \dfrac{1}{\text{Period}} = \dfrac{1}{2\pi/3} = \dfrac{3}{2\pi}$

23. The graph of $y = 4 \sin \pi x - 3 \cos \pi x$ is shown in the figure. We use the zoom feature or the built-in approximation routine to locate the x intercepts in this interval at $x = -0.80$ and $x = 0.20.$ The phase shift for $y = 5 \sin(\pi x - 0.64),$ as determined earlier (Problem 19), is 0.20.

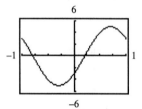

25. The graph of $y = -5 \sin 3x + 3 \cos 3x$ is shown in the figure. We use the zoom feature or the built-in approximation routine to locate the x intercepts in this interval at $x = -0.87$ and $x = 0.18.$ The phase shift for $y = \sqrt{34} \sin(3t + 2.60),$ as determined earlier (Problem 21), is $-0.87.$

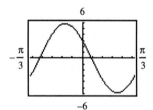

27. $M = -3$ and $N = -4$

Locate $P(M, N) = P(-3, -4)$ to determine C: $R = |R((-3)^2 + (-4)^2)| = 5;$ $\sin C = -\dfrac{4}{5} = -0.8;$

$\cos C = -\dfrac{3}{5} = -0.6.$ Find the reference angle α. Then $C = \alpha + \pi.$ $\sin \alpha = \dfrac{4}{5} = 0.8.$

$\alpha = \sin^{-1}(0.8) \approx 0.93.$ Thus, $C \approx 0.93 + \pi \approx 4.07.$ We can now write

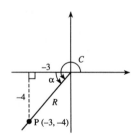

y = –3 sin 8t – 4 cos 8t = 5 sin(8t + 4.07)

Amplitude = | 5 | = 5 cm

Period and Phase Shift:

$$8t + 4.07 = 0 \qquad 8t + 4.07 = 2\pi$$

$$t = -\frac{4.07}{8} \qquad t = -\frac{4.07}{8} + \frac{\pi}{4}$$

$$\text{Period} = \frac{\pi}{4} \text{ sec} \quad \text{Phase Shift} = -\frac{4.07}{8} \approx -0.51 \text{ sec}$$

$$\text{Frequency} = \frac{1}{\text{Period}} = \frac{1}{(\pi/4) \text{ sec}} = \frac{4}{\pi} \text{ Hz}$$

CHAPTER 4   REVIEW EXERCISE

1. $\dfrac{1}{\csc x}$ 2. $\dfrac{1}{\cos x}$ 3. $\dfrac{\cos x}{\sin x}$ 4. $\dfrac{\sin x}{\cos x}$

5. $1 - \cos^2 x$ 6. $\cos x$

7. $\csc x \sin x = \dfrac{1}{\sin x} \sin x$ Reciprocal identity

$\qquad\qquad\quad = 1$ Algebra

$\qquad\qquad\quad = \dfrac{1}{\cos x} \cos x$ Algebra

$\qquad\qquad\quad = \sec x \cos x$ Reciprocal identity

8. $\cot x \sin x = \dfrac{\cos x}{\sin x} \sin x$ Quotient identity

$\qquad\qquad\quad = \cos x$ Algebra

9. $\tan x = \dfrac{\sin x}{\cos x}$ Quotient identity

$\qquad\quad = \dfrac{-\sin(-x)}{\cos(-x)}$ Identities for negatives

10. $\dfrac{\sin^2 x}{\cos x} = \dfrac{1 - \cos^2 x}{\cos x}$ Pythagorean identity

$\qquad\quad = \dfrac{1}{\cos x} - \dfrac{\cos^2 x}{\cos x}$ Algebra

$\qquad\quad = \dfrac{1}{\cos x} - \cos x$ Algebra

$\qquad\quad = \sec x - \cos x$ Reciprocal identity

Chapter 4   Identities

11. $\dfrac{\csc x}{\cos x} = \dfrac{\dfrac{1}{\sin x}}{\cos x}$ Reciprocal identity

$= \dfrac{1}{\sin x} \div \cos x$ Algebra

$= \dfrac{1}{\sin x} \cdot \dfrac{1}{\cos x}$ Algebra

$= \dfrac{1}{\sin x \cos x}$ Algebra

$= \dfrac{\sin^2 x + \cos^2 x}{\sin x \cos x}$ Pythagorean identity

$= \dfrac{\sin^2 x}{\sin x \cos x} + \dfrac{\cos^2 x}{\sin x \cos x}$ Algebra

$= \dfrac{\sin x}{\cos x} + \dfrac{\cos x}{\sin x}$ Algebra

$= \tan x + \cot x$ Quotient identities

12. $\cos^2 x(1 + \cot^2 x) = \cos^2 x \csc^2 x$ Pythagorean identity

$= \cos^2 x \dfrac{1}{\sin^2 x}$ Reciprocal identity

$= \dfrac{\cos^2 x}{\sin^2 x}$ Algebra

$= \cot^2 x$ Quotient identity

13. $\dfrac{\sin \alpha \csc \alpha}{\cot \alpha} = \dfrac{\sin \alpha \cdot \dfrac{1}{\sin \alpha}}{\cot \alpha}$ Reciprocal identity

$= \dfrac{1}{\cot \alpha}$ Algebra

$= \tan \alpha$ Reciprocal identity

14. $\dfrac{\sin^2 u - \cos^2 u}{\sin u \cos u} = \dfrac{\sin^2 u}{\sin u \cos u} - \dfrac{\cos^2 u}{\sin u \cos u}$ Algebra

$= \dfrac{\sin u}{\cos u} - \dfrac{\cos u}{\sin u}$ Algebra

$= \tan u - \cot u$ Quotient identities

15. $\dfrac{\sec \theta - \csc \theta}{\sec \theta \csc \theta} = \dfrac{\sec \theta}{\sec \theta \csc \theta} - \dfrac{\csc \theta}{\sec \theta \csc \theta}$ Algebra

$= \dfrac{1}{\csc \theta} - \dfrac{1}{\sec \theta}$ Algebra

$= \sin \theta - \cos \theta$ Reciprocal identities

16. $\cos(x + y) = \cos x \cos y - \sin x \sin y$

$\cos(x + 2\pi) = \cos x \cos 2\pi - \sin x \sin 2\pi$

$= \cos x(1) - \sin x(0)$

$= \cos x$

17. $\sin(x + y) = \sin x \cos y + \cos x \sin y$

$\sin(x + y) = \sin x \cos \pi + \cos x \sin \pi$

$= \sin x(-1) + \cos x(0)$

$= -\sin x$

18. $\cos 2x = \cos 2(30°) = \cos 60° = \dfrac{1}{2}$

$1 - 2 \sin^2 x = 1 - 2 \sin^2(30°)$

$= 1 - 2(\sin 30°)^2$

$= 1 - 2\left(\dfrac{1}{2}\right)^2 = 1 - \dfrac{1}{2}$

$= \dfrac{1}{2}$

19. $\sin \dfrac{x}{2} = \sin \dfrac{\pi/2}{2} = \sin \dfrac{\pi}{4} = \dfrac{1}{\sqrt{2}}$

Since $\dfrac{\pi}{4}$ is in the first quadrant, the sign of the square root is chosen to be positive.

$$\sqrt{\dfrac{1 - \cos x}{2}} = \sqrt{\dfrac{1 - \cos \pi/2}{2}} = \sqrt{\dfrac{1 - 0}{2}} = \sqrt{\dfrac{1}{2}} = \dfrac{1}{\sqrt{2}}$$

20. $\sin x \sin y = \dfrac{1}{2}[\cos(x - y) - \cos(x + y)]$ Let $x = 8t$ and $y = 5t$

$\sin 8t \sin 5t = \dfrac{1}{2}[\cos(8t - 5t) - \cos(8t + 5t)] = \dfrac{1}{2}(\cos 3t - \cos 13t) = \dfrac{1}{2}\cos 3t - \dfrac{1}{2}\cos 13t$

21. $\sin x + \sin y = 2 \sin \dfrac{x + y}{2} \cos \dfrac{x - y}{2}$ Let $x = w$ and $y = 5w$

$\sin w + \sin 5w = 2 \sin \dfrac{w + 5w}{2} \cos \dfrac{w - 5w}{2} = 2 \sin 3w \cos (-2w) = 2 \sin 3w \cos 2w$

22. $M = -\sqrt{3}$ and $N = -1$

Locate $P(M, N) = P(-\sqrt{3}, -1)$ to determine

C: $R = \sqrt{(-\sqrt{3})^2 + (-1)^2} = 2$; $\sin C = -\dfrac{1}{2}$;

$\cos = -\dfrac{\sqrt{3}}{2}$; $C = \pi + \dfrac{\pi}{6} = \dfrac{7\pi}{6}$.

(Reference triangle is a special $30° - 60°$ triangle.)

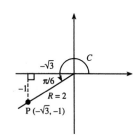

Thus, $y = -\sqrt{3}\, \sin t - \cos t = 2\, \sin\left(t + \dfrac{7\pi}{6}\right)$

Amplitude $= |2| = 2$.

Period and Phase Shift:

$$t + \frac{7\pi}{6} = 0 \qquad\qquad t + \frac{7\pi}{6} = 2\pi$$

$$t = -\frac{7\pi}{6} \qquad\qquad t = -\frac{7\pi}{6} + 2\pi$$

Period $= 2\pi$ Phase Shift $= -\dfrac{7\pi}{6}$

Frequency $= \dfrac{1}{\text{Period}} = \dfrac{1}{2\pi}$

23. $\dfrac{1 - \cos^2 t}{\sin^3 t} = \dfrac{\sin^2 t}{\sin^3 t}$ Pythagorean identity

    $= \dfrac{1}{\sin t}$ Algebra

    $= \csc t$ Reciprocal identity

24. $\dfrac{(\cos \alpha - 1)^2}{\sin^2 \alpha} = \dfrac{(\cos \alpha - 1)^2}{1 - \cos^2 \alpha}$ Pythagorean identity

    $= \dfrac{(\cos \alpha - 1)(\cos \alpha - 1)}{(1 - \cos \alpha)(1 + \cos \alpha)}$ Algebra

    $= \dfrac{(-1)(\cos \alpha - 1)}{1 + \cos \alpha}$ Algebra

    $= \dfrac{1 - \cos \alpha}{1 + \cos \alpha}$ Algebra

Key algebraic steps: $\dfrac{(b - 1)^2}{1 - b^2} = \dfrac{(b - 1)(b - 1)}{(1 - b)(1 + b)} = \dfrac{(-1)(b - 1)}{1 + b} = \dfrac{1 - b}{1 + b}$

25. $\dfrac{1 - \tan^2 x}{1 - \tan^4 x} = \dfrac{1 - \tan^2 x}{(1)^2 - (\tan^2 x)^2}$ Algebra

    $= \dfrac{1 - \tan^2 x}{(1 - \tan^2 x)(1 + \tan^2 x)}$ Algebra

    $= \dfrac{1}{1 + \tan^2 x}$ Algebra

    $= \dfrac{1}{\sec^2 x}$ Pythagorean identity

    $= \left(\dfrac{1}{\sec x}\right)^2$ Algebra

    $= \cos^2 x$ Reciprocal identity

Key algebraic steps: $\dfrac{1 - c^2}{1 - c^4} = \dfrac{1 - c^2}{(1)^2 - (c^2)^2} = \dfrac{1 - c^2}{(1 - c^2)(1 + c^2)} = \dfrac{1}{1 + c^2}$

26. $\cot^2 x \cos^2 x = (\csc^2 x - 1)\cos^2 x$ Pythagorean identity

$= \csc^2 x \cos^2 x - \cos^2 x$ Algebra

$= \left(\dfrac{1}{\sin x}\right)^2 \cos^2 x - \cos^2 x$ Reciprocal identity

$= \left(\dfrac{\cos x}{\sin x}\right)^2 - \cos^2 x$ Algebra

$= \cot^2 x - \cos^2 x$ Quotient identity

27. $\dfrac{\sin x}{1 - \cos x} = \dfrac{\sin x}{1 - \cos x}\dfrac{1 + \cos x}{1 + \cos x}$ Algebra

$= \dfrac{\sin x(1 + \cos x)}{1 - \cos^2 x}$ Algebra

$= \dfrac{\sin x(1 + \cos x)}{\sin^2 x}$ Pythagorean identity

$= \dfrac{\sin x}{\sin^2 x}(1 + \cos x)$ Algebra

$= \dfrac{1}{\sin x}(1 + \cos x)$ Algebra

$= \csc x(1 + \cos x)$ Reciprocal identity

28. $\dfrac{1 - \tan^2 x}{1 - \cot^2 x} = \dfrac{1 - \tan^2 x}{1 - \dfrac{1}{\tan^2 x}}$ Reciprocal identity

$= \dfrac{\tan^2 x(1 - \tan^2 x)}{\tan^2 x\left(1 - \dfrac{1}{\tan^2 x}\right)}$ Algebra

$= \dfrac{\tan^2 x(1 - \tan^2 x)}{\tan^2 x - 1}$ Algebra

$= \dfrac{-\tan^2 x(\tan^2 x - 1)}{\tan^2 x - 1}$ Algebra

$= -\tan^2 x$ Algebra

$= -(\sec^2 x - 1)$ Pythagorean identity

$= 1 - \sec^2 x$ Algebra

Key algebraic steps: $\dfrac{1 - a^2}{1 - \dfrac{1}{a^2}} = \dfrac{a^2(1 - a^2)}{a^2\left(1 - \dfrac{1}{a^2}\right)} = \dfrac{a^2(1 - a^2)}{a^2 - 1} = \dfrac{-a^2(a^2 - 1)}{a^2 - 1} = -a^2$

29. $\tan(x + \pi) = \dfrac{\tan x + \tan \pi}{1 - \tan x \tan \pi}$ Sum identity

$= \dfrac{\tan x + 0}{1 - \tan x \cdot 0}$ Known values

$= \tan x$ Algebra

Chapter 4   Identities

30. $\quad 1 - (\cos \beta - \sin \beta)^2$ $= \quad 1 - (\cos^2\beta - 2 \sin \beta \cos \beta + \sin^2\beta)$ $\qquad$ Algebra

$\qquad\qquad\qquad\qquad = \quad 1 - (\cos^2\beta + \sin^2\beta - 2 \sin \beta \cos \beta)$ $\qquad$ Algebra

$\qquad\qquad\qquad\qquad = \quad 1 - (1 - 2 \sin \beta \cos \beta)$ $\qquad$ Pythagorean identity

$\qquad\qquad\qquad\qquad = \quad 1 - 1 + 2 \sin \beta \cos \beta$ $\qquad$ Algebra

$\qquad\qquad\qquad\qquad = \quad 2 \sin \beta \cos \beta$ $\qquad$ Algebra

$\qquad\qquad\qquad\qquad = \quad \sin 2\beta$ $\qquad$ Double-angle identity

31. $\quad \dfrac{\sin 2x}{\cot x} = \dfrac{2 \sin x \cos x}{\cot x}$ $\qquad$ Double-angle identity

$\qquad\qquad = \dfrac{2 \sin x \cos x}{\dfrac{\cos x}{\sin x}}$ $\qquad$ Quotient identity

$\qquad\qquad = 2 \sin x \cos x + \dfrac{\cos x}{\sin x}$ $\qquad$ Algebra

$\qquad\qquad = 2 \sin x \cos x \cdot \dfrac{\sin x}{\cos x}$ $\qquad$ Algebra

$\qquad\qquad = 2 \sin^2 x$ $\qquad$ Algebra

$\qquad\qquad = 2 \sin^2 x - 1 + 1$ $\qquad$ Algebra

$\qquad\qquad = 1 + (2 \sin^2 x - 1)$ $\qquad$ Algebra

$\qquad\qquad = 1 - (1 - 2 \sin^2 x)$ $\qquad$ Algebra

$\qquad\qquad = 1 - \cos 2x$ $\qquad$ Double-angle identity

32. $\quad \dfrac{2 \tan x}{1 + \tan^2 x} = \dfrac{2 \dfrac{\sin x}{\cos x}}{1 + \dfrac{\sin^2 x}{\cos^2 x}}$ $\qquad$ Quotient identity

$\qquad\qquad = \dfrac{\cos^2 x \cdot 2 \dfrac{\sin x}{\cos x}}{\cos^2 x \cdot 1 + \cos^2 x \cdot \dfrac{\sin^2 x}{\cos^2 x}}$ $\qquad$ Algebra

$\qquad\qquad = \dfrac{2 \sin x \cos x}{\cos^2 x + \sin^2 x}$ $\qquad$ Algebra

$\qquad\qquad = \dfrac{2 \sin x \cos x}{1}$ $\qquad$ Pythagorean identity

$\qquad\qquad = 2 \sin x \cos x$ $\qquad$ Algebra

$\qquad\qquad = \sin 2x$ $\qquad$ Double-angle identity

33. $\quad 2 \csc 2x = \dfrac{2}{\sin 2x}$ $\qquad$ Reciprocal identity

$\qquad\qquad = \dfrac{2}{2 \sin x \cos x}$ $\qquad$ Double-angle identity

$\qquad\qquad = \dfrac{1}{\sin x \cos x}$ $\qquad$ Algebra

$$= \frac{\sin^2 x + \cos^2 x}{\sin x \cos x} \qquad \text{Pythagorean identity}$$

$$= \frac{\sin^2 x}{\sin x \cos x} + \frac{\cos^2 x}{\sin x \cos x} \qquad \text{Algebra}$$

$$= \frac{\sin x}{\cos x} + \frac{\cos x}{\sin x} \qquad \text{Algebra}$$

$$= \tan x + \cot x \qquad \text{Quotient identities}$$

34. $\dfrac{\cot \frac{x}{2}}{1 + \cos x} = \cot \dfrac{x}{2} \cdot \dfrac{1}{1 + \cos x} \qquad$ Algebra

$$= \frac{1}{\tan \frac{x}{2}} \cdot \frac{1}{1 + \cos x} \qquad \text{Reciprocal identity}$$

$$= \frac{1}{\dfrac{\sin x}{1 + \cos x}} \cdot \frac{1}{1 + \cos x} \qquad \text{Half-angle identity}$$

$$= \frac{1 + \cos x}{\sin x} \cdot \frac{1}{1 + \cos x} \qquad \text{Algebra}$$

$$= \frac{1}{\sin x} \qquad \text{Algebra}$$

$$= \csc x \qquad \text{Reciprocal identity}$$

35. $\dfrac{\sin(x - y)}{\sin(x + y)} = \dfrac{\sin x \cos y - \cos x \sin y}{\sin x \cos y + \cos x \sin y} \qquad$ Sum and difference identities

$$= \frac{\dfrac{\sin x \cos y}{\cos x \cos y} - \dfrac{\cos x \sin y}{\cos x \cos y}}{\dfrac{\sin x \cos y}{\cos x \cos y} + \dfrac{\cos x \sin y}{\cos x \cos y}} \qquad \text{Algebra}$$

$$= \frac{\dfrac{\sin x}{\cos x} - \dfrac{\sin y}{\cos y}}{\dfrac{\sin x}{\cos x} + \dfrac{\sin y}{\cos y}} \qquad \text{Algebra}$$

$$= \frac{\tan x - \tan y}{\tan x + \tan y} \qquad \text{Quotient identity}$$

36. $\csc 2x = \dfrac{1}{\sin 2x} \qquad$ Reciprocal identity

$$= \frac{1}{2 \sin x \cos x} \qquad \text{Double-angle identity}$$

$$= \frac{\sin^2 x + \cos^2 x}{2 \sin x \cos x} \qquad \text{Pythagorean identity}$$

$$= \frac{\sin^2 x}{2 \sin x \cos x} + \frac{\cos^2 x}{2 \sin x \cos x} \qquad \text{Algebra}$$

$$= \frac{\sin x}{2 \cos x} + \frac{\cos x}{2 \sin x} \qquad \text{Algebra}$$

$$= \frac{1}{2} \frac{\sin x}{\cos x} + \frac{1}{2} \frac{\cos x}{\sin x} \qquad \text{Algebra}$$

$$= \frac{1}{2} \tan x + \frac{1}{2} \cot x \qquad \text{Quotient identities}$$

$$= \frac{\tan x + \cot x}{2} \qquad \text{Algebra}$$

37. $\dfrac{2 - \sec^2 x}{\sec^2 x} = \dfrac{1 - \dfrac{1}{\cos^2 x}}{\dfrac{1}{\cos^2 x}} \qquad \text{Reciprocal identity}$

$$= \frac{\cos^2 x \cdot 2 - \cos^2 x \cdot \dfrac{1}{\cos^2 x}}{\cos^2 x \cdot \dfrac{1}{\cos^2 x}} \qquad \text{Algebra}$$

$$= \frac{2 \cos^2 x - 1}{1} \qquad \text{Algebra}$$

$$= 2 \cos^2 x - 1 \qquad \text{Algebra}$$

$$= \cos 2x \qquad \text{Double-angle identity}$$

38. $\tan \dfrac{x}{2} = \dfrac{1 - \cos x}{\sin x} \qquad \text{Half-angle identity}$

$$= \frac{\dfrac{1}{\cos x} - \dfrac{\cos x}{\cos x}}{\dfrac{\sin x}{\cos x}} \qquad \text{Algebra}$$

$$= \frac{\sec x - \dfrac{\cos x}{\cos x}}{\dfrac{\sin x}{\cos x}} \qquad \text{Reciprocal identity}$$

$$= \frac{\sec x - 1}{\dfrac{\sin x}{\cos x}} \qquad \text{Algebra}$$

$$= \frac{\sec x - 1}{\tan x} \qquad \text{Quotient identity}$$

39. $\dfrac{\sin t + \sin 5t}{\cos t + \cos 5t} = \dfrac{2 \sin \dfrac{t + 5t}{2} \cos \dfrac{t - 5t}{2}}{2 \cos \dfrac{t + 5t}{2} \cos \dfrac{t - 5t}{2}} \qquad \text{Sum-product identities}$

$$= \frac{2 \sin 3t \cos(-2t)}{2 \cos 3t \cos(-2t)} \qquad \text{Algebra}$$

$$= \frac{\sin 3t}{\cos 3t} \qquad \text{Algebra}$$

$$= \tan 3t \qquad \text{Quotient identity}$$

40. $\dfrac{\sin x + \sin y}{\cos x - \cos y} = \dfrac{2\sin\frac{x+y}{2}\cos\frac{x-y}{2}}{-2\sin\frac{x+y}{2}\sin\frac{x-y}{2}}$ Sum-product identities

$$= -\dfrac{\cos\frac{x-y}{2}}{\sin\frac{x-y}{2}}$$ Algebra

$$= -\cot\frac{x-y}{2}$$ Quotient identity

41. $\dfrac{\cos x - \cos y}{\cos x + \cos y} = \dfrac{-2\sin\frac{x+y}{2}\sin\frac{x-y}{2}}{2\cos\frac{x+y}{2}\cos\frac{x-y}{2}}$ Sum-product identities

$$= -\dfrac{\sin\frac{x+y}{2}}{\cos\frac{x+y}{2}}\dfrac{\sin\frac{x-y}{2}}{\cos\frac{x-y}{2}}$$ Algebra

$$= -\tan\frac{x+y}{2}\tan\frac{x-y}{2}$$ Quotient identity

42. $\sin x \sin y = \frac{1}{2}[\cos(x-y) - \cos(x+y)]$ Let $x = 165°$ and $y = 15°$

$\sin 165° \sin 15° = \frac{1}{2}[\cos(165° - 15°) - \cos(165° + 15°)] = \frac{1}{2}[\cos 150° - \cos 180°]$

$$= \frac{1}{2}\left[-\frac{\sqrt{3}}{2} - (-1)\right] = -\frac{\sqrt{3}}{4} + \frac{1}{2} \text{ or } \frac{1}{2} - \frac{\sqrt{3}}{4}$$

43. $\cos x - \cos y = -2\sin\frac{x+y}{2}\sin\frac{x-y}{2}$ Let $x = 165°$ and $y = 75°$

$\cos 165° - \cos 15° = -2\sin\dfrac{165° + 75°}{2}\sin\dfrac{165° - 75°}{2} = -2\sin 120° \sin 45°$

$$= -2\left(\frac{\sqrt{3}}{2}\right)\left(\frac{\sqrt{2}}{2}\right) = -\frac{\sqrt{6}}{2}$$

44. *Find sin x:* We start with the Pythagorean identity $\sin^2 x + \cos^2 x = 1$ and solve for sin x:

$$\sin x = \pm\sqrt{1 - \cos^2 x}$$

Since cos x and tan x are negative, x is associated with the second quadrant, where sin x is positive;

hence, $\sin x = \sqrt{1 - \cos^2 x} = \sqrt{1 - \left(-\frac{2}{3}\right)^2} = \sqrt{\frac{5}{9}} = \frac{\sqrt{5}}{3}$

Find tan x: $\tan x = \dfrac{\sin x}{\cos x} = \dfrac{\frac{\sqrt{5}}{3}}{-\frac{2}{3}} = -\dfrac{\sqrt{5}}{2}$ *Find cot x:* $\cot x = \dfrac{1}{\tan x} = \dfrac{1}{-\frac{\sqrt{5}}{2}} = -\dfrac{2}{\sqrt{5}}$

Find sec x: $\sec x = \dfrac{1}{\cos x} = \dfrac{1}{-\frac{2}{3}} = -\dfrac{3}{2}$ *Find csc x:* $\csc x = \dfrac{1}{\sin x} = \dfrac{1}{\frac{\sqrt{5}}{3}} = \dfrac{3}{\sqrt{5}}$

Chapter 4   Identities

45. First draw a reference triangle in the first quadrant and find

sin x and cos x: $R = \sqrt{3^2 + 4^2} = 5$; $\sin x = \frac{4}{5}$, $\cos x = \frac{3}{5}$.

Now use the double-angle identities:

$$\sin 2x \;=\; 2 \sin x \cos x = 2\left(\frac{4}{5}\right)\left(\frac{3}{5}\right) = \frac{24}{25}$$

$$\cos 2x \;=\; 2 \cos^2 x - 1 = 2\left(\frac{3}{5}\right)^2 - 1 = \frac{18}{25} - 1 = -\frac{7}{25}$$

$$\tan 2x \;=\; \frac{2 \tan x}{1 - \tan^2 x} = \frac{2\left(\frac{4}{3}\right)}{1 - \left(\frac{4}{3}\right)^2} = \frac{\frac{8}{3}}{1 - \left(\frac{16}{9}\right)} = -\frac{24}{7}$$

46. We are given cos x. We can find $\sin\frac{x}{2}$ and $\cos\frac{x}{2}$ from the half-angle identities, after determining their sign, as follows: if $-\pi < x < -\frac{\pi}{2}$, than $-\frac{\pi}{2} < \frac{x}{2} < -\frac{\pi}{4}$. Thus, $\frac{x}{2}$ is in the fourth quadrant, where sine is negative and cosine is positive. Using half-angle identities, we obtain:

$$\sin\frac{x}{2} \;=\; -\sqrt{\frac{1 - \cos x}{2}} = -\sqrt{\frac{1 - \left(-\frac{5}{13}\right)}{2}} = -\sqrt{\frac{9}{13}} \;\text{or}\; -\frac{3}{\sqrt{13}}$$

$$\cos\frac{x}{2} \;=\; \sqrt{\frac{1 + \cos x}{2}} = \sqrt{\frac{1 + \left(-\frac{5}{13}\right)}{2}} = \sqrt{\frac{4}{13}} \;\text{or}\; \frac{2}{\sqrt{13}}$$

$$\tan\frac{x}{2} \;=\; \frac{\sin\left(\frac{x}{2}\right)}{\cos\left(\frac{x}{2}\right)} = \frac{-\frac{3}{\sqrt{13}}}{\frac{2}{\sqrt{13}}} = -\frac{3}{2}$$

47. $M = -1$ and $N = 1$

Locate $P(M, N) = P(-1, 1)$ to determine C.

$R = \sqrt{(-1)^2 + 1^2} = \sqrt{2}$

$\sin C = \frac{1}{\sqrt{2}}$ $\cos C = \frac{-1}{\sqrt{2}}$ $C = \frac{3\pi}{4}$

Thus, $y = -\sin \pi t + \cos \pi t$

$= \sqrt{2} \sin\left(\pi t + \frac{3\pi}{4}\right)$

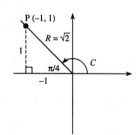

Amplitude $= |\sqrt{2}| = \sqrt{2}$

Period and Phase Shift:

$$\pi t + \frac{3\pi}{4} = 0 \qquad\qquad \pi t + \frac{3\pi}{4} = 2\pi$$

$$t = -\frac{3}{4} \qquad\qquad\qquad t = -\frac{3}{4}$$

$+\ 2$

Period $= 2$ \qquad Phase Shift $= -\frac{3}{4}$

Frequency $= \dfrac{1}{\text{Period}} = \dfrac{1}{2}$

Graph:

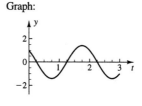

48. M $= 1$ and N $= \sqrt{3}$

Locate P(M, N) $=$ P$(1, \sqrt{3})$ to determine C:

$$R = \sqrt{1^2 + \left(\sqrt{3}\right)^2} = 2$$

$$\sin C = \frac{\sqrt{3}}{2} \qquad \cos C = \frac{1}{2} \qquad C = \frac{\pi}{3}$$

Thus, $y = \sin 2\pi t + \sqrt{3} \cos 2\pi t$

$$= 2 \sin\left(2\pi t + \frac{\pi}{3}\right)$$

Amplitude $= |2| = 2$

Period and Phase Shift:

$$2\pi t + \frac{\pi}{3} = 0 \qquad\qquad 2\pi t + \frac{\pi}{3} = 2\pi$$

$$t = -\frac{1}{6} \qquad\qquad\qquad t = -\frac{1}{6} + 1$$

Period $= 1$ \qquad Phase Shift $= -\frac{1}{6}$

Frequency $= \dfrac{1}{\text{Period}} = 1$

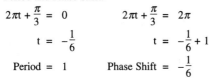

Graph:

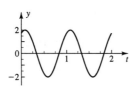

49. To obtain sin x and cos x from sec 2x, we use the half-angle identities with x replaced by 2x. Thus,

$$\sin \frac{x}{2} = \pm\sqrt{\frac{1 - \cos x}{2}} \text{ becomes } \sin \frac{2x}{2} = \pm\sqrt{\frac{1 - \cos 2x}{2}} \text{ or } \sin x = \pm\sqrt{\frac{1 - \cos 2x}{2}}$$

$$\cos \frac{x}{2} = \pm\sqrt{\frac{1 + \cos x}{2}} \text{ becomes } \cos \frac{2x}{2} = \pm\sqrt{\frac{1 + \cos 2x}{2}} \text{ or } \cos x = \pm\sqrt{\frac{1 + \cos 2x}{2}}$$

Chapter 4   Identities

To obtain cos 2x from sec 2x, we use the reciprocal identity

$$\sec 2x = \frac{1}{\cos 2x} \qquad \cos 2x = \frac{1}{\sec 2x} = \frac{1}{-\dfrac{13}{12}} = -\frac{12}{13}$$

Since $-\dfrac{\pi}{2} < x < 0$, x is in the fourth quadrant, where sine is negative and cosine is positive. Thus,

$$\sin x = -\sqrt{\frac{1 - \cos 2x}{2}} = -\sqrt{\frac{1 - \left(-\dfrac{12}{13}\right)}{2}} = -\sqrt{\frac{25}{26}} \text{ or } -\frac{5}{\sqrt{26}}$$

$$\cos x = \sqrt{\frac{1 + \cos 2x}{2}} = \sqrt{\frac{1 + \left(-\dfrac{12}{13}\right)}{2}} = \sqrt{\frac{1}{26}} \text{ or } \frac{1}{\sqrt{26}}$$

$$\tan x = \frac{\sin x}{\cos x} = \frac{-\dfrac{5}{\sqrt{26}}}{\dfrac{1}{\sqrt{26}}} = -5$$

50. $\dfrac{\cot x}{\csc x + 1} = \dfrac{(\csc x - 1)\cot x}{(\csc x - 1)(\csc x + 1)}$ Algebra

$\qquad\qquad = \dfrac{(\csc x - 1)\cot x}{\csc^2 x - 1}$ Algebra

$\qquad\qquad = \dfrac{(\csc x - 1)\cot x}{\cot^2 x}$ Pythagorean identity

$\qquad\qquad = \dfrac{\csc x - 1}{\cot x}$ Algebra

51. $\cot 3x = \dfrac{1}{\tan 3x}$ Reciprocal identity

$\qquad\quad = 1 \div \tan 3x$ Algebra

$\qquad\quad = 1 \div \tan(2x + x)$ Algebra

$\qquad\quad = 1 \div \dfrac{\tan 2x + \tan x}{1 - \tan 2x \tan x}$ Sum identity for tangent

$\qquad\quad = \dfrac{1 - \tan 2x \tan x}{\tan 2x + \tan x}$ Algebra

$\qquad\quad = \dfrac{1 - \dfrac{2 \tan x}{1 - \tan^2 x} \cdot \tan x}{\dfrac{2 \tan x}{1 - \tan^2 x} + \tan x}$ Double-angle identity

$\qquad\quad = \dfrac{(1 - \tan^2 x) \cdot 1 - \dfrac{(1 - \tan^2 x)}{1} \cdot \dfrac{2 \tan x}{1 - \tan^2 x} \cdot \tan x}{\dfrac{2 \tan x}{1 - \tan^2 x} \cdot \dfrac{(1 - \tan^2 x)}{1} + \tan x(1 - \tan^2 x)}$ Algebra

$\qquad\quad = \dfrac{1 - \tan^2 x - 2 \tan x \cdot \tan x}{2 \tan x + \tan x - \tan^3 x}$ Algebra

$$= \frac{1 - \tan^2 x - 2 \tan^2 x}{3 \tan x - \tan^3 x}$$ Algebra

$$= \frac{1 - 3 \tan^2 x}{3 \tan x - \tan^3 x}$$ Algebra

$$= \frac{(-1)(1 - 3 \tan^2 x)}{(-1)(3 \tan x - \tan^3 x)}$$ Algebra

$$= \frac{3 \tan^2 x - 1}{\tan^3 x - 3 \tan x}$$ Algebra

52. The definitions of the circular functions involved a point (a, b) on a unit circle. Recall:

$$\sin x = b \qquad \cos x = a \qquad \tan x = \frac{b}{a} \qquad a \neq 0$$

Thus, $\tan x = \dfrac{b}{a} = \dfrac{\sin x}{\cos x}$

53. We are to show $\cos(x + 2k\pi) = \cos x$. But,

$\cos(x + 2k\pi) = \cos x \cos 2k\pi - \sin x \sin 2k\pi$

by the sum identity for cosine. A quick

sketch shows (a, b) = (1, 0), R = 1, $\cos 2k\pi = 1$,

and $\sin 2k\pi = 0$. Hence,

$\cos(x + 2k\pi) = \cos x \cdot 1 - \sin x \cdot 0$

$\cos(x + 2k\pi) = \cos x$

54. We are to show $\cot(x + k\pi) = \cot x$. But,

$$\cot(x + k\pi) = \frac{1}{\tan(x + k\pi)}$$

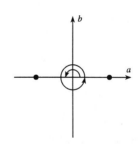

$$= \frac{1}{\dfrac{\tan x + \tan k\pi}{1 - \tan x \tan k\pi}} \qquad \text{Sum identity}$$

$$= \frac{1 - \tan x \tan k\pi}{\tan x + \tan k\pi} \qquad \text{Algebra}$$

A quick sketch shows (a, b) = (±1, 0), R = 1,

$\tan k\pi = \dfrac{0}{\pm 1} = 0$. Hence, $\cot(x + k\pi) = \dfrac{1 - \tan x \cdot 0}{\tan x + 0}$

$$= \frac{1}{\tan x}$$

Algebra

$$= \cot x$$

Reciprocal identity

Chapter 4   Identities

55. If we let $x + y = u$, $x - y = v$, and solve the system, we obtain $2x = u + v$, $2y = u - v$; that is,
$x = \dfrac{u + v}{2}$, $y = \dfrac{u - v}{2}$. Hence,

$$\sin x \sin y = \frac{1}{2}[\cos(x - y) - \cos(x + y)] \text{ becomes } \sin\frac{u + v}{2} \sin\frac{u - v}{2} = \frac{1}{2}[\cos v - \cos u]$$

upon substitution. Solving for the quantity in the brackets, we obtain

$$\frac{1}{2}[\cos v - \cos u] = \sin\frac{u + v}{2} \sin\frac{u - v}{2}; \quad \cos v - \cos u = 2\sin\frac{u + v}{2} \sin\frac{u - v}{2}$$

56.
$\sin 3x = \sin(2x + x)$	Algebra
$= \sin 2x \cos x + \cos 2x \sin x$	Sum identity
$= 2\sin x \cos x \cos x + (1 - 2\sin^2 x)\sin x$	Double-angle identities
$= 2\sin x \cos^2 x + \sin x - 2\sin^3 x$	Algebra
$= 2\sin x(1 - \sin^2 x) + \sin x - 2\sin^3 x$	Pythagorean identity
$= 2\sin x - 2\sin^3 x + \sin x - 2\sin^3 x$	Algebra
$= 3\sin x - 4\sin^3 x$	Algebra

57. $M = 1.6$ and $N = -1.2$

Locate $P(M, N) = P(1.6, -1.2)$ to determine

C: $R = \sqrt{(1.6)^2 + (-1.2)^2} = 2$; $\sin C = \dfrac{-1.2}{2} = -0.6$;

$\cos C = \dfrac{1.6}{2} = 0.8$. Find the reference angle α.

Then $C = -|\alpha|$.
$\sin \alpha = 0.6$, $\alpha = \sin^{-1}(0.6) \approx 0.64$.

Thus, $C = -|\alpha| \approx 0.64$. $|C|$ is minimum for this choice of C.

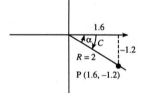

We can now write:
$y = 1.6 \sin 4t - 1.2 \cos 4t = 2\sin(4t - 0.64)$
Amplitude $= |2| = 2$

Period and Phase Shift:

$$4t - 0.64 = 0 \qquad 4t - 0.64 = 2\pi$$
$$t = 0.16 \qquad\qquad t = 0.16 + \frac{\pi}{2}$$

Period $= \dfrac{\pi}{2}$ Phase Shift $= 0.16$

Frequency $= \dfrac{1}{\text{Period}} = \dfrac{1}{\dfrac{\pi}{2}} = \dfrac{2}{\pi}$

58. The graph of f(x) is shown in the figure. The graph appears to be a basic cosine curve with period 2π, amplitude $= \frac{1}{2}$ (y max − y min) $= \frac{1}{2}$ (6 − 2) = 2, displaced upward by k = 4 units. It appears that g(x) = 4 + 2 cos x would be an appropriate choice. We verify f(x) = g(x) as follows:

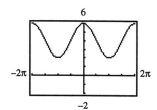

$$f(x) = \frac{3 \sin^2 x}{1 - \cos x} + \frac{\tan^2 x \cos^2 x}{1 + \cos x}$$

$$= \frac{3 \sin^2 x}{1 - \cos x} + \frac{\dfrac{\sin^2 x}{\cos^2 x} \cos^2 x}{1 + \cos x} \qquad \text{Quotient identity}$$

$$= \frac{3 \sin^2 x}{1 - \cos x} + \frac{\sin^2 x}{1 + \cos x} \qquad \text{Algebra}$$

$$= \frac{3(1 - \cos^2 x)}{1 - \cos x} + \frac{1 - \cos^2 x}{1 + \cos x} \qquad \text{Pythagorean identity}$$

$$= \frac{3(1 + \cos x)(1 - \cos x)}{1 - \cos x} + \frac{(1 - \cos x)(1 + \cos x)}{1 + \cos x} \qquad \text{Algebra}$$

$$= 3 + 3 \cos x + 1 - \cos x \qquad \text{Algebra}$$

$$= 4 + 2 \cos x = g(x)$$

59. The graph of f(x) is shown in the figure. The graph appears to have vertical asymptotes $x = -\frac{\pi}{4}$ and $x = \frac{\pi}{4}$, x intercepts $-\frac{\pi}{2}$, 0, and $\frac{\pi}{2}$, and period $\frac{\pi}{2}$. It appears that g(x) = tan 2x would be an appropriate choice. We verify f(x) = g(x) as follows:

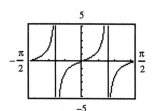

$$f(x) = \frac{\sin x}{\cos x - \sin x} + \frac{\sin x}{\cos x + \sin x}$$

$$= \frac{\sin x(\cos x + \sin x)}{(\cos x - \sin x)(\cos x + \sin x)} + \frac{\sin x(\cos x - \sin x)}{(\cos x + \sin x)(\cos x - \sin x)} \qquad \text{Algebra}$$

$$= \frac{\sin x(\cos x + \sin x) + \sin x(\cos x - \sin x)}{(\cos x - \sin x)(\cos x + \sin x)} \qquad \text{Algebra}$$

$$= \frac{\sin x \cos x + \sin^2 x + \sin x \cos x - \sin^2 x}{\cos^2 x - \sin^2 x} \qquad \text{Algebra}$$

$$= \frac{2 \sin x \cos x}{\cos^2 x - \sin^2 x} \qquad \text{Algebra}$$

$$= \frac{\sin 2x}{\cos 2x} \qquad \text{Double-angle identities}$$

$$= \tan 2x = g(x) \qquad \text{Quotient identity}$$

Chapter 4   Identities

Key algebraic steps: $\dfrac{a}{b-a}+\dfrac{a}{b+a} = \dfrac{a(b+a)}{(b-a)(b+a)}+\dfrac{a(b-a)}{(b+a)(b-a)} = \dfrac{a(b+a)+a(b-a)}{(b-a)(b+a)}$

$$= \dfrac{ab+a^2+ab-a^2}{b^2-a^2} = \dfrac{2ab}{b^2-a^2}$$

60. The graph of $f(x)$ is shown in the figure. The graph appears to be an upside down cosine curve

    with period π, amplitude $= \dfrac{1}{2}$ (y max $-$ y min) $= \dfrac{1}{2}(3-1) = 1$,

    displaced upward by $k = 2$ units.
    It appears that $g(x) = 2 - \cos 2x$ would be an appropriate choice.
    We verify that $f(x) = g(x)$ as follows:

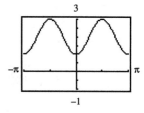

$$
\begin{aligned}
f(x) &= 3\sin^2 x + \cos^2 x \\
&= 3\sin^2 x + 1 - \sin^2 x &&\text{Pythagorean identity} \\
&= 2\sin^2 x + 1 &&\text{Algebra} \\
&= 2 - (1 - 2\sin^2 x) &&\text{Algebra} \\
&= 2 - \cos 2x &&\text{Double-angle identity}
\end{aligned}
$$

61. The graph of $f(x)$ is shown in the figure. The graph appears to have vertical asymptotes $x = -\dfrac{3\pi}{4}$,

    $-\dfrac{\pi}{4}, \dfrac{\pi}{4}$, and $\dfrac{3\pi}{4}$, and period π. It appears to have high and low points with y coordinates -3 and -1, respectively. It appears that $g(x) = -2 + \sec 2x$ would be an appropriate choice. We verify $f(x) = g(x)$ as follows:

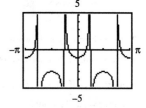

$$
\begin{aligned}
f(x) &= \frac{3 - 4\cos^2 x}{1 - 2\sin^2 x} \\[2mm]
&= \frac{1 - (-2 + 4\cos^2 x)}{1 - 2\sin^2 x} &&\text{Algebra} \\[2mm]
&= \frac{1 - 2(2\cos^2 x - 1)}{1 - 2\sin^2 x} &&\text{Algebra} \\[2mm]
&= \frac{1 - 2\cos 2x}{\cos 2x} &&\text{Double-angle identities} \\[2mm]
&= \frac{1}{\cos 2x} - \frac{2\cos 2x}{\cos 2x} &&\text{Algebra} \\[2mm]
&= \sec 2x - 2 \ \text{ or } \ -2 + \sec 2x = g(x) &&\text{Reciprocal identity, Algebra}
\end{aligned}
$$

62. The graph of f(x) is shown in the figure. The graph appears to have vertical asymptotes $x = -2\pi$,

    $x = 0$, and $x = 2\pi$, and period 2π. Its x intercepts are difficult to determine, but since there appears to be symmetry with respect to the points where the curve crosses the line $y = 2$, it appears to be a cotangent curve displaced upward by $k = 2$ units. It appears that $g(x) = 2 + \cot\dfrac{x}{2}$ would be an appropriate choice. We verify $f(x) = g(x)$ as follows:

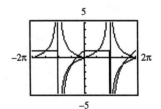

$$f(x) \;=\; \frac{2 + \sin x - 2\cos x}{1 - \cos x}$$

$$=\; \frac{2 - 2\cos x + \sin x}{1 - \cos x} \qquad \text{Algebra}$$

$$=\; \frac{2 - 2\cos x}{1 - \cos x} + \frac{\sin x}{1 - \cos x} \qquad \text{Algebra}$$

$$=\; 2 + 1 \div \frac{1 - \cos x}{\sin x} \qquad \text{Algebra}$$

$$=\; 2 + 1 \div \tan\frac{x}{2} \qquad \text{Half-angle identity}$$

$$=\; 2 + \cot\frac{x}{2} \qquad \text{Reciprocal identity}$$

Key algebraic steps: $\dfrac{2 + a - 2b}{1 - b} = \dfrac{2 - 2b + a}{1 - b} = \dfrac{2 - 2b}{1 - b} + \dfrac{a}{1 - b} = 2 + 1 \div \dfrac{1 - b}{a}$

63. The graphs are shown in the figure. The graphs of y1 and y2 coincide only on the intervals for which the graph shows a horizontal straight line (for y3). These are the intervals:
    $-2\pi \le x < -\pi$ and $0 \le x < \pi$

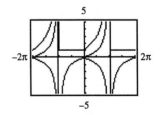

64. The graphs are shown in the figure. The graphs of y1 and y2 coincide only on the intervals for which the graph shows a horizontal straight line (for y3). These are the intervals:
    $-\pi < x \le 0$ and $\pi < x \le 2\pi$

Chapter 4   Identities

65.

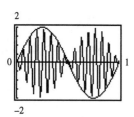

66. $\cos x \sin y = \dfrac{1}{2}[\sin(x + y) - \sin(x - y)]$

Let $x = 30\pi X$ and $y = 2\pi X$

$2 \cos 30\pi X \sin 2\pi X$

$\quad = 2\left(\dfrac{1}{2}\right)[\sin(30\pi X + 2\pi X) - \sin(30\pi X - 2\pi X)]$

$\quad = \sin 32\pi X - \sin 28\pi X$

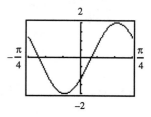

67. The graph of $y = 1.6 \sin 4t - 1.2 \cos 4t$ is shown
in the figure. We use the zoom feature or the built-
in approximation routine to locate the t intercepts
in this interval at $t = -0.62$ and $t = 0.16$. The phase
shift for $y = 2 \sin(4t - 0.64)$, as determined earlier
(Problem 57) is 0.16.

68. $\begin{aligned}
\sqrt{u^2 - a^2} &= \sqrt{(a \sec x)^2 - a^2} && \text{Using the given substitution}\\
&= \sqrt{a^2 \sec^2 x - a^2} && \text{Algebra}\\
&= \sqrt{a^2(\sec^2 x - 1)} && \text{Algebra}\\
&= \sqrt{a^2 \tan^2 x} && \text{Pythagorean identity}\\
&= |a||\tan x| && \text{Algebra}\\
&= a \tan x && \text{Since } a > 0 \text{ and x is in quadrant I or IV}
\end{aligned}$

$\left(\text{given } -\dfrac{\pi}{2} < x < \dfrac{\pi}{2}\right)$, thus, $\tan x > 0$.

69. In Problem 57, Exercise 4.3, the formula
$$\tan(\theta_2 - \theta_1) = \frac{m_2 - m_1}{1 + m_1 m_2}$$
was derived. Since the given lines have slopes $4 = m_2$ and $\dfrac{1}{3} = m_1$, we can write

$$\tan(\theta_2 - \theta_1) = \frac{4 - \dfrac{1}{3}}{1 + \left(\dfrac{1}{3}\right)(4)} = \frac{3 \cdot 4 - 3 \cdot \dfrac{1}{3}}{3 + 3\left(\dfrac{1}{3}\right)4} = \frac{12 - 1}{3 + 4} = \frac{11}{7}$$

$$\theta_2 - \theta_1 = \tan^{-1}\left(\frac{11}{7}\right)$$

$$\approx 57.5°$$

70. We note that $\tan \theta = \frac{5}{8}$ and $\tan 2\theta = \frac{5 + x}{8}$ (see figure). Then,

$$\tan 2\theta = \frac{2 \tan \theta}{1 - \tan^2\theta},$$

$$\theta = \tan^{-1}\frac{5}{8} \approx 32.005°$$

$$\frac{5 + x}{8} = \frac{2\left(\frac{5}{8}\right)}{1 - \left(\frac{5}{8}\right)^2} = \frac{\frac{5}{4}}{1 - \frac{25}{64}} = \frac{\frac{5}{4}}{\frac{39}{64}} = \frac{80}{39}$$

$$8 \cdot \frac{5 + x}{8} = 8 \cdot \frac{80}{39}$$

$$5 + x = \frac{640}{39}$$

$$x = \frac{445}{39} \approx 11.410$$

71. Sum $= 0.3 \cos 120\pi t - 0.3 \cos 140\pi t$.

$\cos x - \cos y = -2 \sin\frac{x + y}{2} \sin\frac{x - y}{2}$. Let $x = 120\pi t$ and $y = 140\pi t$.

$$0.3 \cos 120\pi t - 0.3 \cos 140\pi t = 0.3(\cos 120\pi t - \cos 140\pi t)$$

$$= 0.3(-2) \sin\frac{120\pi t + 140\pi t}{2} \sin\frac{120\pi t - 140\pi t}{2}$$

$$= -0.6 \sin 130\pi t \sin(-10\pi t)$$

$$= 0.6 \sin 130\pi t \sin 10\pi t$$

To find the beat frequency, we note:

Period of first tone $= \frac{2\pi}{B_1} = \frac{2\pi}{120\pi} = \frac{1}{60}$; Frequency of first tone $= \frac{1}{\text{Period}} = \frac{1}{\frac{1}{60}} = 60$ Hz

Period of second tone $= \frac{2\pi}{B_2} = \frac{2\pi}{140\pi} = \frac{1}{70}$; Frequency of second tone $= \frac{1}{\text{Period}} = \frac{1}{\frac{1}{70}} = 70$ Hz

Beat frequency = Frequency of second tone – Frequency of first tone

$$f_b = 70 \text{ Hz} - 60 \text{ Hz} = 10 \text{ Hz}$$

72. (A)

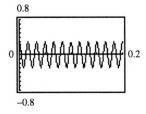

(B)

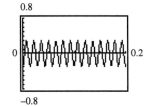

(C) y3 = 0.3 cos 120πt – 0.3 cos 140πt

0.8

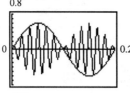

0.2

–0.8

(D) y3 = 0.6 sin 130πt sin 10πt

0.8

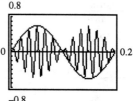

0.2

–0.8

73. M = –8 and N = –6.
Locate P(M, N) = P(–8 ,–6) to determine C: R = \R((–8)² +

(–6)²) = 10, sin C = $\frac{-6}{10}$ = –0.6,

tan C = $\frac{-6}{-8}$ = 0.75. Find the reference angle α. Then C = π +

α.

tan α = 0.75, α = tan⁻¹(0.75) = 0.6435. Thus, C ≈ 3.79. We

can now write:
y = –8 sin 3t – 6 cos 3t = 10 sin(3t + 3.79)

Amplitude = | 10| = 10

Period and Phase Shift:

3t + 3.79 = 0   3t + 3.79 = 2π

t = –1.26   t = –1.26 + $\frac{2\pi}{3}$

Period = $\frac{2\pi}{3}$ Phase Shift ≈ –1.26

Frequency = $\frac{1}{\text{Period}}$ = $\frac{1}{\frac{3}{2\pi}}$ = $\frac{2\pi}{3}$

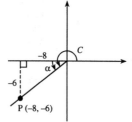

74. The graph of y = –8 sin 3t – 6 cos 3t is shown in the

figure. We use the zoom feature or the built-in

approximation routine to locate the t intercepts in this

interval at t = –1.26, –0.21, 0.83, and 1.88. The phase
shift for y = 10 sin(3t + 3.79) is –1.26.

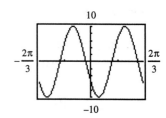

Chapter 5   Inverse Trigonometric Functions;
              Trigonometric Equations

EXERCISE 5.1 Inverse Sine, Cosine, and Tangent Functions

1. $y = \sin^{-1} 0$ is equivalent to $\sin y = 0$.

    No reference triangle can be drawn, but the
    only y between $-\dfrac{\pi}{2}$ and $\dfrac{\pi}{2}$ which has sine
    equal to 0 is $y = 0$. Thus, $\sin^{-1} 0 = 0$.

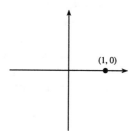

3. $y = \arccos \dfrac{\sqrt{3}}{2}$ is equivalent to $\cos y = \dfrac{\sqrt{3}}{2}$.

    What y between 0 and π has cosine equal to $\dfrac{\sqrt{3}}{2}$?
    y must be associated with a reference triangle in
    the first quadrant. Reference triangle is a special
    $30° - 60°$ triangle.

    $y = \dfrac{\pi}{6}$ $\arccos \dfrac{\sqrt{3}}{2} = \dfrac{\pi}{6}$

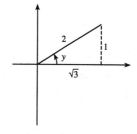

5. $y = \tan^{-1} 1$ is equivalent to $\tan y = 1$.
    What y between $-\dfrac{\pi}{2}$ and $\dfrac{\pi}{2}$ has tangent equal to 1?
    y must be associated with a reference triangle in
    the first quadrant. Reference triangle is a special
    $45°$ triangle.

    $y = \dfrac{\pi}{4}$ $\tan^{-1} 1 = \dfrac{\pi}{4}$

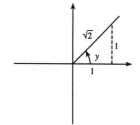

7. $y = \cos^{-1} \dfrac{1}{2}$ is equivalent to $\cos y = \dfrac{1}{2}$.

    What y between 0 and π has cosine equal to $\dfrac{1}{2}$?
    y must be associated with a reference triangle in
    the first quadrant. Reference triangle is a special
    $30° - 60°$ triangle.

    $y = \dfrac{\pi}{3}$ $\cos^{-1} \dfrac{1}{2} = \dfrac{\pi}{3}$

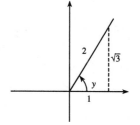

Chapter 5   Inverse Trigonometric Functions; Trigonometric Equations

9. $y = \arctan \dfrac{1}{\sqrt{3}}$ is equivalent to $\tan y = \dfrac{1}{\sqrt{3}}$. What y between $-\dfrac{\pi}{2}$ and $\dfrac{\pi}{2}$ has tangent equal to $\dfrac{1}{\sqrt{3}}$?

From the figure in Problem 3, we see that $y = \dfrac{\pi}{6}$. Thus, $\arctan \dfrac{1}{\sqrt{3}} = \dfrac{\pi}{6}$.

11. $y = \tan^{-1} 0$ is equivalent to $\tan y = 0$. What y between $-\dfrac{\pi}{2}$ and $\dfrac{\pi}{2}$ has tangent equal to 0? From the figure in Problem 1, we see that $y = 0$. Thus, $\tan^{-1} 0 = 0$.

13. Calculator in radian mode: $\cos^{-1} 0.4038 = 1.155$

15. Calculator in radian mode: $\tan^{-1} 43.09 = 1.548$

17. 1.131 is not in the domain of the inverse sine function. $-1 \le 1.131 \le 1$ is false. arcsin 1.131 is not defined.

19. $y = \arccos\left(-\dfrac{1}{2}\right)$ is equivalent to $\cos y = -\dfrac{1}{2}$.

What y between 0 and π has cosine equal to $-\dfrac{1}{2}$?
y must be associated with a reference triangle in the the second quadrant. Reference triangle is a special $30° - 60°$ triangle.

$y = \dfrac{2\pi}{3} \qquad \arccos\left(-\dfrac{1}{2}\right) = \dfrac{2\pi}{3}$

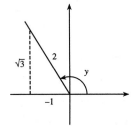

21. $y = \tan^{-1}(-1)$ is equivalent to $\tan y = -1$.
What y between $-\dfrac{\pi}{2}$ and $\dfrac{\pi}{2}$ has tangent equal to -1?
y must be negative and associated with a reference triangle in the fourth quadrant. Reference triangle is a special $45°$ triangle.

$y = -\dfrac{\pi}{4} \qquad \tan^{-1}(-1) = -\dfrac{\pi}{4}$

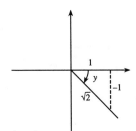

23. $y = \sin^{-1}\left(-\dfrac{\sqrt{3}}{2}\right)$ is equivalent to $\sin y = -\dfrac{\sqrt{3}}{2}$.

What y between $-\dfrac{\pi}{2}$ and $\dfrac{\pi}{2}$ has sine equal to $-\dfrac{\sqrt{3}}{2}$?
y must be negative and associated with a reference triangle in the fourth quadrant. Reference triangle is a special $30° - 60°$ triangle.

$y = -\dfrac{\pi}{3} \qquad \sin^{-1}\left(-\dfrac{\sqrt{3}}{2}\right) = -\dfrac{\pi}{3}$

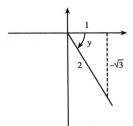

25. $y = \cos^{-1}\left(-\dfrac{\sqrt{3}}{2}\right)$ is equivalent to $\cos y = -\dfrac{\sqrt{3}}{2}$.

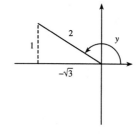

What y between 0 and π has cosine equal to $-\dfrac{\sqrt{3}}{2}$?
y must be associated with a reference triangle in the
second quadrant. Reference triangle is a special
$30° - 60°$ triangle.

$y = \dfrac{5\pi}{6}$ $\cos^{-1}\left(-\dfrac{\sqrt{3}}{2}\right) = \dfrac{5\pi}{6}$

27. $\sin[\sin^{-1}(-0.6)] = -0.6$ from the sine-inverse sine identity.

29. $\tan^{-1}[\tan(-1.5)] = -1.5$ from the tangent-inverse tangent identity.

31. Let $y = \cos^{-1}\dfrac{1}{2}$, then $\cos y = \dfrac{1}{2}$, $0 \le y \le \pi$. From the reference triangle (see Problem 7), we see
that $\tan\left(\cos^{-1}\dfrac{1}{2}\right) = \tan y = \sqrt{3}$.

33. Let $y = \sin^{-1}\left(-\dfrac{\sqrt{2}}{2}\right)$, then $\sin y = -\dfrac{\sqrt{2}}{2}$, $-\dfrac{\pi}{2} \le y \le \dfrac{\pi}{2}$.

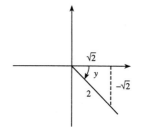

Draw the reference triangle associated with y.
Then, $\cos y = \cos\left[\sin^{-1}\left(-\dfrac{\sqrt{2}}{2}\right)\right]$ can be determined
directly from the triangle or by recognizing that $y = -\dfrac{\pi}{4}$.

$\cos\left[\sin^{-1}\left(-\dfrac{\sqrt{2}}{2}\right)\right] = \cos y = \dfrac{\sqrt{2}}{2}$

or $= \cos\left(-\dfrac{\pi}{4}\right) = \dfrac{\sqrt{2}}{2}$

35. Let $y = \cos^{-1}\left(-\dfrac{1}{2}\right)$, then $\cos y = -\dfrac{1}{2}$, $0 \le y \le \pi$. Draw the reference triangle associated with y (see
Problem 19). Then, $\cot y = \cot\left[\cos^{-1}\left(-\dfrac{1}{2}\right)\right]$ can be determined directly from the triangle, or by
recognizing that $y = \dfrac{2\pi}{3}$.

$\cot\left[\cos^{-1}\left(-\dfrac{1}{2}\right)\right] = \cot y = -\dfrac{1}{\sqrt{3}}$

or $= \cot\dfrac{2\pi}{3} = -\dfrac{1}{\sqrt{3}}$

37. Calculator in radian mode: $\tan^{-1}(-4.038) = -1.328$

39. Calculator in radian mode: $\sec[\sin^{-1}(-0.0399)] = \dfrac{1}{\cos[\sin^{-1}(-0.0399)]} = 1.001$

41. Calculator in radian mode: $\csc[\tan^{-1}(-4.118)] = \dfrac{1}{\sin[\tan^{-1}(-4.118)]} = -1.029$

Chapter 5   Inverse Trigonometric Functions; Trigonometric Equations

43. Calculator in radian mode: $\sqrt{2} + \tan^{-1}\left(\sqrt[3]{5}\right) = 2.456$

45.

x	−1.0	−0.8	−0.6	−0.4	−0.2	0
$\sin^{-1}x$	−1.57	−0.92	−0.64	−0.41	−0.20	0

x	0.2	0.4	0.6	0.8	1.0
$\sin^{-1}x$	0.20	0.41	0.64	0.92	1.57

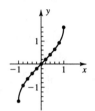

47. $\theta = \arccos\left(-\dfrac{1}{2}\right)$ is equivalent to $\cos\theta = -\dfrac{1}{2}$; $0° \leq \theta \leq 180°$. Thus, $\theta = 120°$.

49. $\theta = \tan^{-1}(-1)$ is equivalent to $\tan\theta = -1$; $-90° < \theta < 90°$. Thus, $\theta = -45°$.

51. $\theta = \sin^{-1}\left(-\dfrac{\sqrt{3}}{2}\right)$ is equivalent to $\sin\theta = -\dfrac{\sqrt{3}}{2}$; $-90° \leq \theta \leq 90°$. Thus, $\theta = -60°$.

53. Calculator in degree mode: $\theta = \tan^{-1} 3.0413 = 71.80°$

55. Calculator in degree mode: $\theta = \arcsin(-0.8107) = -54.16°$

57. Calculator in degree mode: $\theta = \arctan(-17.305) = -86.69°$

59. Let $u = \cos^{-1}\left(-\dfrac{4}{5}\right)$, $v = \sin^{-1}\left(-\dfrac{3}{5}\right)$. Then we are asked to evaluate $\sin(u + v)$, which is

$\sin u \cos v + \cos u \sin v$ from the sum identity. We know $\sin v = \sin\left[\sin^{-1}\left(-\dfrac{3}{5}\right)\right] = -\dfrac{3}{5}$ and

$\cos u = \cos\left[\cos^{-1}\left(-\dfrac{4}{5}\right)\right] = -\dfrac{4}{5}$ from the function-inverse function identities. It remains to find

$\cos v$ and $\sin u$. Note: $0 \leq u \leq \pi$ and $-\dfrac{\pi}{2} \leq v \leq \dfrac{\pi}{2}$.

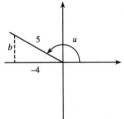

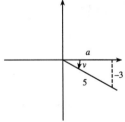

$b = \sqrt{5^2 - (-4)^2} = 3$; $\sin u = \dfrac{3}{5}$ $a = \sqrt{5^2 - (-3)^2} = 4$; $\cos v = \dfrac{4}{5}$

Then, $\sin\left[\cos^{-1}\left(-\frac{4}{5}\right) + \sin^{-1}\left(-\frac{3}{5}\right)\right] = \sin(u+v) = \sin u \cos v + \cos u \sin v$

$$= \left(\frac{3}{5}\right)\left(\frac{4}{5}\right) + \left(-\frac{4}{5}\right)\left(-\frac{3}{5}\right) = \frac{12}{25} + \frac{12}{25} = \frac{24}{25}$$

61. We could proceed as in Problem 59. Alternatively, we can shorten the process by recognizing $\arccos\frac{1}{2} = \frac{\pi}{3}$ and $\arcsin(-1) = -\frac{\pi}{2}$. Then,

$$\sin\left[\arccos\frac{1}{2} + \arcsin(-1)\right] = \sin\left(\frac{\pi}{3} + -\frac{\pi}{2}\right) = \sin\frac{\pi}{3}\cos\left(-\frac{\pi}{2}\right) + \cos\frac{\pi}{3}\sin\left(-\frac{\pi}{3}\right)$$

$$= \frac{\sqrt{3}}{2}(0) + \left(\frac{1}{2}\right)(-1) = -\frac{1}{2}$$

63. Let $y = \sin^{-1}\left(-\frac{4}{5}\right)$. Then, $\sin y = -\frac{4}{5}$. We are asked to evaluate $\sin(2y)$, which is $2\sin y \cos y$ by the double-angle identity. Draw a reference triangle associated with y; then, $\cos y = \cos\left[\sin^{-1}\left(-\frac{4}{5}\right)\right]$ can be determined directly from the triangle.

$\sin y = \sin\left[\sin^{-1}\left(-\frac{4}{5}\right)\right] = -\frac{4}{5}$ by the sine-inverse sine identity.

Note: $-\frac{\pi}{2} \le y \le \frac{\pi}{2}$ $a = \sqrt{5^2 - (-4)^2} = 3$ $\cos y = \frac{3}{5}$.

Thus, $\sin\left[2\sin^{-1}\left(-\frac{4}{5}\right)\right] =$

$2\sin\left[\sin^{-1}\left(-\frac{4}{5}\right)\right]\cos\left[\sin^{-1}\left(-\frac{4}{5}\right)\right] =$

$2\left(-\frac{4}{5}\right)\left(\frac{3}{5}\right) = -\frac{24}{25}$

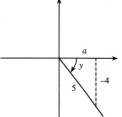

65. Let $y = \cos^{-1}\left(\frac{1}{5}\right)$. Then, $\cos y = \frac{1}{5}$. We are asked to evaluate $\sin\left(\frac{1}{2}y\right)$, which is $\pm\sqrt{\frac{1-\cos y}{2}}$ by the half-angle identity. Since $0 \le y \le \pi$, we have $0 \le \frac{1}{2}y \le \frac{\pi}{2}$. Thus we choose the positive sign of the square root. Hence,

$$\sin\frac{1}{2}y = \sqrt{\frac{1-\cos y}{2}} \qquad \sin\left[\frac{\cos^{-1}\left(\frac{1}{5}\right)}{2}\right] = \sqrt{\frac{1-\frac{1}{5}}{2}} = \sqrt{\frac{2}{5}} = \frac{2}{\sqrt{10}}$$

67.　Let $y = \cos^{-1}x$　　($-1 \le x \le 1$ corresponds to $0 \le y \le \pi$)

or, equivalently, $\cos y = x$　　$0 \le y \le \pi$

Geometrically,

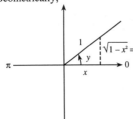　　or　　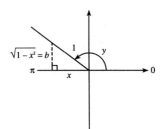

In either case, $b = \sqrt{1 - x^2}$. Thus, $\sin(\cos^{-1} x) = \sin y = \dfrac{b}{R} = \sqrt{1 - x^2}$

69.　Let $y = \arcsin x$　$\left(-1 \le x \le 1 \text{ corresponds to } -\dfrac{\pi}{2} \le y \le \dfrac{\pi}{2}\right)$

or, equivalently, $\sin y = x$　　$-\dfrac{\pi}{2} \le y \le \dfrac{\pi}{2}$

Geometrically,

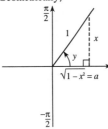　　or　　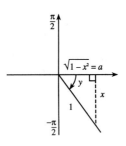

In either case, $a = \sqrt{1 - x^2}$. Thus, $\tan(\arcsin x) = \tan y = \dfrac{b}{a} = \dfrac{x}{\sqrt{1 - x^2}}$

71.　Let $y = \tan^{-1} x$　　　$-\dfrac{\pi}{2} < y < \dfrac{\pi}{2}$　　or, equivalently,

　　　$x = \tan y$　　　$-\dfrac{\pi}{2} < y < \dfrac{\pi}{2}$

Geometrically,

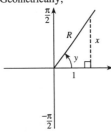　　or　　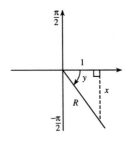

In either case, $R = \sqrt{x^2 + 1}$.

$$\sin(2 \tan^{-1} x) = \sin(2y) = 2 \sin y \cos y = 2 \cdot \frac{b}{R} \cdot \frac{a}{R} = 2 \cdot \frac{1}{\sqrt{x^2 + 1}} \cdot \frac{x}{\sqrt{x^2 + 1}} = \frac{2x}{x^2 + 1}$$

73. Let $u = \cos^{-1} x$ and $v = \sin^{-1} x$ $-1 \le x \le 1$

or, equivalently, $x = \cos u$ $0 \le u \le \pi$ $x = \sin v$ $-\dfrac{\pi}{2} \le v \le \dfrac{\pi}{2}$

Then, $\sin(\cos^{-1} x - \sin^{-1} x) = \sin(u - v) = \sin u \cos v - \cos u \sin v$.

From Problem 67, we have $\sin u = \sin(\cos^{-1} x) = \sqrt{1 - x^2}$

From Problem 69, we have $\cos v = \cos(\sin^{-1} x) = \sqrt{1 - x^2}$ (see figure, Problem 69)

From the sine-inverse sine identity, we have $\sin v = \sin(\sin^{-1} x) = x$

From the cosine-inverse cosine identity, we have $\cos u = \cos(\cos^{-1} x) = x$

Thus, $\sin(\cos^{-1} x - \sin^{-1} x) = \sin u \cos v - \cos u \sin v = \sqrt{1 - x^2} \cdot \sqrt{1 - x^2} - x \cdot x$
$$= 1 - x^2 - x^2 = 1 - 2x^2$$

75. Since $y = a \sin(bx + c)$, $-\dfrac{\pi}{2} \le bx + c \le \dfrac{\pi}{2}$, $\dfrac{y}{a} = \sin(bx + c)$

This is equivalent to: $bx + c = \sin^{-1}\left(\dfrac{y}{a}\right)$, $-1 \le \dfrac{y}{a} \le 1$. Hence, $bx = \sin^{-1}\left(\dfrac{y}{a}\right) - c$,

$x = \dfrac{1}{b} \sin^{-1}\left(\dfrac{y}{a}\right) - \dfrac{c}{b}$. Since $a > 0$, the restriction $-1 \le \dfrac{y}{a} \le 1$ can be written as $-a \le y \le a$.

77.
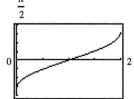

79.
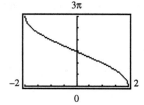

81.
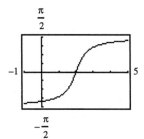

83. (A)
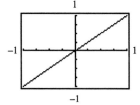

(B)
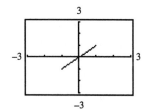

Since $\sin^{-1} x$ is defined only for $-1 \le x \le 1$, no graph results from using values outside this interval.

Some calculators may give an error message in this situation.

Chapter 5   Inverse Trigonometric Functions;
Trigonometric Equations

85. (A)

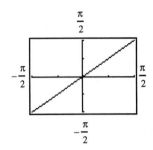

(B)

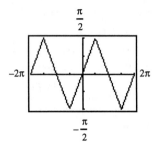

For any real number x, sin x is a real number between −1 and 1, thus sin⁻¹(sin x) is defined for all real numbers x. Enlarging the graphing interval extends the graph of this periodic function.

87. (A) If $\sin\dfrac{\theta}{2} = \dfrac{1}{M}$, then $\dfrac{\theta}{2} = \sin^{-1}\left(\dfrac{1}{M}\right)$ $-1 \le \dfrac{1}{M} \le 1$.

Thus, $\theta = 2\sin^{-1}\left(\dfrac{1}{M}\right)$. Since M must be positive, $\dfrac{1}{M} \le 1$, or $M \ge 1$.

(B) Calculator in degree mode:

For M = 1.7, $\theta = 2\sin^{-1}\left(\dfrac{1}{1.7}\right) = 72°$. For M = 2.3, $\theta = 2\sin^{-1}\left(\dfrac{1}{2.3}\right) = 52°$

89. (A) The volume of the fuel is clearly given by

Volume = (height)(cross-sectional area)

with L = height.

To determine the cross-sectional area (see figure),

we reason that

Area of segment AEBF

= Area of sector ACBF − Area of triangle

ABC

Area of sector ACBF

= 2(area of sector CFB) = $2\left(\dfrac{1}{2}R^2\theta\right) = R^2\theta$

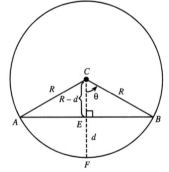

Since triangle CEB is a right triangle, we have

$\cos\theta = \dfrac{EC}{BC} = \dfrac{R-d}{R}$ $\qquad \theta = \cos^{-1}\dfrac{R-d}{R}$

Therefore,

Area of sector ACBF = $R^2 \cos^{-1}\dfrac{R-d}{R}$

Area of triangle ABC = $\dfrac{1}{2}$(base)(altitude) = $\dfrac{1}{2}$(AB)(CE) = $\dfrac{1}{2}$(2 · EB)(CE) = (EB)(R − d)

By the Pythagorean theorem applied to triangle CEB, we have EB² + (R − d)² = R².

EB = $\sqrt{R^2 - (R-d)^2}$. Therefore, Area of triangle ABC = $(R-d)\sqrt{R^2-(R-d)^2}$.

Finally,

$$\text{Area of segment AEBF} = R^2 \cos^{-1}\frac{R-d}{R} - (R-d)\sqrt{R^2 - (R-d)^2}$$

$$\text{Volume} = \left[R^2 \cos^{-1}\frac{R-d}{R} - (R-d)\sqrt{R^2 - (R-d)^2}\right]L$$

(B) Substituting the given values, we have $R = 3$, $d = 2$, $R - d = 1$, $L = 30$

$$\text{Volume} = \left[3^2 \cos^{-1}\frac{1}{3} - 1\sqrt{3^2 - 1^2}\right]30$$

$$= \left[9 \cos^{-1}\frac{1}{3} - \sqrt{8}\right]30 \qquad \text{(Calculator in radian mode)}$$

$$\approx 248 \text{ ft}^3$$

91. The graphs of y1 and y2 are shown in the figure.
     We use the zoom feature or the built-in approxi-
     mation routine to locate the x coordinate of the
     point of intersection at x = 2.6.
     The depth is 2.6 feet.

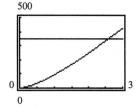

93. From the figure, the following should be clear:

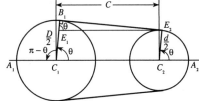

Length of belt = 2[arc A_1B_1 + B_1E_2 + arc E_2A_2]; arc $A_1B_1 = \frac{D}{2}(\pi - \theta)$; arc $E_2A_2 = \frac{d}{2}\theta$.

To find B_1E_2, we note: C_1C_2 has length C. E_1E_2 is constructed parallel to C_1C_2. E_1C_1 is parallel
to E_2C_2. Hence, $E_1C_1C_2E_2$ is a parallelogram. E_1E_2 has length C.
$B_1E_2E_1$ is a right triangle. Thus,

$$(1) \ \cos\theta = \frac{B_1E_1}{E_1E_2} = \frac{\frac{D}{2}-\frac{d}{2}}{C} = \frac{D-d}{2C} \quad \text{and} \quad (2) \ \sin\theta = \frac{B_1E_2}{E_1E_2}, \text{ so } B_1E_2 = E_1E_2 \sin\theta = C\sin\theta$$

Finally, Length of belt = 2[arc A_1B_1 + B_1E_2 + arc E_2A_2] = $2\left[\frac{D}{2}(\pi - \theta) + C\sin\theta + \frac{d}{2}\theta\right]$

$$= D(\pi - \theta) + 2C\sin\theta + d\theta$$

$$L = \pi D + (d - D)\theta + 2C\sin\theta$$

and, from (1) above, $\theta = \cos^{-1}\frac{D-d}{2C}$. Substituting the given values, we have

$$D = 6, d = 4, C = 10 \qquad \text{(Calculator in radian mode)}$$

$$\theta = \cos^{-1}\frac{6-4}{2\cdot 10} = \cos^{-1}\frac{1}{10}$$

$$L = 6\left(\pi - \cos^{-1}\frac{1}{10}\right) + 2\cdot 10\sin\left(\cos^{-1}\frac{1}{10}\right) + 4\cos^{-1}\frac{1}{10} \approx 35.8 \text{ inches}$$

Chapter 5   Inverse Trigonometric Functions; Trigonometric Equations

95. The graphs of y1 and y2 are shown in the figure. We use the zoom feature or the built-in approxima-tion routine to locate the x coordinate of the point of intersection at x = 12.1. The length is 12.1 feet.

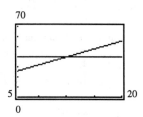

97. (A) Following the hint, we draw AC. Then, since the central angle in a circle subtended by an arc is twice any inscribed angle subtended by the same arc, angle ACB has measure 2θ. Thus, $d = R \cdot 2\theta = 2R\theta$. In triangle ECP, $\tan\theta = \dfrac{x}{R}$, hence,

$$\theta = \tan^{-1}\frac{x}{R} \qquad d = 2R\,\tan^{-1}\frac{x}{R}$$

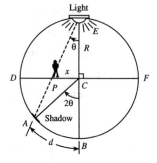

(B) Substituting the given values, we have R = 50, x = 25.

$$d = 2 \cdot 50\,\tan^{-1}\frac{25}{50} = 100\,\tan^{-1}\frac{1}{2}\text{ (Calculator in radian mode)}$$

$$\approx 46.4 \text{ ft.}$$

EXERCISE 5.2 Inverse Cotangent, Secant, and Cosecant Functions

1. $y = \cot^{-1}\sqrt{3}$ is equivalent to $\cot y = \sqrt{3}$ and $0 < y < \pi$.
   What number between 0 and π has cotangent equal to $\sqrt{3}$?
   y must be in the first quadrant.
   $$\cot y = \sqrt{3} = \frac{\sqrt{3}}{1}, y = \frac{\pi}{6}.$$
   Thus, $\cot^{-1}\sqrt{3} = \dfrac{\pi}{6}$

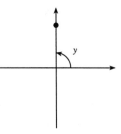

3. $y = \operatorname{arccsc} 1$ is equivalent to $\csc y = 1$ and $-\dfrac{\pi}{2} \le y \le \dfrac{\pi}{2}$, $y \ne 0$.
   What number between $-\dfrac{\pi}{2}$ and $\dfrac{\pi}{2}$ has cosecant equal to 1?
   No reference triangle can be drawn, but from the diagram we see (a, b) = (0, 1)   R = 1 $\csc y = \dfrac{1}{1} = 1$ $y = \dfrac{\pi}{2}$

   Thus, $\operatorname{arccsc} 1 = \dfrac{\pi}{2}$

5. $y = \sec^{-1} \sqrt{2}$ is equivalent to $\sec y = \sqrt{2}$ and $0 \le y \le \pi$, $y \ne \dfrac{\pi}{2}$.

What number between 0 and π has secant equal to $\sqrt{2}$?

y must be in the first quadrant.

$\sec y = \sqrt{2} = \dfrac{\sqrt{2}}{1}$ $y = \dfrac{\pi}{4}$

Thus, $\sec^{-1} \sqrt{2} = \dfrac{\pi}{4}$

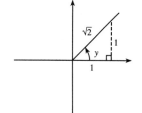

7. Let $y = \cot^{-1} 0$, then $\cot y = 0$, $0 < y < \pi$.

No reference triangle can be drawn, but from the diagram we

see $(a, b) = (0, 1)$ $R = 1$ $\cot y = \dfrac{0}{1} = 0$ $y = \dfrac{\pi}{2}$

Thus, $\sin(\cot^{-1} 0) = \sin\dfrac{\pi}{2} = 1$

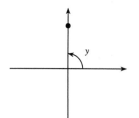

9. Let $y = \csc^{-1} \dfrac{5}{4}$, then $\csc y = \dfrac{5}{4}$, $-\dfrac{\pi}{2} \le y \le \dfrac{\pi}{2}$, $y \ne 0$.

y is in the first quadrant. Draw a reference triangle, find

the third side, then determine tan y from the triangle.

$a = \sqrt{5^2 - 4^2} = 3$

Thus, $\tan\left(\csc^{-1} \dfrac{5}{4}\right) = \tan y = \dfrac{b}{a} = \dfrac{4}{3}$

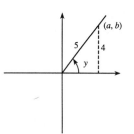

11. $y = \cot^{-1}(-1)$ is equivalent to $\cot y = -1$ and $0 < y < \pi$.

What number between 0 and π has cotangent equal to -1?

y must be positive and in the second quadrant.

$\cot y = -1 = \dfrac{-1}{1}$ $\alpha = \dfrac{\pi}{4}$ $y = \dfrac{3\pi}{4}$

Thus, $\cot^{-1}(-1) = \dfrac{3\pi}{4}$

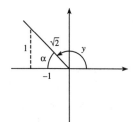

13. $y = \text{arcsec}(-2)$ is equivalent to $\sec y = -2$ and $0 \le y \le \pi$, $y \ne \dfrac{\pi}{2}$.

What number between 0 and π has secant equal to -2? y must

be positive and in the second quadrant.

$\sec y = -2 = \dfrac{2}{-1}$ $\alpha = \dfrac{\pi}{3}$ $y = \dfrac{2\pi}{3}$

Thus, $\text{arcsec}(-2) = \dfrac{2\pi}{3}$

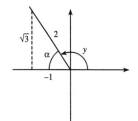

Chapter 5   Inverse Trigonometric Functions; Trigonometric Equations

15. $y = \text{arccsc}(-2)$ is equivalent to $\csc y = -2$ and $-\frac{\pi}{2} \le y \le \frac{\pi}{2}$, $y \ne 0$.

What number between $-\frac{\pi}{2}$ and $\frac{\pi}{2}$ has cosecant equal to -2?

y must be negative and in the fourth quadrant.

$\csc y = -2 = \frac{2}{-1}$ $\alpha = \frac{\pi}{6}$ $y = -\frac{\pi}{6}$

Thus, $\text{arccsc}(-2) = -\frac{\pi}{6}$

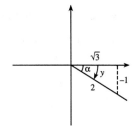

17.   If $x = \frac{1}{2}$, neither $x \le -1$ nor $x \ge 1$ is a true statement. Thus, x is not in the domain of the inverse

cosecant function. $\csc^{-1}\frac{1}{2}$ is not defined.

19.   Let $y = \csc^{-1}\left(-\frac{5}{3}\right)$; then, $\csc y = -\frac{5}{3}$, $-\frac{\pi}{2} \le y \le \frac{\pi}{2}$, $y \ne 0$.

y is negative and in the fourth quadrant. Draw a reference

triangle, find the third side, and then determine cos y from

the triangle.

$a = \sqrt{5^2 - (-3)^2} = 4$ $\left(\csc y = -\frac{5}{3} = \frac{5}{-3}\right)$

Thus, $\cos\left[\csc^{-1}\left(-\frac{5}{3}\right)\right] = \cos y = \frac{a}{R} = \frac{4}{5}$

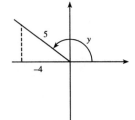

21.   Let $y = \sec^{-1}\left(-\frac{5}{4}\right)$, then, $\sec y = -\frac{5}{4}$, $0 \le y \le \pi$, $y \ne \frac{\pi}{2}$.

y is positive and in the second quadrant. Draw a reference

triangle, find the third side, and then determine cot y from

the triangle.

$b = \sqrt{5^2 - (-4)^2} = 3$ $\left(\sec y = -\frac{5}{4} = \frac{5}{-4}\right)$

Thus, $\cot\left[\sec^{-1}\left(-\frac{5}{4}\right)\right] = \cot y = \frac{a}{b} = -\frac{4}{3}$

23.   From the inverse trigonometric identities, we see $\sec^{-1} x = \cos^{-1}\frac{1}{x}$, $x \ge 1$ or $x \le -1$.

Thus, $\cos(\sec^{-1} x) = \cos\left(\cos^{-1}\frac{1}{x}\right) = \frac{1}{x}$ if $\frac{1}{x}$ is in the domain of the inverse cosine function.

Hence, $\cos[\sec^{-1}(-2)] = \cos\left[\cos^{-1}\left(\frac{1}{-2}\right)\right] = \frac{1}{-2}$ or $-\frac{1}{2}$, since $-\frac{1}{2}$ is in the domain of the inverse

cosine function.

25. We note: $\cot^{-1} x = \tan^{-1} \dfrac{1}{x}$ for $x > 0$.

Thus, $\cot(\cot^{-1} x) = \cot\left(\tan^{-1}\dfrac{1}{x}\right)$

$= \dfrac{1}{\tan\left(\tan^{-1}\dfrac{1}{x}\right)}$ Reciprocal identity

$= \dfrac{1}{\dfrac{1}{x}}$ Tangent-inverse tangent identity

$= x$ Algebra

Since $33.4 > 0$, the above reasoning applies; thus $\cot(\cot^{-1} 33.4) = 33.4$

27. We note: $\csc^{-1} x = \sin^{-1} \dfrac{1}{x}$ for $x \geq 1$ or $x \leq -1$.

Thus, $\csc(\csc^{-1} x) = \csc\left(\sin^{-1}\dfrac{1}{x}\right)$

$= \dfrac{1}{\sin\left(\sin^{-1}\dfrac{1}{x}\right)}$ Reciprocal identity

$= \dfrac{1}{\dfrac{1}{x}}$ $\begin{cases}\text{Sine-inverse sine identity. [If } x \geq 1 \text{ or}\\ x \leq -1 \text{ then } 1/x \text{ will be in the domain}\\ \text{of the inverse sine function.]}\end{cases}$

$= x$ Algebra

Since $-4 \leq -1$, the above reasoning applies; thus, $\csc[\csc^{-1}(-4)] = -4$

29. $\cot^{-1} 3.065 = \tan^{-1} \dfrac{1}{3.065}$ (Calculator in radian mode)

$= 0.315$

31. $\sec^{-1}(-1.963) = \cos^{-1} \dfrac{1}{-1.963}$ (Calculator in radian mode)

$= 2.105$

33. $\csc^{-1} 1.172 = \sin^{-1} \dfrac{1}{1.172}$ (Calculator in radian mode)

$= 1.022$

35. We note: $\cot^{-1} x = \pi + \tan^{-1} \dfrac{1}{x}$ if $x < 0$.

Thus, $\cot^{-1}(-5.104) = \pi + \tan^{-1} \dfrac{1}{-5.104}$ (Calculator in radian mode)

$= 2.948$

37. $\theta = \text{arcsec}(-2)$ is equivalent to $\sec\theta = -2$ $0 \leq \theta \leq 180°$, $\theta \neq 90°$

$\cos\theta = -\dfrac{1}{2}$ $0 \leq \theta \leq 180°$

Thus, $\theta = 120°$

Chapter 5   Inverse Trigonometric Functions; Trigonometric Equations

39. $\theta = \cot^{-1}(-1)$ is equivalent to $\cot\theta = -1$ $0 < \theta < 180°$

$$\tan\theta = \frac{1}{-1} = -1 \quad 0 < \theta < 180°$$

Thus, $\theta = 135°$

41. $\theta = \csc^{-1}\left(-\dfrac{2}{\sqrt{3}}\right)$ is equivalent to $\csc\theta = -\dfrac{2}{\sqrt{3}}$ $-90° \le \theta \le 90°, \ \theta \ne 0°$

$$\sin\theta = -\frac{\sqrt{3}}{2} \quad\quad -90° \le \theta \le 90°$$

Thus, $\theta = -60°$

43.   Calculator in degree mode: $\theta = \cot^{-1}0.3288 = \tan^{-1}\dfrac{1}{0.3288} = 71.80°$

45.   Calculator in degree mode: $\theta = \mathrm{arccsc}(-1.2336) = \arcsin\dfrac{1}{-1.2336} = -54.16°$

47.   Calculator in degree mode: $\theta = \mathrm{arccot}(-0.0578) = 180° + \tan^{-1}\left(\dfrac{1}{-0.0578}\right) = 93.31°$

49.   Let $u = \csc^{-1}\left(-\dfrac{5}{3}\right)$ and $v = \tan^{-1}\dfrac{1}{4}$. Then we are asked to evaluate $\tan(u + v)$, which is

$\dfrac{\tan u + \tan v}{1 + \tan u\ \tan v}$ by the sum identity for tangent. We know $\tan v = \tan\left(\tan^{-1}\dfrac{1}{4}\right) = \dfrac{1}{4}$ by

the tangent-inverse tangent identity. It remains to find $\tan u = \tan\left[\csc^{-1}\left(-\dfrac{5}{3}\right)\right]$. See sketch,

Problem 19. $\tan\left[\csc^{-1}\left(-\dfrac{5}{3}\right)\right] = \dfrac{b}{a} = \dfrac{-3}{4}$ from the reference triangle. Hence,

$$\tan\left[\csc^{-1}\left(-\frac{5}{3}\right) + \tan^{-1}\frac{1}{4}\right] = \tan(u+v) = \frac{\tan u + \tan v}{1 - \tan u\ \tan v}$$

$$= \frac{-\dfrac{3}{4} + \dfrac{1}{4}}{1 - \left(-\dfrac{3}{4}\right)\left(\dfrac{1}{4}\right)} = \frac{-\dfrac{2}{4}}{1 + \dfrac{3}{16}} = \frac{-8}{19}$$

51.   Let $y = \cot^{-1}\left(-\dfrac{3}{4}\right)$. Then we are asked to evaluate

$\tan(2y)$, which is $\dfrac{2\tan y}{1 - \tan^2 y}$ from the double-angle

identity. Draw a reference triangle associated with y.

Then, $\tan y = \tan\left[\cot^{-1}\left(-\dfrac{3}{4}\right)\right]$ can be determined

directly from the triangle. y is positive and in the

second quadrant. $\tan y = \dfrac{4}{-3} = -\dfrac{4}{3}$. Thus,

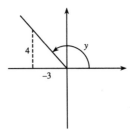

$$\tan\left[2\cot^{-1}\left(-\frac{3}{4}\right)\right] = \tan 2y = \frac{2\left(-\dfrac{4}{3}\right)}{1 - \left(-\dfrac{4}{3}\right)^2} = \frac{-\dfrac{8}{3}}{1 - \left(\dfrac{16}{9}\right)} = \frac{24}{7}$$

53. Let $y = \cot^{-1} x$ $0 < y < \pi$ or, equivalently,

      $x = \cot y$ $0 < y < \pi$

Geometrically,

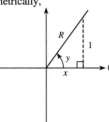

or

In either case, $R = \sqrt{x^2 + 1}$. Thus, $\sin(\cot^{-1} x) = \sin y = \dfrac{b}{R} = \dfrac{1}{\sqrt{x^2 + 1}}$.

55. Let $y = \sec^{-1} x$ $0 \le y \le \pi$ $y \ne \dfrac{\pi}{2}$ or, equivalently,

      $x = \sec y$ $0 \le y \le \pi$

Geometrically,

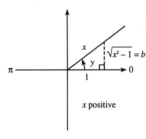

or

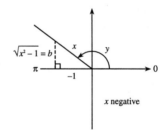

x positive

x negative

In either case, $R = \sqrt{x^2 - 1}$. Thus, if $x > 0$, $\csc(\sec^{-1} x) = \csc y = \dfrac{R}{b} = \dfrac{x}{\sqrt{x^2 - 1}}$.

If $x < 0$, $\csc(\sec^{-1} x) = \csc y = \dfrac{R}{b} = \dfrac{-x}{\sqrt{x^2 - 1}}$. A convenient notation for the quantity x if $x > 0$,

$-x$ if $x < 0$, is $|x|$. Hence, we can write $\csc(\sec^{-1} x) = \dfrac{|x|}{\sqrt{x^2 - 1}}$.

57. Let $y = \cot^{-1} x$ $0 < y < \pi$ or, equivalently,

      $x = \cot y$ $0 < y < \pi$

Then, $\sin(2 \cot^{-1} x) = \sin(2y) = 2 \sin y \cos y$. From Problem 53, we have

$$\sin y = \sin(\cot^{-1} x) = \frac{1}{\sqrt{x^2 + 1}} .$$

From the figure in Problem 53, we also have $\cos y = \cos(\cot^{-1} x) = \dfrac{a}{R} = \dfrac{x}{\sqrt{x^2 + 1}}$.

Thus, $\sin(2 \cot^{-1} x) = 2 \cdot \dfrac{1}{\sqrt{x^2 + 1}} \cdot \dfrac{x}{\sqrt{x^2 + 1}} = \dfrac{2x}{x^2 + 1}$.

59. Let $y = \sec^{-1} x$ $x \leq -1$ or $x \geq 1$

Then, $\sec y = x$ $0 \leq y \leq \pi, \, y \neq \dfrac{\pi}{2}$ Definition of $\sec^{-1}$

$\dfrac{1}{\cos y} = x$ $0 \leq y \leq \pi, \, y \neq \dfrac{\pi}{2}$ Reciprocal identity

$\cos y = \dfrac{1}{x}$ $0 \leq y \leq \pi, \, y \neq \dfrac{\pi}{2}$ Algebra

$y = \cos^{-1} \dfrac{1}{x}$ $0 \leq y \leq \pi, \, y \neq \dfrac{\pi}{2}$ Definition of $\cos^{-1}$

Thus, $\sec^{-1} x = \cos^{-1} \dfrac{1}{x}$ for $x \leq -1$ and $x \geq 1$

61.

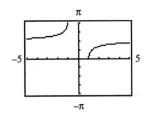

63.

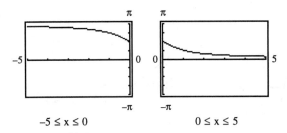

$-5 \leq x \leq 0$ $0 \leq x \leq 5$

EXERCISE 5.3 Trigonometric Equations I

1. Sine is positive in the first
    and

    second quadrants. Sketch the

    reference triangles and angles

    corresponding to x.

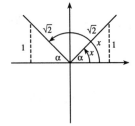

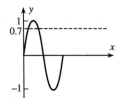

$\alpha = \sin^{-1} \dfrac{1}{\sqrt{2}} = \dfrac{\pi}{4}$

$$x = \begin{cases} \alpha \\ \pi - \alpha \end{cases} = \begin{cases} \sin^{-1} \dfrac{1}{\sqrt{2}} \\ \pi - \sin^{-1} \dfrac{1}{\sqrt{2}} \end{cases} = \begin{cases} \dfrac{\pi}{4} \\ \pi - \dfrac{\pi}{4} \end{cases} = \begin{cases} \dfrac{\pi}{4} \\ \dfrac{3\pi}{4} \end{cases}$$

3. Cosine is negative in the

    second and third quadrants.

    Sketch the reference

    triangles and angles

    corresponding to x.

    $\alpha = \cos^{-1}\dfrac{1}{2} = \dfrac{\pi}{3}$

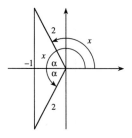

$$x = \begin{cases} \pi - \alpha \\ \pi + \alpha \end{cases} = \begin{cases} \pi - \cos^{-1}\dfrac{1}{2} \\ \pi + \cos^{-1}\dfrac{1}{2} \end{cases} = \begin{cases} \pi - \dfrac{\pi}{3} \\ \pi + \dfrac{\pi}{3} \end{cases} = \begin{cases} \dfrac{2\pi}{3} \\ \dfrac{4\pi}{3} \end{cases}$$

5. Tangent is negative in the

    fourth quadrant. Sketch the

    reference triangle and angle
    corresponding to x.

    $\alpha = \tan^{-1}\sqrt{3} = \dfrac{\pi}{3}$

    $x = -\alpha = -\tan^{-1}\sqrt{3} = -\dfrac{\pi}{3}$

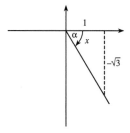

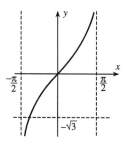

7. Cosine is positive in the

    first and fourth quadrants.

    Sketch the reference triangles

    and angles corresponding to θ.

    $\alpha = \cos^{-1}\dfrac{1}{2} = 60°.$

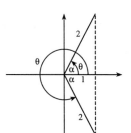

$$\theta = \begin{cases} \alpha \\ 360° - \alpha \end{cases} = \begin{cases} \cos^{-1}\dfrac{1}{2} \\ 360° - \cos^{-1}\dfrac{1}{2} \end{cases} = \begin{cases} 60° \\ 300° \end{cases}$$

Chapter 5 Inverse Trigonometric Functions; Trigonometric Equations

9. Sine is negative in the third and fourth quadrants. Sketch the reference triangles and angles corresponding to θ.

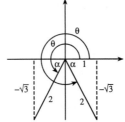 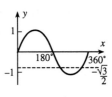

$\alpha = \sin^{-1} \dfrac{\sqrt{3}}{2} = 60°$

$$\theta = \begin{cases} 180° + \alpha \\ 360° - \alpha \end{cases} = \begin{cases} 180° + \sin^{-1} \dfrac{\sqrt{3}}{2} \\ 360° - \sin^{-1} \dfrac{\sqrt{3}}{2} \end{cases} = \begin{cases} 180° + 60° \\ 360° - 60° \end{cases} = \begin{cases} 240° \\ 300° \end{cases}$$

11. Tangent is negative in the fourth quadrant. Sketch the reference triangle and angle corresponding to θ.

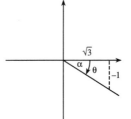 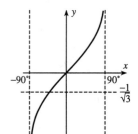

$\alpha = \tan^{-1} \dfrac{1}{\sqrt{3}} = 30°$

$\theta = -\alpha = -\tan^{-1} \dfrac{1}{\sqrt{3}} = -30°$

13. We have found all solutions of $\sin x = \dfrac{1}{\sqrt{2}}$ over one period in Problem 1.

They are $x = \dfrac{\pi}{4}$ and $x = \dfrac{3\pi}{4}$. We find all solutions by adding integer multiples of 2π to these.

$$x = \begin{cases} \dfrac{\pi}{4} + 2k\pi \\ \dfrac{3\pi}{4} + 2k\pi \end{cases} \qquad \text{k any integer}$$

15. We have found all solutions of $\cos x = -\dfrac{1}{2}$ over one period in Problem 3.

They are $x = \dfrac{2\pi}{3}$ and $x = \dfrac{4\pi}{3}$. We find all solutions by adding integer multiples of 2π to these.

$$x = \begin{cases} \dfrac{2\pi}{3} + 2k\pi \\ \dfrac{4\pi}{3} + 2k\pi \end{cases} \qquad \text{k any integer}$$

17. We have found all solutions of $\tan x = -\sqrt{3}$ over one period in Problem 5.

The only solution is $x = -\dfrac{\pi}{3}$. We find all solutions by adding integer multiples of π (the period of the tangent function) to this.

$$x = -\dfrac{\pi}{3} + k\pi \qquad \text{k any integer}$$

19. We have found all solutions of $\cos \theta = \frac{1}{2}$ over one period in Problem 7.

    They are $\theta = 60°$ and $\theta = 300°$. We find all solutions by adding integer multiples of $360°$ to these.

    $$\theta = \begin{cases} 60° + k360° \\ 300° + k360° \end{cases} \quad \text{k any integer}$$

21. We have found all solutions of $\sin \theta = -\frac{\sqrt{3}}{2}$ over one period in Problem 9.

    They are $\theta = 240°$ and $\theta = 300°$. We find all solutions by adding integer multiples of $360°$ to these.

    $$\theta = \begin{cases} 240° + k360° \\ 300° + k360° \end{cases} \quad \text{k any integer}$$

23. We have found all solutions of $\tan \theta = -\frac{1}{\sqrt{3}}$ over one period in Problem 11.

    The only solution is $\theta = -30°$. We find all solutions by adding integer multiples of $180°$ (the period of the tangent function) to this.

    $$\theta = -30° + k180° \quad \text{k any integer}$$

25. Sine is positive in the first and second quadrants. Sketch the reference triangles and angles corresponding to x. $\alpha = \sin^{-1} 0.60 = 0.64$

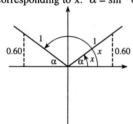 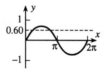

$$x = \begin{cases} \alpha \\ \pi - \alpha \end{cases} = \begin{cases} \sin^{-1} 0.60 \\ \pi - \sin^{-1} 0.60 \end{cases} = \begin{cases} 0.64 \\ 3.14 - 0.64 \end{cases} = \begin{cases} 0.64 \\ 2.50 \end{cases}$$

27. Cosine is negative in the second and third quadrants. Sketch the reference triangles and angles corresponding to x.

    $\alpha = \cos^{-1} 0.46 = 1.09$

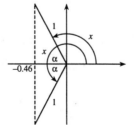

$$x = \begin{cases} \pi - \alpha \\ \pi + \alpha \end{cases} = \begin{cases} \pi - \cos^{-1} 0.46 \\ \pi + \cos^{-1} 0.46 \end{cases} = \begin{cases} 3.14 - 1.09 \\ 3.14 + 1.09 \end{cases} = \begin{cases} 2.05 \\ 4.23 \end{cases}$$

Chapter 5   Inverse Trigonometric Functions; Trigonometric Equations

29. Tangent is negative in the
    fourth quadrant. Sketch the
    reference triangle and angle
    corresponding to x.

    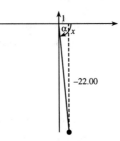
    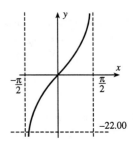

    $\alpha = \tan^{-1} 22.00 = 1.53$

    $x = -\alpha = -\tan^{-1} 22.00 = -1.53$

31. Cosine is positive in the first
    and fourth quadrants. Sketch
    the reference triangles and
    angles corresponding to θ.

    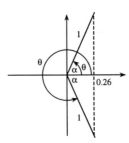
    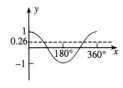

    $\alpha = \cos^{-1} 0.26 = 74.93°$

    $$x = \begin{cases} \alpha \\ 360° - \alpha \end{cases} = \begin{cases} \cos^{-1} 0.26 \\ 360° - \cos^{-1} 0.26 \end{cases} = \begin{cases} 74.93° \\ 360° - 74.93° \end{cases} = \begin{cases} 74.93° \\ 285.07° \end{cases}$$

33. Sine is negative in the third
    and fourth quadrants. Sketch
    the reference triangles and
    angles corresponding to θ.

    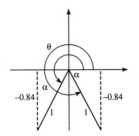
    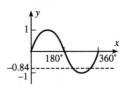

    $\alpha = \sin^{-1} 0.84 = 57.14°$

    $$\theta = \begin{cases} 180° + \alpha \\ 360° - \alpha \end{cases} = \begin{cases} 180° + \sin^{-1} 0.84 \\ 360° - \sin^{-1} 0.84 \end{cases} = \begin{cases} 180° + 57.14° \\ 360° - 57.14° \end{cases} = \begin{cases} 237.14° \\ 302.86° \end{cases}$$

35. $\alpha = \tan^{-1} 9.45 = 83.96°$

    $\theta = -\alpha = -\tan^{-1} 9.45$

    $\quad = -83.96°$

    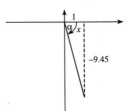
    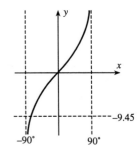

37. We have found all solutions of sin x = 0.60 over one period in Problem 25.

    They are x = 0.64 and x = 2.50. We find all solutions by adding integer multiples of 2π to these.
    $$x = \begin{cases} 0.64 + 2k\pi \\ 2.50 + 2k\pi \end{cases} \quad \text{k any integer}$$

39. We have found all solutions of cos x = –0.46 over one period in Problem 27.

    They are x = 2.05 and x = 4.23. We find all solutions by adding integer multiples of 2π to these.
    $$x = \begin{cases} 2.05 + 2k\pi \\ 4.23 + 2k\pi \end{cases} \quad \text{k any integer}$$

41. We have found all solutions of tan x = –22.00 over one period in Problem 29.

    The only solution is x = –1.53. We find all solutions by adding integer multiples of π (the period of the tangent function) to this.

    $x = -1.53 + k\pi$ k any integer

43. We have found all solutions of cos θ = 0.26 over one period in Problem 31.

    They are θ = 74.93° and θ = 285.07°. We find all solutions by adding integer multiples of 360° to these.
    $$x = \begin{cases} 74.93° + k360° \\ 285.07° + k360° \end{cases} \quad \text{k any integer}$$

45. We have found all solutions of sin θ = –0.84 over one period in Problem 33.

    There are θ = 237.14° and θ = 302.86°. We find all solutions by adding integer multiples of 360° to these
    $$x = \begin{cases} 237.14° + k360° \\ 302.86° + k360° \end{cases} \quad \text{k any integer}$$

47. We have found all solutions of tan θ = –9.45 over one period in Problem 35.

    The only solution is θ = –83.96°. We find all solutions by adding integer multiples of 180° (the period of the tangent function) to this.

    $\theta = -83.96° + k180°$ k any integer

49. There are no reference triangles, but from the graph of y = sin x, we can see that sin x crosses the x-axis (y = 0, sin x = 0) at intervals that are integer multiples of π.

    Solution: x = $k\pi$ k any integer

51. There are no reference triangles, but from the graph of y = cos x, we can see that the solution over one period, $0 \le x \le 2\pi$, is only x = π. We then add integer multiples of 2π.

Solution: x = $\pi + 2k\pi$ k any integer

53. csc x = $2 = \dfrac{2}{1}$

Cosecant is positive in the first and second quadrants. Sketch the reference triangles and angles corresponding to x. The reference triangles are special $30° - 60°$ triangles. $\alpha = \csc^{-1} 2 = \dfrac{\pi}{6}$.

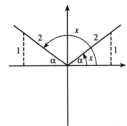

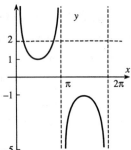

$$ x = \begin{cases} \alpha \\ \pi - \alpha \end{cases} = \begin{cases} \dfrac{\pi}{6} \\ \pi - \dfrac{\pi}{6} \end{cases} = \begin{cases} \dfrac{\pi}{6} \\ \dfrac{5\pi}{6} \end{cases} $$

55. sec $\theta = -5 = \dfrac{5}{-1}$

Secant is negative in the second and third quadrants.

Sketch the reference triangles and angles corresponding to θ.

$\alpha = \sec^{-1} 5$

= 78.463° (from calculator)

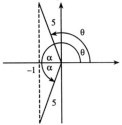

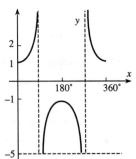

$$ \theta = \begin{cases} 180° - \alpha \\ 180° + \alpha \end{cases} = \begin{cases} 180° - \sec^{-1} 5 \\ 180° + \sec^{-1} 5 \end{cases} = \begin{cases} 180° - 78.463° \\ 180° + 78.463° \end{cases} = \begin{cases} 101.537° \\ 258.463° \end{cases} $$

57. We first find all solutions over one period, $0 \le x \le 2\pi$, then add 2π to these to find solutions $2\pi \le x \le 4\pi$. Sketch the reference triangles and angles corresponding to x. $\sin x = 0.4315$.

Sine is positive in the first and second quadrants. $\alpha = \sin^{-1} 0.4315 = 0.446$.

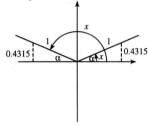
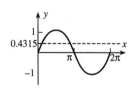

$$x = \begin{cases} \alpha \\ \pi - \alpha \end{cases} = \begin{cases} \sin^{-1} 0.4315 \\ \pi - \sin^{-1} 0.4315 \end{cases} = \begin{cases} 0.446 \\ \pi - 0.446 \end{cases} = \begin{cases} 0.446 \\ 2.695 \end{cases}$$

Also, $x = \begin{cases} 0.446 + 2\pi \\ 2.695 + 2\pi \end{cases} = \begin{cases} 6.729 \\ 8.979 \end{cases}$ Solutions: x = 0.446, 2.695, 6.729, 8.979

59. We first find all solutions over one period, $0 \le x \le 2\pi$, then subtract 2π from these to find solutions $-2\pi \le x \le 0$. Sketch the reference triangles and angles corresponding to x. $\sin x = -0.2243$.

Sine is negative in the third and fourth quadrants. $\alpha = \sin^{-1} 0.2243 = 0.226$.

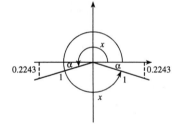

$$x = \begin{cases} \pi + \alpha \\ 2\pi - \alpha \end{cases} = \begin{cases} \pi + \sin^{-1} 0.4315 \\ 2\pi - \sin^{-1} 0.2243 \end{cases} = \begin{cases} \pi + 0.226 \\ 2\pi - 0.226 \end{cases} = \begin{cases} 3.368 \\ 6.057 \end{cases}$$

Also, $x = \begin{cases} 3.368 - 2\pi \\ 6.057 - 2\pi \end{cases} = \begin{cases} -2.915 \\ -0.226 \end{cases}$ Solutions: x = -2.915, -0.226, 3.368, 6.057

Chapter 5   Inverse Trigonometric Functions; Trigonometric Equations

61. We first find all solutions over one period, $-90° \le \theta \le 90°$, then add or subtract multiples of 180° (the period of the tangent function) to find solutions $-360° \le \theta \le 360°$.

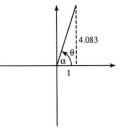 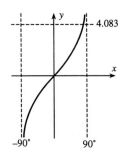

$\tan \theta = 4.083$

Tangent is positive in the first quadrant.

$\theta = \alpha = \tan^{-1} 4.083 = 76.238°$

Also, $\theta = \alpha + 180°$, $\alpha - 180°$, $\alpha - 360°$ are within the range $-360° \le \theta \le 360°$

Solutions: $\theta = \begin{cases} \alpha + 180° \\ \alpha \\ \alpha - 180° \\ \alpha - 360° \end{cases} = \begin{cases} 256.238° \\ 76.238° \\ -103.762° \\ -283.762° \end{cases}$

63. From the graph of the equation y = sin x, $0 \le x \le 2\pi$, we can see that solutions of sin x = a correspond to places where the line y = a crosses the graph of y = sin x. Thus, we can see:

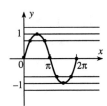

(A) If a > 1 or a < –1, the horizontal line will not cross the graph of y = sin x at all. No solution, if a > 1 or a < –1.

(B) If a = 1 or a = –1, the horizontal line crosses the graph of y = sin x exactly once. One solution, if a = 1 or a = –1.

(C) If –1 < a < 0 or 0 < a < 1, the horizontal line crosses the graph of y = sin x exactly twice. Two solutions, if –1 < a < 0 or 0 < a < 1.

(D) The only horizontal line that crosses the graph more than twice is the x-axis, corresponding to a = 0. More than two solutions, if a = 0.

65. The solutions of the equation sin x = –0.21 are the x coordinates of the points of intersection of the graphs of y1 = sin x and y2 = –0.21 in the graphing calculator figure. Examining the graph, we see that there are two points of intersection. Using the zoom and trace procedures of the calculator, we obtain the following approximations to the x coodinates of these intersection points: $x = \begin{cases} -2.930 \\ -0.212 \end{cases}$

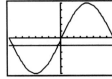

67. The solutions of the equation sec x = 4 are the x coordinates of the points of intersection of the graphs of y1 = sec x and y2 = 4 in the graphing calculator figure.

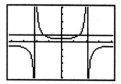

Examining the graph, we see that there are two points of intersection. Using the zoom

and trace procedures of the calculator, we obtain the following approximations to the x coordinates of these intersection points: $x = \begin{cases} -1.318 \\ 1.318 \end{cases}$

69. The solutions of the equation tan x = 2 are the x coordinates of the points of intersection of the graphs of y1 = tan x and y2 = 2 in the graphing calculator figure.

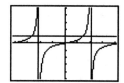

Examining the graph, we see that there are two points of intersection. Using the zoom

and trace procedures of the calculator, we obtain the following approximations to the x coordinates of these intersection points: $x = \begin{cases} -2.034 \\ 1.107 \end{cases}$

EXERCISE 5.4   Trigonometric Equations II

1. $1 + \cos 4x = 0$ $0 \le x \le \dfrac{\pi}{2}$ is equivalent to

     $1 + \cos 4x = 0$ $0 \le 4x \le 2\pi$

     $\cos 4x = -1$ The only solution of $\cos \theta = -1$, $0 \le \theta \le 2\pi$, is $\theta = \pi$.

     $4x = \pi$

     $x = \dfrac{\pi}{4}$

3. $1 + \sqrt{2} \sin \dfrac{\theta}{2} = 0$ $0 \le \theta < 720°$ is equivalent to

$1 + \sqrt{2} \sin \dfrac{\theta}{2} = 0$ $0 \le \dfrac{\theta}{2} < 360°$

$\sqrt{2} \sin \dfrac{\theta}{2} = -1$

$\sin \dfrac{\theta}{2} = -\dfrac{1}{\sqrt{2}}$ $\alpha = \sin^{-1} \dfrac{1}{\sqrt{2}} = 45°$

$\theta = \begin{cases} 180° + \alpha \\ 180° - \alpha \end{cases} = \begin{cases} 180° + 45° \\ 360° - 45° \end{cases} = \begin{cases} 225° \\ 315° \end{cases}$

$\dfrac{\theta}{2} = 225°$ $\dfrac{\theta}{2} = 315°$

$\theta = 450°$ $\theta = 630°$ Solutions: $\theta = 450°, 630°$

5. $\sqrt{3} - \tan 2x = 0$ $-\dfrac{\pi}{4} < x < \dfrac{\pi}{4}$ is equivalent to

$\sqrt{3} - \tan 2x = 0$ $-\dfrac{\pi}{2} < 2x < \dfrac{\pi}{2}$

$\tan 2x = \sqrt{3}$

$2x = \dfrac{\pi}{3}$

$x = \dfrac{\pi}{6}$

7. $4 \cos^2 x - 3 = 0$ $0 \le x \le 2\pi$

$4 \cos^2 x = 3$

$\cos^2 x = \dfrac{3}{4}$

$\cos x = \pm \dfrac{\sqrt{3}}{2}$

$\cos x = \dfrac{\sqrt{3}}{2}$ $0 \le x \le 2\pi$ $\cos x = -\dfrac{\sqrt{3}}{2}$ $0 \le x \le 2\pi$

$x = \dfrac{\pi}{6}, \dfrac{11\pi}{6}$ $x = \dfrac{5\pi}{6}, \dfrac{7\pi}{6}$

9. $\sin^2 \theta = \sin \theta$ $0° \le \theta \le 180°$

$\sin^2 \theta - \sin \theta = 0$

$\sin \theta (\sin \theta - 1) = 0$

$\sin \theta = 0$ $0 \le \theta \le 180°$ $\sin \theta - 1 = 0$ $0 \le \theta \le 180°$

$\theta = 0°, 180°$ $\sin \theta = 1$

$\theta = 90°$

Solutions: $\theta = 0°, 90°, 180°$

11. $2 \sin 2x = \sqrt{3}$ $0 \le x \le \pi$ is equivalent to

 $2 \sin 2x = \sqrt{3}$ $0 \le 2x \le 2\pi$

 $\sin 2x = \dfrac{\sqrt{3}}{2}$

 $2x = \dfrac{\pi}{3}, \dfrac{2\pi}{3}$

 $2x = \dfrac{\pi}{3}$ $2x = \dfrac{2\pi}{3}$

 $x = \dfrac{\pi}{6}$ $x = \dfrac{\pi}{3}$ Solutions: $x = \dfrac{\pi}{6}, \dfrac{\pi}{3}$

13. $\sin x \cos x = \dfrac{1}{2}$ $0 \le x \le \pi$

 $2 \sin x \cos x = 1$

 $\sin 2x = 1$ $0 \le x \le \pi$ is equivalent to

 $\sin 2x = 1$ $0 \le 2x \le 2\pi$

 $2x = \dfrac{\pi}{2}$ is the only solution $0 \le 2x \le 2\pi$

 $x = \dfrac{\pi}{4}$

15. $\tan \dfrac{x}{2} - 1 = 0$ $0 \le x \le 2\pi$ is equivalent to

 $\tan \dfrac{x}{2} - 1 = 0$ $0 \le \dfrac{x}{2} \le \pi$

 $\tan \dfrac{x}{2} = 1$

 $\dfrac{x}{2} = \dfrac{\pi}{4}$

 $x = \dfrac{2\pi}{4} = \dfrac{x}{2}$

17. First solve for x over one period, $0 \le x \le 2\pi$. Then add integer multiples of 2π to find all solutions.

 $\sin 2x = \sin x$ $0 \le x < 2\pi$

 $2 \sin x \cos x = \sin x$

 $2 \sin x \cos x - \sin x = 0$

 $\sin x (2 \cos x - 1) = 0$

 $\sin x = 0$ $2 \cos x - 1 = 0$

 $x = 0, \pi$ $\cos x = \dfrac{1}{2}$

 $x = \dfrac{\pi}{3}, \dfrac{5\pi}{3}$

Chapter 5   Inverse Trigonometric Functions; Trigonometric Equations

Thus, all solutions over one period, $0 \le x < 2\pi$, are $x = 0, \pi, \dfrac{\pi}{3}, \dfrac{5\pi}{3}$.

Thus, the solutions, if x is allowed to range over all real numbers, are

$$x = \begin{cases} 2k\pi \\ \pi + 2k\pi \\ \dfrac{\pi}{3} + 2k\pi \\ \dfrac{5\pi}{3} + 2k\pi \end{cases} \qquad \text{k any integer}$$

19. $\cos^2 \theta = \dfrac{1}{2} \sin 2\theta$ \qquad\qquad all θ

$\cos^2 \theta = \dfrac{1}{2} 2 \sin \theta \cos \theta = \sin \theta \cos \theta$

$\cos^2 \theta - \sin \theta \cos \theta = 0$

$\cos \theta (\cos \theta - \sin \theta) = 0$

$\cos \theta = 0$ \qquad\qquad\qquad $\cos \theta - \sin \theta = 0$

Solutions over $0 \le \theta < 360°$ \qquad Solutions over $0 \le \theta < 360°$

are $\theta = 90°$ or $270°$ \qquad\qquad are $\cos \theta = \sin \theta$

$\qquad\qquad\qquad\qquad\qquad\qquad 1 = \dfrac{\sin \theta}{\cos \theta}$

$\qquad\qquad\qquad\qquad\qquad\qquad 1 = \tan \theta$

$\qquad\qquad\qquad\qquad\qquad\qquad \theta = 45°$ or $225°$

Thus, all solutions over one period, $0 \le \theta < 360°$, are $\theta = 90°, 270°, 45°, 225°$.

Thus, the solutions, if θ is allowed to range over all possible values, are

$$\theta = \begin{cases} 90° + k360° \\ 270° + k360° \\ 45° + k360° \\ 225° + k360° \end{cases} \qquad \text{k any integer}$$

21. $2 \sin^2 x = 3 \sin x - 1$ \qquad $0 \le x < 2\pi$

$2 \sin^2 x - 3 \sin x + 1 = 0$

$(2 \sin x - 1)(\sin x - 1) = 0$

$2 \sin x - 1 = 0$ \qquad\qquad\qquad $\sin x - 1 = 0$

$2 \sin x = 1$ \qquad\qquad\qquad\qquad $\sin x = 1$

$\sin x = \dfrac{1}{2}$ \qquad\qquad\qquad\qquad $x = \dfrac{\pi}{2}$

$x = \dfrac{\pi}{6}, \dfrac{5\pi}{6}$

Solutions: $x = \dfrac{\pi}{6}, \dfrac{5\pi}{6}, \dfrac{\pi}{2}$

23. $\sin^2 \theta + 2 \cos \theta = -2$ $0° \le \theta < 360°$

    $1 - \cos^2 \theta + 2 \cos \theta = -2$

    $-\cos^2 \theta + 2 \cos \theta + 3 = 0$

    $\cos^2 \theta - 2 \cos \theta - 3 = 0$

    $(\cos \theta - 3)(\cos \theta + 1) = 0$

    $\cos \theta - 3 = 0$ $\cos \theta + 1 = 0$

       $\cos \theta = 3$ $\cos \theta = -1$

    No solution $\theta = 180°$ Solution: $\theta = 180°$

25. $\cos 2\theta + \sin^2 \theta = 0$ $0 \le \theta < 360°$

    $1 - 2 \sin^2\theta + \sin^2 \theta = 0$

          $1 - \sin^2 \theta = 0$

            $\sin^2 \theta = 1$

             $\sin \theta = \pm 1$

  $\sin \theta = 1$ $\sin \theta = -1$

    $\theta = 90°$ $\theta = 270°$ Solutions: $\theta = 90°, 270°$

27. $4 \cos^2 2x - 4 \cos 2x + 1 = 0$ $0 \le x \le \pi$

    $(2 \cos 2x - 1)^2 = 0$

      $2 \cos 2x - 1 = 0$ $0 \le x \le \pi$ is equivalent to

      $2 \cos 2x - 1 = 0$ $0 \le 2x \le 2\pi$

  $2 \cos 2x = 1$

    $\cos 2x = \dfrac{1}{2}$

        $2x = \dfrac{\pi}{3}, \dfrac{5\pi}{3}$

         $x = \dfrac{\pi}{6}, \dfrac{5\pi}{6}$

29. $\cos x = \cot x$ $0 \le x \le 2\pi$

  $\cos x = \dfrac{\cos x}{\sin x}$

        $\sin x \cos x = \cos x$ $\sin x \ne 0$

    $\sin x \cos x - \cos x = 0$

      $\cos x(\sin x - 1) = 0$

    $\cos x = 0$ $\sin x - 1 = 0$

       $x = \dfrac{\pi}{2}, \dfrac{3\pi}{2}$ $\sin x = 1$

                            $x = \dfrac{\pi}{2}$ Solutions: $x = \dfrac{\pi}{2}, \dfrac{3\pi}{2}$

Chapter 5   Inverse Trigonometric Functions; Trigonometric Equations

31. $4 \cos^2 \theta = 7 \cos \theta + 2$ $0° \leq \theta \leq 180°$

$$4 \cos^2 \theta - 7 \cos \theta - 2 = 0$$

$$(4 \cos \theta + 1)(\cos \theta - 2) = 0$$

$4 \cos \theta + 1 = 0$ $0° \leq \theta \leq 180°$ $\cos \theta - 2 = 0$

$\qquad 4 \cos \theta = -1$ $\cos \theta = 2$

$\qquad \cos \theta = -\dfrac{1}{4}$ No solution

$\qquad\qquad \theta = 180° - \cos^{-1} \dfrac{1}{4} = 104.5°$

33. $\cos 2x + 10 \cos x = 5$ $0 \leq x < 2\pi$

$$2 \cos^2 x - 1 + 10 \cos x = 5$$

$$2 \cos^2 x + 10 \cos x - 6 = 0$$

$\qquad \cos^2 x + 5 \cos x - 3 = 0$ Quadratic in $\cos x$

$$\cos x = \frac{-5 \pm \sqrt{5^2 - 4(1)(-3)}}{2 \cdot 1} = \frac{-5 \pm \sqrt{37}}{2}$$

$\cos x = -5.5414$ $\cos x = 0.5414$

No solution $x = \begin{cases} \cos^{-1} 0.5414 \\ 2\pi - \cos^{-1} 0.5414 \end{cases} = \begin{cases} 0.9987 \\ 5.284 \end{cases}$

Solutions: $x = 0.9987,\ 5.284$

35.   First, let $u = 2x$ and solve $\sin u = 0.4836$ for all real u. We start by finding all solutions over one period, say $0 \leq u \leq 2\pi$.

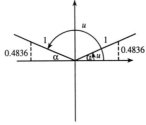

$u = \begin{cases} \alpha + 2k\pi \\ \pi - \alpha + 2k\pi \end{cases}$ k any integer $u = \begin{cases} (\sin^{-1} 0.4836) + 2k\pi \\ (\pi + \sin^{-1} 0.4836) + 2k\pi \end{cases}$

Now replace u with $2x$ and solve for x: $2x = \begin{cases} (\sin^{-1} 0.4836) + 2k\pi \\ (\pi - \sin^{-1} 0.4836) + 2k\pi \end{cases}$

$x = \begin{cases} \dfrac{\sin^{-1} 0.4836}{2} + k\pi \\ \dfrac{\pi - \sin^{-1} 0.4836}{2} + k\pi \end{cases} \approx \begin{cases} 0.2524 + k\pi \\ 1.318 + k\pi \end{cases}$ k any integer

37. First, let u = 4x and solve cos u = −0.6670 for all real u. We start by finding all solutions over one period, say $0 \le u \le 2\pi$.

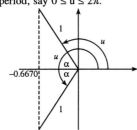

$$u = \begin{cases} \pi - \alpha + 2k\pi \\ \pi + \alpha + 2k\pi \end{cases} \quad \text{k any integer} \qquad u = \begin{cases} (\pi - \cos^{-1} 0.6670) + 2k\pi \\ (\pi + \cos^{-1} 0.6670) + 2k\pi \end{cases}$$

Now replace u with 4x and solve for x: $4x = \begin{cases} \pi - \cos^{-1}(0.6670) + 2k\pi \\ \pi + \cos^{-1}(0.6670) + 2k\pi \end{cases}$

$$x = \begin{cases} \dfrac{\pi - \cos^{-1}(0.6670)}{4} + \dfrac{k\pi}{2} \\[2mm] \dfrac{\pi + \cos^{-1}(0.6670)}{4} + \dfrac{k\pi}{2} \end{cases} \approx \begin{cases} 0.5752 + \dfrac{k\pi}{2} \\[2mm] 0.9956 + \dfrac{k\pi}{2} \end{cases} \quad \text{k any integer}$$

39. $\cos^2 x = 3 - 5 \cos x$

$\cos^2 x + 5 \cos x - 3 = 0$ x any real number

See Problem 33.

$$x = \begin{cases} \cos^{-1} 0.5414 \\ 2\pi - \cos^{-1} 0.5414 \end{cases} = \begin{cases} 0.9987 \\ 5.284 \end{cases} \text{ are the solutions over one period } 0 \le x \le 2\pi.$$

If x can range over all real numbers,

$$x = \begin{cases} 0.9987 + 2k\pi \\ 5.284 \ + 2k\pi \end{cases} \quad \text{or} \quad \begin{cases} 0.9987 + 2k\pi \\ -0.9987 + 2k\pi \end{cases} \quad \text{k any integer}$$

41. $2 \cos 2x = 7 \cos x$ We start by solving for all x over one period, $0 \le x < 2\pi$

$$2(2 \cos^2 x - 1) = 7 \cos x$$
$$4 \cos^2 x - 2 = 7 \cos x$$
$$4 \cos^2 x - 7 \cos x - 2 = 0$$
$$(4 \cos x + 1)(\cos x - 2) = 0$$

$$4 \cos x + 1 = 0 \qquad\qquad \cos x - 2 = 0$$
$$4 \cos x = -1 \qquad\qquad \cos x = 2$$
$$\cos x = -\frac{1}{4} \qquad\qquad \text{No solution}$$

$$x = \begin{cases} \pi - \cos^{-1} \dfrac{1}{4} \\ \pi + \cos^{-1} \dfrac{1}{4} \end{cases} = \begin{cases} 1.823 \\ 4.460 \end{cases} \quad \text{are the solutions over one period } 0 \le x \le 2\pi$$

If x can range over all real numbers $x = \begin{cases} 1.823 + 2k\pi \\ 4.460 + 2k\pi \end{cases}$ k any integer

43.
$$\sin x = \cos x \qquad \text{x any real number}$$
$$\pm\sqrt{1 - \cos^2 x} = \cos x \qquad \text{x any real number}$$
$$1 - \cos^2 x = \cos^2 x \qquad \text{Squaring both sides}$$
$$1 = 2\cos^2 x$$

$$\cos^2 x = \frac{1}{2}$$

$$\cos x = \pm\frac{1}{\sqrt{2}}$$

In squaring both sides, we may have introduced extraneous solutions; hence, it is necessary to check solutions of this equation in the original equation.

$$\cos x = \frac{1}{\sqrt{2}} \qquad\qquad\qquad\qquad \cos x = -\frac{1}{\sqrt{2}}$$

$$x = \begin{cases} \cos^{-1} \dfrac{1}{\sqrt{2}} \\ 2\pi - \cos^{-1} \dfrac{1}{\sqrt{2}} \end{cases} = \begin{cases} \dfrac{\pi}{4} \\ \dfrac{7\pi}{4} \end{cases} \qquad x = \begin{cases} \pi - \cos^{-1}\left(\dfrac{1}{\sqrt{2}}\right) \\ \pi + \cos^{-1}\left(\dfrac{1}{\sqrt{2}}\right) \end{cases} = \begin{cases} \dfrac{3\pi}{4} \\ \dfrac{5\pi}{4} \end{cases}$$

Check:

$$x = \frac{\pi}{4} \qquad\qquad x = \frac{7\pi}{4} \qquad\qquad x = \frac{3\pi}{4} \qquad\qquad x = \frac{5\pi}{4}$$

$$\sin x = \cos x \qquad \sin x = \cos x \qquad \sin x = \cos x \qquad \sin x = \cos x$$

$$\sin\frac{\pi}{4} = \cos\frac{\pi}{4} \qquad \sin\frac{7\pi}{4} = \cos\frac{7\pi}{4} \qquad \sin\frac{3\pi}{4} = \cos\frac{3\pi}{4} \qquad \sin\frac{5\pi}{4} = \cos\frac{5\pi}{4}$$

$$\frac{1}{\sqrt{2}} = \frac{1}{\sqrt{2}} \qquad -\frac{1}{\sqrt{2}} \ne \frac{1}{\sqrt{2}} \qquad \frac{1}{\sqrt{2}} \ne -\frac{1}{\sqrt{2}} \qquad -\frac{1}{\sqrt{2}} = -\frac{1}{\sqrt{2}}$$

A solution   Not a solution   Not a solution   A solution

Thus, the solutions over one period are $x = \dfrac{\pi}{4}, \dfrac{5\pi}{4}$.

Thus, if x can range over all real numbers, $x = \begin{cases} \dfrac{\pi}{4} + 2k\pi \\ \dfrac{5\pi}{4} + 2k\pi \end{cases}$ k any integer

45. $2\sin^2 x + 3\cos x = 3$ all real x

$$2(1 - \cos^2 x) + 3\cos x = 3$$
$$2 - 2\cos^2 x + 3\cos x = 3$$

$$0 = 2\cos^2 x - 3\cos x + 1$$

$$0 = (2\cos x - 1)(\cos x - 1)$$

$$
\begin{array}{ll}
2\cos x - 1 = 0 & \cos x - 1 = 0 \\
\cos x = \dfrac{1}{2} & \cos x = 1 \\
x = \dfrac{\pi}{3}, -\dfrac{\pi}{3} & x = 0
\end{array}
$$

Thus, the solutions over one period are $x = 0, \dfrac{\pi}{3}, -\dfrac{\pi}{3}$.

Thus, if x can range over all real numbers, $x = \begin{cases} 0 + 2k\pi \\ \dfrac{\pi}{3} + 2k\pi \\ -\dfrac{\pi}{3} + 2k\pi \end{cases}$ k any integer

47. $\sin x + \cos x = 1$ $0 \le x < 2x$

$$
\begin{aligned}
\pm\sqrt{1 - \cos^2 x} + \cos x &= 1 \\
\pm\sqrt{1 - \cos^2 x} &= 1 - \cos x \\
1 - \cos^2 x &= (1 - \cos x)^2 \qquad \text{Squaring both sides} \\
1 - \cos^2 x &= 1 - 2\cos x + \cos^2 x \\
0 &= -2\cos x + 2\cos^2 x \\
0 &= 2\cos x(-1 + \cos x)
\end{aligned}
$$

$$
\begin{array}{ll}
2\cos x = 0 & -1 + \cos x = 0 \\
\cos x = 0 & \cos x = 1 \\
x = \dfrac{\pi}{2}, \dfrac{3\pi}{2} & x = 0
\end{array}
$$

In squaring both sides, we may have introduced extraneous solutions; hence, it is necessary to check solutions of these equations in the original equation.

Check:	$x = \dfrac{\pi}{2}$	$x = \dfrac{3\pi}{2}$	$x = 0$
	$\sin x + \cos x = 1$	$\sin x + \cos x = 1$	$\sin x + \cos x = 1$
	$\sin\dfrac{\pi}{2} + \cos\dfrac{\pi}{2} = 1$	$\sin\dfrac{3\pi}{2} + \cos\dfrac{3\pi}{2} = 1$	$\sin 0 + \cos 0 = 1$
	$1 + 0 = 1$	$-1 + 0 \ne 1$	$0 + 1 = 1$
	A solution	Not a solution	A solution

Solutions: $x = 0, \dfrac{\pi}{2}$

49. $\sec x + \tan x = 1$ $0 \le x < 2\pi$

$$
\begin{aligned}
\pm\sqrt{1 + \tan^2 x} + \tan x &= 1 \\
\pm\sqrt{1 + \tan^2 x} &= 1 - \tan x \qquad \text{Squaring both sides} \\
1 + \tan^2 x &= (1 - \tan x)^2 \qquad \text{Squaring both sides} \\
1 + \tan^2 x &= 1 - 2\tan x + \tan^2 x \\
0 &= -2\tan x
\end{aligned}
$$

$$\tan x = 0$$

$$x = 0, \pi$$

In squaring both sides, we may have introduced extraneous solutions; hence, it is necessary to check solutions of these equations in the original equation.

$x = 0$	$x = \pi$
$\sec x + \tan x = 1$	$\sec x + \tan x = 1$
$\sec 0 + \tan 0 = 1$	$\sec \pi + \tan \pi = 1$
$1 + 0 = 1$	$-1 + 0 \neq 1$

A solution   Not a solution   Solution: $x = 0$

51. $\sin 3x + \sin x = 0$ $0 \leq x \leq \pi$

Apply the sum-product identity: $\sin A + \sin B = 2 \sin\dfrac{A + B}{2} \cos\dfrac{A - B}{2}$

$$\sin 3x + \sin x = 2 \sin\dfrac{3x + x}{2} \cos\dfrac{3x - x}{2}$$

The equation becomes: $2 \sin\dfrac{3x + x}{2} \cos\dfrac{3x - x}{2} = 0$ $0 \leq x \leq \pi$

$$2 \sin 2x \cos x = 0 \qquad 0 \leq x \leq \pi$$

$$\sin 2x = 0 \quad \text{or} \quad \cos x = 0 \qquad 0 \leq x \leq \pi$$

This is equivalent to: $\sin 2x = 0$ $0 \leq 2x \leq 2\pi$ or $\cos x = 0$ $0 \leq x \leq \pi$

$$2x = 0, \pi, 2\pi \qquad\qquad\qquad x = \dfrac{\pi}{2}$$

$$x = 0, \dfrac{\pi}{2}, \pi$$

Solutions: $x = 0, \dfrac{\pi}{2}, \pi$

53. $\sin 5x - \sin 3x = \sin x$ $0 \leq x \leq \dfrac{\pi}{2}$

Apply the sum-product identity: $\sin A - \sin B = 2 \cos\dfrac{A + B}{2} \sin\dfrac{A - B}{2}$

$$\sin 5x - \sin 3x = 2 \cos\dfrac{5x + 3x}{2} \sin\dfrac{5x - 3x}{2}$$

The equation becomes: $2 \cos\dfrac{5x + 3x}{2} \sin\dfrac{5x - 3x}{2} = \sin x$ $0 \leq x \leq \dfrac{\pi}{2}$

$$2 \cos 4x \sin x = \sin x$$

$$2 \cos 4x \sin x - \sin x = 0$$

$$\sin x(2 \cos 4x - 1) = 0 \qquad\qquad 0 \leq x \leq \dfrac{\pi}{2}$$

$\sin x = 0$ or $2 \cos 4x - 1 = 0$ $0 \leq x \leq \dfrac{\pi}{2}$

$\sin x = 0$ $0 \leq x \leq \dfrac{\pi}{2}$ $2 \cos 4x - 1 = 0$ $0 \leq x \leq \dfrac{\pi}{2}$

$\quad x = 0$ This is equivalent to:

$$2 \cos 4x - 1 = 0 \qquad 0 \leq 4x \leq 2\pi$$

$$\cos 4x = \dfrac{1}{2}$$

$$4x = \dfrac{\pi}{3}, \dfrac{5\pi}{3}$$

Solutions: $x = 0, \dfrac{\pi}{12}, \dfrac{5\pi}{12}$ $x = \dfrac{\pi}{12}, \dfrac{5\pi}{12}$

55. To transform sin t + cos t using the given identity, we note: M = 1 and N = 1.

Locate P(M, N) = P(1, 1) to determine C: $R = \sqrt{1^2 + 1^2} = \sqrt{2}$, $\sin C = \dfrac{1}{\sqrt{2}}$, $\tan C = 1$, $C = \dfrac{\pi}{4}$

Thus, $\sin t + \cos t = \sqrt{2}\sin\left(t + \dfrac{\pi}{4}\right)$. Therefore, the equation sin t + cos t = 1 becomes

$$\sqrt{2}\sin\left(t + \frac{\pi}{4}\right) = 1$$

$$\sin\left(t + \frac{\pi}{4}\right) = \frac{1}{\sqrt{2}}$$

$$t + \frac{\pi}{4} = \frac{\pi}{4}, \frac{3\pi}{4}$$

$$t = 0, \frac{\pi}{2}$$

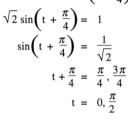 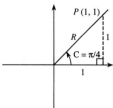

Thus, all solutions over one period, $0 \le t \le 2\pi$, are $t = 0, \dfrac{\pi}{2}$.

Thus, all solutions are given by $t = \begin{cases} 0 + 2k\pi \\ \dfrac{\pi}{2} + 2k\pi \end{cases}$ k any integer

57. To transform sin πt – cos πt using the given identity, we note: M = 1 and N = –1.

Locate P(M ,N) = P(1, –1) to determine C: $R = \sqrt{1^2 + 1^2} = \sqrt{2}$, $\sin C = \dfrac{-1}{\sqrt{2}}$, $\tan C = -1$,

$C = -\dfrac{\pi}{4}$. Thus, $\sin \pi t - \cos \pi t = \sin\left(\pi t - \dfrac{\pi}{4}\right)$. Therefore, the equation sin πt – cos πt = 1

becomes $\sqrt{2}\sin\left(\pi t - \dfrac{\pi}{4}\right) = 1$

$$\sin\left(\pi t - \frac{\pi}{4}\right) = \frac{1}{\sqrt{2}}$$

$$\pi t - \frac{\pi}{4} = \frac{\pi}{4}, \frac{3\pi}{4}$$

$$\pi t = \frac{\pi}{2}, \pi$$

$$t = \frac{1}{2}, 1$$

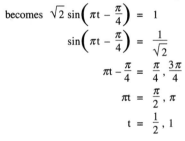 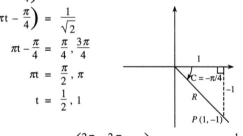

Thus, all solutions over one period, $\left(\dfrac{2\pi}{B} = \dfrac{2\pi}{\pi} = 2\right)$ $0 \le t < 2$, are $t = \dfrac{1}{2}$, 1.

Thus, all solutions are given by $t = \begin{cases} \dfrac{1}{2} + 2k \\ 1 + 2k \end{cases}$ k any integer

59. From the figure, we see that y1 = cos x and y2 = x

intersect only once on the interval $-2\pi \le x \le 2\pi$.

Using zoom and trace procedures, the solution is

found to be 0.739. Since |cos x| ≤ 1, while |x| > 1

for real x not shown, there can be no other solutions.

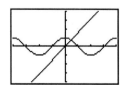

Chapter 5   Inverse Trigonometric Functions; Trigonometric Equations

61. From the figure, we see that $y1 = \sin 2x + 2 \cos x$
and $y2 = 1$ intersect twice on the indicated interval.
Using zoom and trace procedures, the solutions are
found to be 0.257 and 2.029.

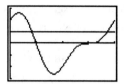

63. From the figure, we see that $y1 = \sin x \sin 2x$ and
$y2 = -0.5$ intersect four times on the indicated
interval.
Using zoom and trace procedures, the solutions are
found to be 1.844, 2.564, 3.720, and 4.439.

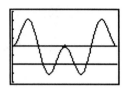

65. $I = 30 \sin 120\pi t$ $I = 25$

$$25 = 30 \sin 120\pi t$$
$$\sin 120\pi t = \frac{25}{30}$$
$$120\pi t = \sin^{-1}\frac{25}{30} \text{ will yield the least positive solution of the equation}$$
$$t = \frac{1}{120\pi} \sin^{-1}\frac{25}{30} = 0.002613 \text{ sec}$$

67. Following the hint, we solve:

$$I \cos^2 \theta = 0.70I \qquad 0° \le \theta \le 180°$$

$$\cos^2 \theta = 0.70$$
$$\cos \theta = \pm\sqrt{0.70}$$
$$\theta = \cos^{-1}\sqrt{0.70} \text{ will yield the least positive solution of the equation}$$
$$\theta = 33.21°$$

69. We use the given formula $A = \frac{1}{2}r^2(\theta - \sin \theta)$
with $r = 10$ and $A = 40$.
$$40 = \frac{1}{2}(10)^2(\theta - \sin \theta) = 50(\theta - \sin \theta)$$
$$\theta - \sin \theta = 0.8$$
We graph $y1 = x - \sin x$ and $y2 = 0.8$ on the
interval from 0 to π. From the figure, we see that
$y1$ and $y2$ intersect once on the given interval.
Using the zoom and trace procedures, the solution is found to
be 1.78 radians.

71. The doorway consists of a rectangle of area

4 × 8 = 32 square feet plus a segment ABCDA

(see Problem 69) as shown.

To find the area of the segment, we need to

find θ and r. In triangle OCD, we have $\sin\dfrac{\theta}{2} = \dfrac{2}{r}$.

Since s = rθ, we can also write $r = \dfrac{s}{\theta} = \dfrac{5}{\theta}$.

Therefore, θ must satisfy the relation

$\sin\dfrac{\theta}{2} = 2 \div r = 2 \div \dfrac{5}{\theta} = \dfrac{2\theta}{5}$; $\sin\dfrac{\theta}{2} = \dfrac{2\theta}{5}$

We graph $y1 = \sin\dfrac{x}{2}$ and $y2 = \dfrac{2x}{5}$. From the figure, we

see that y1 and y2 intersect once on the interval 0 to π.

Using the zoom and trace procedures, the solution

is found to be θ = 2.262205 radians. Then,

$r = \dfrac{5}{\theta} = 2.21023$ ft. Finally,

$$A = \dfrac{1}{2}r^2(\theta - \sin\theta)$$

$$= \dfrac{1}{2}(2.21023)^2(2.262205 - \sin 2.262205)$$

$$= 3.64 \text{ square feet.}$$

Hence, the area of the doorway = 32 + 3.64 = 35.64 square feet.

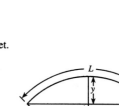

73. (A) Analyzing triangle OAB, we see the following:

$$OA = r - y$$

$$OA^2 + AB^2 = r^2$$

$$(r - y)^2 + x^2 = r^2$$

$$r^2 - 2ry + y^2 + x^2 = r^2$$

$$y^2 + x^2 = 2ry$$

$$r = \dfrac{y^2 + x^2}{2y} = \dfrac{2.5^2 + 5.5^2}{2(2.5)} = 7.3$$

$$\sin\theta = \dfrac{x}{r}$$

$$\theta = \sin^{-1}\dfrac{x}{r} = \sin^{-1}\dfrac{5.5}{7.3} = 0.85325 \text{ rad.}$$

$$L = 2\theta \cdot r = 2(0.85325)(7.3) = 12.4575 \text{ mm}$$

(B) From part (A), we can write $r = \dfrac{y^2 + x^2}{2y}$, $\theta = \sin^{-1}\dfrac{x}{r} = \sin^{-1}\dfrac{2xy}{x^2 + y^2}$,

$$L = 2r\theta = \dfrac{y^2 + x^2}{y}\sin^{-1}\dfrac{2xy}{x^2 + y^2}$$

We want to find y if L = 12.4575 and x = 5.4. Substituting, we find that y must satisfy

$$12.4575 = \dfrac{y^2 + 29.16}{y}\sin^{-1}\dfrac{10.8y}{29.16 + y^2}$$

We graph

$$y1 = \frac{y^2 + 29.16}{y} \sin^{-1} \frac{10.8}{29.16 + y^2} - 12.4575.$$

From the figure, we see that y1 has one zero on the
interval from 0 to 4.

Using the zoom and trace features, the solution is
found to be y = 2.6495 mm.

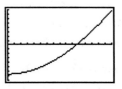

75. We are to solve $3.78 \times 10^7 = \frac{3.44 \times 10^7}{1 - 0.206 \cos \theta}$. For convenience, we can divide both sides of this

equation by 10^7: $3.78 = \frac{3.44}{1 - 0.206 \cos \theta}$

$$3.78(1 - 0.206 \cos \theta) = 3.44$$

$$3.78 - (3.78)(0.206) \cos \theta = 3.44$$

$$-(3.78)(0.206) \cos \theta = 3.44 - 3.78$$

$$\cos \theta = \frac{3.44 - 3.78}{-(3.78)(0.206)}$$

$$\theta = \cos^{-1} \frac{3.44 - 3.78}{-(3.78)(0.206)} = 64.1°$$

77. $r = 2 \sin \theta$ $\qquad$ $0° \le \theta \le 360°$

$r = 2(1 - \sin \theta)$

We solve this system of equations by equating the right sides:

$2 \sin \theta = 2(1 - \sin \theta) = 2 - 2 \sin \theta$

$4 \sin \theta = 2$

$\sin \theta = \frac{1}{2}$, $\theta = 30°$ or $150°$

If we substitute these values of θ in either of the original equations, we obtain

$r = 2 \sin 30° = 1$ $\qquad\qquad$ $r = 2 \sin 150° = 1$

Thus, the solutions of the system of equations are

$(r, \theta) = (1, 30°)$ and $(r, \theta) = (1, 150°)$

79. $xy = -2$

$(u \cos \theta - v \sin \theta)(u \sin \theta + v \cos \theta) = -2$ $\qquad$ Substitution

$u \cos \theta u \sin \theta + u \cos \theta v \cos \theta - v \sin \theta u \sin \theta - v \sin \theta v \cos \theta = -2$ $\qquad$ Multiplication

$u^2 \cos \theta \sin \theta + uv \cos^2 \theta - uv \sin^2 \theta - v^2 \sin \theta \cos \theta = -2$

$u^2 \sin \theta \cos \theta + uv(\cos^2 \theta - \sin^2 \theta) - v^2 \sin \theta \cos \theta = -2$

Chapter 5   Inverse Trigonometric Functions; Trigonometric Equations

We are to find the least positive θ so that the coefficient of the uv term will be zero.

$$\cos^2\theta - \sin^2\theta = 0$$
$$\cos 2\theta = 0$$
$$2\theta = \cos^{-1} 0 \quad \text{yields the least positive } \theta$$
$$2\theta = 90°$$
$$\theta = 45°$$

CHAPTER 5   REVIEW EXERCISE

1. $y = \cos^{-1} \dfrac{\sqrt{3}}{2}$ is equivalent to $\cos y = \dfrac{\sqrt{3}}{2}$.

What y between 0 and π has cosine equal to $\dfrac{\sqrt{3}}{2}$?
y must be associated with a reference triangle in
the first quadrant. Reference triangle is a special
$30° - 60°$ triangle.

$$y = \frac{\pi}{6} \qquad \cos^{-1} \frac{\sqrt{3}}{2} = \frac{\pi}{6}$$

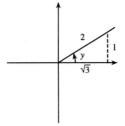

2. $y = \arcsin \dfrac{1}{2}$ is equivalent to $\sin y = \dfrac{1}{2}$. What y between $-\dfrac{\pi}{2}$ and $\dfrac{\pi}{2}$ has sine equal to $\dfrac{1}{2}$? y must be

associated with a reference triangle in the first quadrant. See graph in Problem 1. $\arcsin \dfrac{1}{2} = \dfrac{\pi}{6}$

3. $y = \sin^{-1}(-1)$ is equivalent to $\sin y = -1$.
No reference triangle can be drawn, but the
only y between $-\dfrac{\pi}{2}$ and $\dfrac{\pi}{2}$ that has sine equal
to -1 is $y = -\dfrac{\pi}{2}$. Thus, $\sin^{-1}(-1) = -\dfrac{\pi}{2}$.

4. $y = \cos^{-1}(-1)$ is equivalent to $\cos y = -1$.
No reference triangle can be drawn, but the
only y between 0 and π that has cosine equal
to -1 is $y = \pi$. Thus, $\cos^{-1}(-1) = \pi$.

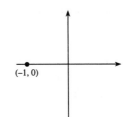

(0, −1)

(−1, 0)

Chapter 5   Inverse Trigonometric Functions; Trigonometric Equations

5. $y = \sin^{-1}\left(-\dfrac{\sqrt{3}}{2}\right)$ is equivalent to

    $\sin y = -\dfrac{\sqrt{3}}{2}$. What y between $-\dfrac{\pi}{2}$ and

    $\dfrac{\pi}{2}$ has sine equal to $-\dfrac{\sqrt{3}}{2}$? y must be

    fourth quadrant. Reference triangle is a

    special 30° – 60° triangle.

    $$y = -\dfrac{\pi}{3} \qquad \sin^{-1}\left(-\dfrac{\sqrt{3}}{2}\right) = -\dfrac{\pi}{3}$$

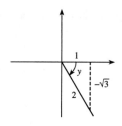

6. $y = \arccos\left(-\dfrac{1}{2}\right)$ is equivalent to $\cos y = -\dfrac{1}{2}$

    What y between 0 and π has cosine equal to

    $-\dfrac{1}{2}$? y must be associated with a reference

    associated with a reference triangle in the

    triangle in the second quadrant. Reference

    triangle is a special 30° – 60° triangle.

    $$y = \dfrac{2\pi}{3} \qquad \arccos\left(-\dfrac{1}{2}\right) = \dfrac{2\pi}{3}$$

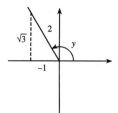

7. $y = \arctan 1$ is equivalent to $\tan y = 1$.

    What y between $-\dfrac{\pi}{2}$ and $\dfrac{\pi}{2}$ has tangent equal

    to 1? y must be associated with a reference

    triangle in the first quadrant. Reference

    triangle is a special 45° triangle.

    $$y = \dfrac{\pi}{4} \qquad \arctan 1 = \dfrac{\pi}{4}$$

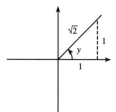

8. $y = \tan^{-1}\left(-\sqrt{3}\right)$ is equivalent to $\tan y = -\sqrt{3}$. What y between $-\dfrac{\pi}{2}$ and $\dfrac{\pi}{2}$ has tangent equal to $-\sqrt{3}$?

    y must be associated with a reference triangle in the fourth quadrant. See graph in Problem 5.

    $$\tan^{-1}\left(-\sqrt{3}\right) = -\dfrac{\pi}{3}$$

9. $y = \arccos 0$ is equivalent to $\cos y = 0$. No

    reference triangle can be drawn, but the only

    y between 0 and π that has cosine equal to

    0 is $y = \dfrac{\pi}{2}$. Thus, $\arccos 0 = \dfrac{\pi}{2}$

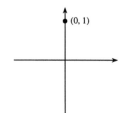

10. 2 is not in the domain of the inverse sine function. $-1 \leq 2 \leq 1$ is false. $\sin^{-1} 2$ is not defined.

11. $y = \cot^{-1}\left(-\dfrac{1}{\sqrt{3}}\right)$ is equivalent to $\cot y = -\dfrac{1}{\sqrt{3}}$. What y between 0 and π has cotangent

equal to $-\dfrac{1}{\sqrt{3}}$? y must be associated with a reference triangle in the second quadrant.

See graph in Problem 6. $y = \dfrac{2\pi}{3}$, $\cot^{-1}\left(-\dfrac{1}{\sqrt{3}}\right) = \dfrac{2\pi}{3}$

12. $y = \text{arcsec}(-2)$ is equivalent to $\sec y = -2$. What y between 0 and π has secant equal to -2?

y must be associated with a reference triangle in the second quadrant.

See graph in Problem 6. $y = \dfrac{2\pi}{3}$, $\sec^{-1}(-2) = \dfrac{2\pi}{3}$

13. $y = \text{arccsc}(-1)$ is equivalent to $\csc y = -1$.

No reference triangle can be drawn, but the only

y between $-\dfrac{\pi}{2}$ and $\dfrac{\pi}{2}$ that has cosecant equal to

-1 is $y = -\dfrac{\pi}{2}$. Thus, $\text{arccsc}(-1) = -\dfrac{\pi}{2}$

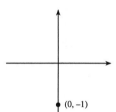

14. $y = \sec^{-1}\left(-\sqrt{2}\right)$ is equivalent to $\sec y = -\sqrt{2}$.

What y between 0 and π has secant equal to $-\sqrt{2}$?

y must be associated with a reference triangle in

the second quadrant. Reference triangle is a

special 45° triangle.

$y = \dfrac{3\pi}{4}$ $\sec^{-1}\left(-\sqrt{2}\right) = \dfrac{3\pi}{4}$

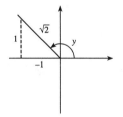

15. Calculator in radian mode: $\sin^{-1}(-0.8277) = -0.9750$

16. Calculator in radian mode: $\cos^{-1}(0.6068) = 0.9188$

17. -1.328 is not in the domain of the inverse cosine function. $-1 \leq -1.328 \leq 1$ is false.

arccos(-1.328) is not defined.

18. Calculator in radian mode: $\tan^{-1}(75.14) = 1.557$

19. Calculator in radian mode: $\cot^{-1} 5.632 = \tan^{-1} \dfrac{1}{5.632} = 0.1757$

20. Calculator in radian mode: $\text{arccsc}(-3.548) = \arcsin \dfrac{1}{-3.548} = -0.2857$

21. $2 \cos x - \sqrt{3} = 0 \qquad 0 \le x \le 2\pi$

$$\cos x = \frac{\sqrt{3}}{2}$$

Cosine is positive in the first and fourth quadrants. Sketch the reference triangles and angles

corresponding to x. $\alpha = \cos^{-1} \dfrac{\sqrt{3}}{2} = \dfrac{\pi}{6}$

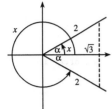 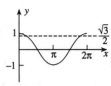

$$x = \begin{cases} \alpha \\ 2\pi - \alpha \end{cases} = \begin{cases} \cos^{-1}\dfrac{\sqrt{3}}{2} \\ 2\pi - \cos^{-1}\dfrac{\sqrt{3}}{2} \end{cases} = \begin{cases} \dfrac{\pi}{6} \\ 2\pi - \dfrac{\pi}{6} \end{cases} = \begin{cases} \dfrac{\pi}{6} \\ \dfrac{11\pi}{6} \end{cases}$$

22. $2 \sin^2 \theta = \sin \theta \qquad 0° \le \theta < 360°$

$2 \sin^2 \theta - \sin \theta = 0$

$\sin \theta (2 \sin \theta - 1) = 0$

$\begin{array}{ll} \sin \theta = 0 & 2 \sin \theta - 1 = 0 \\ \theta = 0°, 180° & \sin \theta = \dfrac{1}{2} \\ & \theta = 30°, 150° \qquad \text{Solutions: } \theta = 0°, 30°, 150°, 180° \end{array}$

23. $4 \cos^2 x - 3 = 0 \qquad 0 \le x < 2\pi$

$$\cos^2 x = \frac{3}{4}$$

$$\cos x = \pm\sqrt{\frac{3}{4}}$$

$$\cos x = \pm\frac{\sqrt{3}}{2}$$

$\begin{array}{ll} \cos x = \pm\dfrac{\sqrt{3}}{2} & \cos x = -\dfrac{\sqrt{3}}{2} \\ x = \dfrac{\pi}{6}, \dfrac{11\pi}{6} & x = \dfrac{5\pi}{6}, \dfrac{7\pi}{6} \qquad \text{Solutions: } x = \dfrac{\pi}{6}, \dfrac{5\pi}{6}, \dfrac{7\pi}{6}, \dfrac{11\pi}{6} \end{array}$

24. $2 \cos^2 \theta + 3 \cos \theta + 1 = 0 \qquad 0° \le \theta < 360°$

$(2 \cos \theta + 1)(\cos \theta + 1) = 0$

$\begin{array}{ll} 2 \cos \theta + 1 = 0 & \cos \theta + 1 = 0 \\ \cos \theta = -\dfrac{1}{2} & \cos \theta = -1 \\ \theta = 120°, 240° & \theta = 180° \qquad \text{Solutions: } \theta = 120°, 180°, 240° \end{array}$

25. $\sqrt{2} \sin 4x - 1 = 0$ $0 \le x < \dfrac{\pi}{2}$ is equivalent to

    $\sqrt{2} \sin 4x - 1 = 0$ $0 \le 4x < 2\pi$

$$\sin 4x = \dfrac{1}{\sqrt{2}}$$

$$4x = \dfrac{\pi}{4}, \dfrac{3\pi}{4}$$

$$x = \dfrac{\pi}{16}, \dfrac{3\pi}{16}$$

26. $\tan \dfrac{\theta}{2} + \sqrt{3} = 0$ $-180° < \theta < 180°$ is equivalent to

    $\tan \dfrac{\theta}{2} + \sqrt{3} = 0$ $-90° < \dfrac{\theta}{2} < 90°$

$$\tan \dfrac{\theta}{2} = -\sqrt{3}$$

$$\dfrac{\theta}{2} = -60°$$

$$\theta = -120°$$

27. $\cos(\cos^{-1} 0.315) = 0.315$ from the cosine-inverse cosine identity.

28. $\tan^{-1}[\tan(-1.5)] = -1.5$ from the tangent-inverse tangent identity.

29.   Let $y = \tan^{-1}\left(-\dfrac{3}{4}\right)$, then $\tan y = -\dfrac{3}{4}$, $-\dfrac{\pi}{2} < y < \dfrac{\pi}{2}$.

    Draw the reference triangle associated with y, then

    $\sin y = \sin\left[\tan^{-1}\left(-\dfrac{3}{4}\right)\right]$ can be determined directly

    from the triangle.

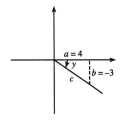

    $a^2 + b^2 = c^2$ $c = \sqrt{4^2 + (-3)^2} = 5$

    $\sin\left[\tan^{-1}\left(-\dfrac{3}{4}\right)\right] = \sin y = \dfrac{-3}{5} = -\dfrac{3}{5}$

30.   Let $y = \arccos\left(-\dfrac{2}{3}\right)$, then $\cos y = -\dfrac{2}{3}$, $0 \le y \le \pi$.

    Sketch the reference triangle associated with y, then

    $\cot y = \cot\left[\cos^{-1}\left(-\dfrac{2}{3}\right)\right]$ can be determined directly

    from the triangle.

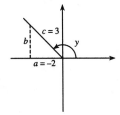

    $a^2 + b^2 = c^2$ $b = \sqrt{3^2 - (-2)^2} = \sqrt{5}$

    $\cot\left[\arccos\left(-\dfrac{2}{3}\right)\right] = \cot y = \dfrac{-2}{\sqrt{5}} = -\dfrac{2}{\sqrt{5}}$

31. Let $y = \sin^{-1}\left(-\dfrac{1}{\sqrt{5}}\right)$, then $\sin y = -\dfrac{1}{\sqrt{5}}$,

$-\dfrac{\pi}{2} \le y \le \dfrac{\pi}{2}$. Sketch the reference triangle

associated with y, then $\sec y = \sec\left[\sin^{-1}\left(-\dfrac{1}{\sqrt{5}}\right)\right]$

can be determined directly from the triangle.

$a^2 + b^2 = c^2 \qquad a = \sqrt{(\sqrt{5})^2 - (-1)^2} = 2$

$\sec\left[\sin^{-1}\left(-\dfrac{1}{\sqrt{5}}\right)\right] = \sec y = \dfrac{\sqrt{5}}{2}$

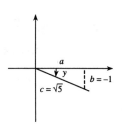

32. We would expect that $\sec(\sec^{-1} 4)$ equals 4 since 4 is in the domain of the inverse secant function.

Using identities from the text, we can prove that our guess is correct as follows:

$\sec(\sec^{-1} 4) = \dfrac{1}{\cos(\sec^{-1} 4)}$ Reciprocal identity

$= \dfrac{1}{\cos\left(\cos^{-1}\dfrac{1}{4}\right)}$ Inverse trigonometric identity

$= \dfrac{1}{\dfrac{1}{4}}$ Cosine-inverse cosine identity

$= 4$

33. Let $y = \cot^{-1}\left(-\dfrac{1}{3}\right)$, then $\cot y = -\dfrac{1}{3}$, $0 < y < \pi$.

Draw the reference triangle associated with y, then

$\csc y = \csc\left[\cot^{-1}\left(-\dfrac{1}{3}\right)\right]$ can be determined

directly from the triangle.

$a^2 + b^2 = c^2 \qquad c = \sqrt{(-1)^2 + 3^2} = \sqrt{10}$

$\csc\left[\cot^{-1}\left(-\dfrac{1}{3}\right)\right] = \csc y = \dfrac{\sqrt{10}}{3}$

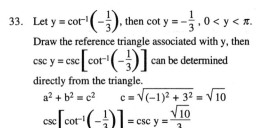
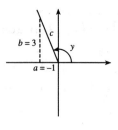

34. Let $y = \text{arccsc } 5$, then $\csc y = 5$, $-\dfrac{\pi}{2} \le y \le \dfrac{\pi}{2}$.

Draw the reference triangle associated with y,

then $\cos y = \cos(\text{arccsc } 5)$ can be determined

directly from the triangle.

$a^2 + b^2 = c^2 \qquad a = \sqrt{5^2 + 1^2} = \sqrt{24} = 2\sqrt{6}$

$\cos(\text{arccsc } 5) = \cos y = \dfrac{2\sqrt{6}}{5}$

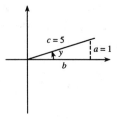

35. Calculator in radian mode: $\sin^{-1}(\cos 22.37) = -1.192$

Chapter 5   Inverse Trigonometric Functions; Trigonometric Equations

36. Calculator in radian mode: $\sin^{-1}(\tan 1.345) = \text{error}$

    $\sin^{-1}(\tan 1.345) = \sin^{-1}(4.353)$ is not defined.

    4.353 is not in the domain of the inverse sine function.

37. Calculator in radian mode: $\sin[\tan^{-1}(-14.00)] = -0.9975$

38. Calculator in radian mode: $\csc[\cos^{-1}(-0.4081)] = \dfrac{1}{\sin[\cos^{-1}(-0.4081)]} = 1.095$

39. Calculator in radian mode: $\sin^{-1}(\sin 2.000) = 1.142$

40. Calculator in radian mode: $\sin[\sec^{-1}(-2.987)] = \sin\left[\cos^{-1}\dfrac{1}{-2.987}\right] = 0.9423$

41. Calculator in radian mode: $\cos(\cot^{-1} 6.823) = \cos\left(\tan^{-1}\dfrac{1}{6.823}\right) = 0.9894$

42. Calculator in radian mode:

    $\sec[\operatorname{arccsc}(-25.52)] = \dfrac{1}{\cos[\operatorname{arccsc}(-25.52)]} = \dfrac{1}{\cos\left[\arcsin\left(\dfrac{1}{-25.52}\right)\right]} = 1.001$

43. $\theta = \cos^{-1}\left(-\dfrac{1}{\sqrt{2}}\right)$ is equivalent to $\cos\theta = -\dfrac{1}{\sqrt{2}}$ $0° \le \theta \le 180°$. Thus, $\theta = 135°$

44. $\theta = \tan^{-1}\left(\dfrac{1}{\sqrt{3}}\right)$ is equivalent to $\tan\theta = \dfrac{1}{\sqrt{3}}$ $-90° < \theta < 90°$. Thus, $\theta = 30°$

45. $\theta = \arcsin(-1)$ is equivalent to $\sin\theta = -1$ $-90° \le \theta \le 90°$. Thus, $\theta = -90°$

46. Calculator in degree mode: $\theta = \cos^{-1} 0.3456 = 69.78°$

47. Calculator in degree mode: $\theta = \arctan(-12.45) = -85.41°$

48. Calculator in degree mode: $\theta = \sin^{-1}(0.0025) = 0.14°$

49. $\sin^2\theta = -\cos 2\theta$ $0° \le \theta \le 360°$

    $\sin^2\theta = -(1 - 2\sin^2\theta) = -1 + 2\sin^2\theta$

    $\qquad 0 = -1 + \sin^2\theta$

    $\sin^2\theta = 1$

    $\sin\theta = \pm 1$

    $\sin\theta = 1 \qquad\qquad \sin\theta = -1$

    $\quad \theta = 90° \qquad\qquad \theta = 270° \qquad$ Solutions: $\theta = 90°, 270°$

Chapter 5   Inverse Trigonometric Functions; Trigonometric Equations

50. $\sin 2x = \dfrac{1}{2}$ $0 \le x < \pi$ is equivalent to

$\sin 2x = \dfrac{1}{2}$ $0 \le 2x < 2\pi$

$2x = \dfrac{\pi}{6}, \dfrac{5\pi}{6}$

$x = \dfrac{\pi}{12}, \dfrac{5\pi}{12}$

51. $2 \cos x + 2 = -\sin^2 x$ $-\pi \le x < \pi$

$2 \cos x + 2 = -(1 - \cos^2 x) = -1 + \cos^2 x$

$0 = \cos^2 x - 2 \cos x - 3$

$0 = (\cos x - 3)(\cos x + 1)$

$\cos x - 3 = 0$ $\cos x + 1 = 0$

$\cos x = 3$ $\cos x = -1$

No solution $x = -\pi$ Soluition: $x = -\pi$

52. $2 \sin^2 \theta - \sin \theta = 0$ All θ

$\sin \theta(2 \sin \theta - 1) = 0$

$\sin \theta = 0$ $2 \sin \theta - 1 = 0$

$\theta = 0°, 180°$ $\sin \theta = \dfrac{1}{2}$

$\theta = 30°, 150°$

Thus, the solutions over one period are 0°, 180°, 30°, 150°.

Thus, if θ can range over all degree measures, $\theta = \begin{cases} 0° + k360° \\ 180° + k360° \\ 30° + k360° \\ 150° + k360° \end{cases}$ k any integer

53. $\sin 2x = \sqrt{3} \sin x$ All real x

$2 \sin x \cos x = \sqrt{3} \sin x$

$2 \sin x \cos x - \sqrt{3} \sin x = 0$

$\sin x(2 \cos x - \sqrt{3}) = 0$

$\sin x = 0$ $2 \cos x - \sqrt{3} = 0$

$x = 0, \pi$ $\cos x = \dfrac{\sqrt{3}}{2}$

$x = \dfrac{\pi}{6}, -\dfrac{\pi}{6}$

Thus, the solutions over one period, $-\pi < x \leq \pi$, are $0, \pi, \dfrac{\pi}{6}, -\dfrac{\pi}{6}$.

Thus, if x can range over all real numbers, $x = \begin{cases} 0 + 2k\pi \\ \pi + 2k\pi \\ \dfrac{\pi}{6} + 2k\pi \\ -\dfrac{\pi}{6} + 2k\pi \end{cases}$ k any integer

54. $2 \sin^2 \theta + 5 \cos \theta + 1 = 0$ $0° \leq \theta < 360°$

$2(1 - \cos^2 \theta) + 5 \cos \theta + 1 \ = \ 0$

$2 - 2 \cos^2 \theta + 5 \cos \theta + 1 \ = \ 0$

$-2 \cos^2 \theta + 5 \cos \theta + 3 \ = \ 0$

$2 \cos^2 \theta - 5 \cos \theta - 3 \ = \ 0$

$(2 \cos \theta + 1)(\cos \theta - 3) \ = \ 0$

$2 \cos \theta + 1 \ = \ 0$	$\cos \theta - 3 \ = \ 0$
$\cos \theta \ = \ -\dfrac{1}{2}$	$\cos \theta \ = \ 3$
$\theta \ = \ 120°, 240°$	No solution

Solutions: $\theta = 120°, 240°$

55. $3 \sin 2x \ = \ -2 \cos^2 2x$ $0 \leq x \leq \pi$ is equivalent to

$3 \sin 2x \ = \ -2 \cos^2 2x$ $0 \leq 2x \leq 2\pi$

$3 \sin 2x \ = \ -2(1 - \sin^2 2x) = -2 + 2 \sin^2 2x$

$0 \ = \ 2 \sin^2 2x - 3 \sin 2x - 2$

$0 \ = \ (2 \sin 2x + 1)(\sin 2x - 2)$

$2 \sin 2x + 1 \ = \ 0$	$\sin 2x - 2 \ = \ 0$
$\sin 2x \ = \ -\dfrac{1}{2}$ $0 \leq 2x \leq 2\pi$	$\sin 2x \ = \ 2$
$2x \ = \ \dfrac{7\pi}{6}, \dfrac{11\pi}{6}$	No solution
$x \ = \ \dfrac{7\pi}{12}, \dfrac{11\pi}{12}$	Solutions: $x = \dfrac{7\pi}{12}, \dfrac{11\pi}{12}$

56. $\sin x = 0.7088$. Sine is positive in the first and second quadrants.

$x = \begin{cases} \sin^{-1} 0.7088 \\ \pi - \sin^{-1} 0.7088 \end{cases} = \begin{cases} 0.7878 \\ 2.354 \end{cases}$ are the solutions over one period $0 \leq x < 2\pi$.

If x can range over all real numbers, $x = \begin{cases} 0.7878 + 2k\pi \\ 2.354 + 2k\pi \end{cases}$ k any integer

57. $\cos x = -0.1187$. Cosine is negative in the second and third quadrants.

$$x = \begin{cases} \pi - \cos^{-1} 0.1187 \\ \pi + \cos^{-1} 0.1187 \end{cases} = \begin{cases} \pi - 1.452 \\ \pi + 1.452 \end{cases} = \begin{cases} 1.690 \\ 4.593 \end{cases} \text{ are the solutions over one period } 0 \le x < 2\pi.$$

If x can range over all real numbers, $x = \begin{cases} 1.690 + 2k\pi \\ 4.593 + 2k\pi \end{cases}$ k any integer

58. $\tan x = -4.318$. Tangent is negative in the fourth quadrant, $-\dfrac{\pi}{2} < x < \dfrac{\pi}{2}$.

  $x = -\tan^{-1}(4.318)$ is the only solution $-\dfrac{\pi}{2} < x < \dfrac{\pi}{2}$

  $x = -1.343$

If x can range over all real numbers, $x = -1.343 + k\pi$, k any integer.

59. $\csc x = 4.138$

  $\dfrac{1}{\sin x} = 4.138$

  $\sin x = \dfrac{1}{4.138} = 0.2417$

Sine is positive in the first and second quadrants.

$$x = \begin{cases} \sin^{-1} 0.2417 \\ \pi - \sin^{-1} 0.2417 \end{cases} = \begin{cases} 0.2441 \\ 2.898 \end{cases} \text{ are the solutions over one period } 0 \le x < 2\pi.$$

If x can range over all real numbers, $x = \begin{cases} 0.2441 + 2k\pi \\ 2.898 \ + 2k\pi \end{cases}$ k any integer

60. $\sin^2 x + 2 = 4 \sin x$

  $\sin^2 x - 4 \sin x + 2 = 0$ Quadratic in sin x

  $\sin x = \dfrac{-(-4) \pm \sqrt{(-4)^2 - 4(1)(2)}}{2(1)} = \dfrac{4 \pm \sqrt{8}}{2}$

  $\sin x = 3.4142$ $\sin x = 0.5858$

  No solution $x = \begin{cases} \sin^{-1} 0.5858 \\ \pi - \sin^{-1} 0.5858 \end{cases} = \begin{cases} 0.6259 \\ 2.516 \end{cases} \begin{array}{l} \text{are the solutions over} \\ \text{one period } 0 \le x < 2\pi. \end{array}$

If x can range over all real numbers, $x = \begin{cases} 0.6259 + 2k\pi \\ 2.516 \ + 2k\pi \end{cases}$ k any integer

61. $\tan^2 x = 2 \tan x + 1$

  $\tan^2 x - 2 \tan x - 1 = 0$ Quadratic in tan x

  $\tan x = \dfrac{-(-2) \pm \sqrt{(-2)^2 - 4(1)(-1)}}{2(1)} = \dfrac{2 \pm \sqrt{8}}{2}$

$$\tan x = 2.4142 \qquad\qquad \tan x = -0.4142$$

$$x = \tan^{-1}(2.4142) = 1.178 \qquad x = \tan^{-1}(-0.4142) = -0.3927$$

$x = 1.178, -0.3927$ are the solutions over one period $-\dfrac{\pi}{2} < x < \dfrac{\pi}{2}$.

If x can range over all real numbers, $x = \begin{cases} 1.178 + 2k\pi \\ -0.3927 + 2k\pi \end{cases}$ k any integer

62. $\cos x = 1 - \sin x \qquad 0 \le x < 2\pi$

$$\cos^2 x = (1 - \sin x)^2 \qquad\qquad \text{Squaring both sides}$$

$$\cos^2 x = 1 - 2\sin x + \sin^2 x$$

$$1 - \sin^2 x = 1 - 2\sin x + \sin^2 x \qquad \text{Pythagorean identity}$$

$$0 = 2\sin^2 x - 2\sin x$$

$$0 = 2\sin x(\sin x - 1)$$

$2\sin x = 0$	$\sin x - 1 = 0$
$\sin x = 0$	$\sin x = 1$
$x = 0, \pi$	$x = \dfrac{\pi}{2}$

In squaring both sides, we may have introduced extraneous solutions; hence, it is necessary to check solutions of these equations in the original equation.

$x = 0$	$x = \pi$	$x = \dfrac{\pi}{2}$
$\cos x = 1 - \sin x$	$\cos x = 1 - \sin x$	$\cos x = 1 - \sin x$
$\cos 0 = 1 - \sin 0$	$\cos \pi = 1 - \sin \pi$	$\cos \dfrac{\pi}{2} = 1 - \sin \dfrac{\pi}{2}$
$1 = 1 - 0$	$-1 \ne 1 - 0$	$0 = 1 - 1$
A solution	Not a solution	A solution

Solutions: $x = 0, \dfrac{\pi}{2}$

63. $\cos^2 2x = \cos 2x + \sin^2 2x \qquad 0 \le x < \pi$ is equivalent to

$\cos^2 2x = \cos 2x + \sin^2 2x \qquad 0 \le 2x < 2\pi$

$\cos^2 2x = \cos 2x + 1 - \cos^2 2x \qquad \text{Pythagorean identity}$

$$2\cos^2 2x - \cos 2x - 1 = 0$$

$$(2\cos 2x + 1)(\cos 2x - 1) = 0$$

$2\cos 2x + 1 = 0 \qquad 0 \le 2x < 2\pi$	$\cos 2x - 1 = 0 \qquad 0 \le 2x < 2\pi$
$\cos 2x = -\dfrac{1}{2}$	$\cos 2x = 1$
$2x = \dfrac{2\pi}{3}, \dfrac{4\pi}{3}$	$2x = 0$
$x = \dfrac{\pi}{3}, \dfrac{2\pi}{3}$	$x = 0$

Solutions: $x = 0, \dfrac{\pi}{3}, \dfrac{2\pi}{3}$

64. $\sin 7x - \sin x = \sin 3x \qquad 0 \le x \le \dfrac{\pi}{2}$

Apply the sum-product identity: $\sin A - \sin B = 2 \cos \dfrac{A + B}{2} \sin \dfrac{A - B}{2}$

$$\sin 7x - \sin x = 2 \cos \dfrac{7x + x}{2} \sin \dfrac{7x - x}{2} = 2 \cos 4x \sin 3x$$

The equation becomes $\qquad 2 \cos 4x \sin 3x = \sin 3x \qquad 0 \le x \le \dfrac{\pi}{2}$

$$2 \cos 4x \sin 3x - \sin 3x = 0$$

$$\sin 3x(2 \cos 4x - 1) = 0$$

$\sin 3x = 0 \qquad 0 \le x \le \dfrac{\pi}{2}$	$2 \cos 4x - 1 = 0 \qquad 0 \le x \le \dfrac{\pi}{2}$
is equivalent to	is equivalent to
$\sin 3x = 0 \qquad 0 \le 3x \le \dfrac{3\pi}{2}$	$2 \cos 4x - 1 = 0 \qquad 0 \le 4x \le 2\pi$
$3x = 0, \pi$	$\cos 4x = \dfrac{1}{2}$
$x = 0, \dfrac{\pi}{3}$	$4x = \dfrac{\pi}{3}, \dfrac{5\pi}{3}$
	$x = \dfrac{\pi}{12}, \dfrac{5\pi}{12}$

Solutions: $x = 0, \dfrac{\pi}{12}, \dfrac{5\pi}{12}, \dfrac{\pi}{3}$

65. $2 + 2 \sin x = 1 + 2 \cos^2 x \qquad\qquad 0 \le x \le 2\pi$

$2 + 2 \sin x = 1 + 2(1 - \sin^2 x) \qquad\qquad$ Pythagorean identity

$2 + 2 \sin x = 1 + 2 - 2 \sin^2 x$

$2 \sin^2 x + 2 \sin x - 1 = 0 \qquad\qquad$ Quadratic in sin x

$$\sin x = \dfrac{-2 \pm \sqrt{(2)^2 - 4(2)(-1)}}{2(2)} = \dfrac{-2 \pm \sqrt{12}}{4}$$

$\sin x = -1.3660 \qquad \sin x = 0.3660$

No solution

$$x = \begin{cases} \sin^{-1} 0.3660 \\ \pi - \sin^{-1} 0.3660 \end{cases} = \begin{cases} 0.375 \\ \pi - 0.375 \end{cases} = \begin{cases} 0.375 \\ 2.77 \end{cases}$$

Solutions: $x = 0.375, 2.77$

66. Let $y = \tan^{-1}\left(-\frac{3}{4}\right)$. Then, $\tan y = -\frac{3}{4}$. We are asked to evaluate

$\sin(2y)$, which is $2 \sin y \cos y$ from the double-angle identity.

Sketch a reference triangle associated with y; then,

$\cos y = \cos\left[\tan^{-1}\left(-\frac{3}{4}\right)\right]$ and $\sin y = \sin\left[\tan^{-1}\left(-\frac{3}{4}\right)\right]$ can

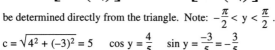

be determined directly from the triangle. Note: $-\frac{\pi}{2} < y < \frac{\pi}{2}$.

$c = \sqrt{4^2 + (-3)^2} = 5 \qquad \cos y = \frac{4}{5} \qquad \sin y = \frac{-3}{5} = -\frac{3}{5}$

Thus, $\sin\left[2 \tan^{-1}\left(-\frac{3}{4}\right)\right] = 2 \sin\left[\tan^{-1}\left(-\frac{3}{4}\right)\right] \cos\left[\tan^{-1}\left(-\frac{3}{4}\right)\right] = 2\left(-\frac{3}{5}\right)\left(\frac{4}{5}\right) = -\frac{24}{25}$

67. Let $u = \sin^{-1}\left(\frac{3}{5}\right)$ and $v = \cos^{-1}\left(\frac{4}{5}\right)$. Then we are asked to evaluate $\sin(u + v)$, which is

$\sin u \cos v + \cos u \sin v$ from the sum identity. We know $\sin u = \sin\left[\sin^{-1}\left(\frac{3}{5}\right)\right] = \frac{3}{5}$ and

$\cos v = \cos\left[\cos^{-1}\left(\frac{4}{5}\right)\right] = \frac{4}{5}$ from the function-inverse function identities. It remains to find

$\cos u$ and $\sin v$. Note: $-\frac{\pi}{2} \le u \le \frac{\pi}{2}$ and $0 \le v \le \pi$.

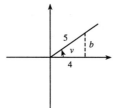

$a = \sqrt{5^2 - 3^2} = 4 \qquad \cos u = \frac{4}{5} \qquad\qquad b = \sqrt{5^2 - 4^2} = 3 \qquad \sin v = \frac{3}{5} \quad (u = v)$

Then, $\sin\left[\sin^{-1}\left(\frac{3}{5}\right) + \cos^{-1}\left(\frac{4}{5}\right)\right] = \sin(u + v) = \sin u \cos v + \cos u \sin v$

$$= \left(\frac{3}{5}\right)\left(\frac{4}{5}\right) + \left(\frac{4}{5}\right)\left(\frac{3}{5}\right) = \frac{12}{25} + \frac{12}{25} = \frac{24}{25}$$

68. Let $y = \sec^{-1}\left(\frac{13}{5}\right)$. Then, $\sec y = \frac{13}{5}$, hence $\cos y = \frac{5}{13}$.

We are asked to evaluate $\cot \frac{1}{2} y$, which is $\dfrac{1}{\tan \frac{1}{2} y}$, or

$\dfrac{1 + \cos y}{\sin y}$ from the half-angle identity. Sketch a reference

triangle associated with y; then $\sin y = \sin\left[\sec^{-1}\left(\frac{13}{5}\right)\right]$

can be determined directly from the triangle.

Note: $-\frac{\pi}{2} < y < \frac{\pi}{2}$. $\qquad a = \sqrt{13^2 - 5^2} = 12 \qquad \sin y = \frac{12}{13}$

Thus, $\cot\left[\dfrac{1}{2}\sec^{-1}\left(\dfrac{13}{5}\right)\right] = \cot\dfrac{1}{2}\,y = \dfrac{1+\cos y}{\sin y} = \dfrac{1+\dfrac{5}{13}}{\dfrac{12}{13}} = \dfrac{13+5}{12} = \dfrac{18}{12} = \dfrac{3}{2}$

69. Let $y = \sin^{-1} x \qquad -\dfrac{\pi}{2} \le y \le \dfrac{\pi}{2} \qquad$ or, equivalently, $\qquad \sin y = x \qquad -\dfrac{\pi}{2} \le y \le \dfrac{\pi}{2}$
     Geometrically,

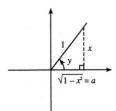   or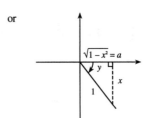

In either case, $a = \sqrt{1 - x^2}$. Thus, $\tan(\sin^{-1} x) = \tan y = \dfrac{b}{a} = \dfrac{x}{\sqrt{1-x^2}}$.

70. Let $y = \tan^{-1} x \qquad -\dfrac{\pi}{2} < y < \dfrac{\pi}{2} \qquad$ or, equivalently $\qquad x = \tan y \qquad -\dfrac{\pi}{2} < y < \dfrac{\pi}{2}$
     Geometrically,

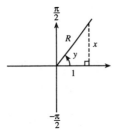   or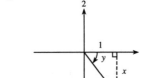

In either case, $R = \sqrt{x^2 + 1}$. Thus, $\cos(\tan^{-1} x) = \cos y = \dfrac{a}{R} = \dfrac{x}{\sqrt{x^2+1}}$.

71. Let $u = \tan^{-1} x$ and $v = \sin^{-1} x \quad -1 < x < 1 \qquad$ or, equivalently,
     $\qquad x = \tan u \qquad -\dfrac{\pi}{2} < u < \dfrac{\pi}{2} \qquad x = \sin v \qquad -\dfrac{\pi}{2} < v < \dfrac{\pi}{2}$
     Then, $\cos(\tan^{-1} x + \sin^{-1} x) = \cos(u + v) = \cos u \cos v - \sin u \sin v$.
     From Problem 70, we have $\cos u = \cos(\tan^{-1} x) = \dfrac{1}{\sqrt{x^2+1}}$
     From the figure in Problem 70, we have $\sin u = \sin(\tan^{-1} x) = \dfrac{x}{\sqrt{x^2+1}}$
     From the figure in Problem 69, we have $\cos v = \cos(\sin^{-1} x) = \sqrt{1-x^2}$
     From the sine-inverse sine identity, we have $\sin v = \sin(\sin^{-1} x) = x$

Thus, $\cos(\tan^{-1} x + \sin^{-1} x) = \cos u \cos v - \sin u \sin v$

$$= \frac{1}{\sqrt{x^2 + 1}} \cdot \sqrt{1 - x^2} - \frac{x}{\sqrt{x^2 + 1}} \cdot x = \frac{\sqrt{1 - x^2} - x^2}{\sqrt{x^2 + 1}}$$

72.   Let $u = \sec^{-1} x$ $0 \le u \le \pi$ or, equivalently, $x = \sec u$ $0 \le u \le \pi$

Geometrically,

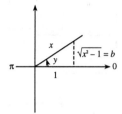   or   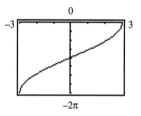

In either case, $\cos(2 \sec^{-1} x) = \cos 2u = 2\cos^2 u - 1 = 2\left(\frac{1}{x}\right)^2 - 1 = \frac{2}{x^2} - 1$ or $\frac{2 - x^2}{x^2}$.

73.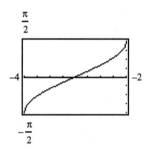

74.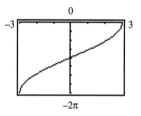

75.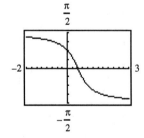

76.   From the figure, we see that $y1 = \sin x$ and
$y2 = 0.25$ intersect twice on the indicated interval.
Using zoom and trace procedures, the solutions are
found to be 0.253 and 2.889.

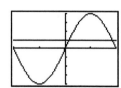

Chapter 5　　Inverse Trigonometric Functions; Trigonometric Equations

77.　From the figure, we see that y1 = cot x and

　　y2 = −4 intersect twice on the indicated interval.

　　Using zoom and trace procedures, the solutions are

　　found to be −0.245 and 2.897.

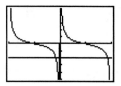

78.　From the figure, we see that y1 = sec x and

　　y2 = 2 intersect twice on the indicated interval.

　　Using zoom and trace procedures, the solutions are

　　found to be −1.047 and 1.047.

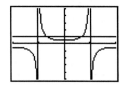

79.　From the figure, we see that y1 = cos x and y2 = x^2

　　intersect twice on the interval from −π to π.

　　Using zoom and trace procedures, the solutions

　　in the interval shown are found to be −0.824

　　and 0.824. Since | cos x | ≤ 1, while x^2 > 1

　　for real x not shown, there can be no other
　　solutions.

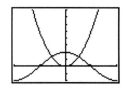

80.　From the figure, we see that the only intersection,

　　0 ≤ x ≤ 2π, of y1 = sin x and y2 = $\sqrt{x}$ is at x = 0.

　　Since | sin x | ≤ 1, while $\sqrt{x}$ > 1 for real x not
　　shown, there can be no other solutions.

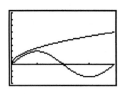

81.　From the figure, we see that y1 = 2 sin x cos 2x

　　and y2 = 1 intersect twice on the indicated interval.

　　Using zoom and trace procedures, the solutions are
　　found to be 4.227 and 5.197.

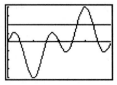

82.　From the figure, we see that y1 = sin $\frac{x}{2}$ + 3 sin x and

　　y2 = 2 intersect four times on the indicated interval.

　　Using zoom and trace procedures, the solutions are
　　found to be 0.604, 2.797, 7.246, and 8.203.

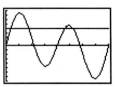

83.　From the figure, we see that

　　y1 = sin x + 2 sin 2x + 3 sin 3x and y2 = 3

　　intersect twice on the indicated interval.

　　Using zoom and trace procedures, the solutions
　　are found to be 0.2278 and 1.008.

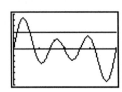

Chapter 5   Inverse Trigonometric Functions; Trigonometric Equations

84. We are to solve $0.05 = 0.08 \sin 880\pi t$.

$$\sin 880\pi t = \frac{0.05}{0.08}$$

$$880\pi t = \sin^{-1} \frac{0.05}{0.08} \quad \text{will yield the smallest positive t}$$

$$t = \frac{1}{880\pi} \sin^{-1} \frac{0.05}{0.08} \qquad \text{(calculator in radian mode)}$$

$$= 0.00024 \text{ sec}$$

85. We are to solve $20 = 30 \sin 120\pi t$

$$\sin 120\pi t = \frac{20}{30}$$

$$120\pi t = \sin^{-1} \frac{20}{30} \quad \text{will yield the smallest positive t}$$

$$t = \frac{1}{120\pi} \sin^{-1} \frac{20}{30} \qquad \text{(calculator in radian mode)}$$

$$= 0.001936 \text{ sec}$$

86. (A) From the figure in the text we can see that $\tan \frac{\theta}{2} = \frac{500}{x}$.

Thus, $\frac{\theta}{2} = \arctan \frac{500}{x} \qquad \theta = 2 \arctan \frac{500}{x}$

(B) We use the above formula with $x = 1200$.

$$\theta = 2 \arctan \frac{500}{1200} = 45.2°$$

87. We redraw the text figure in a side view.

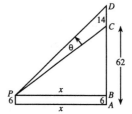

We note:

$\tan BPC = \dfrac{BC}{x} = \dfrac{62 - 6}{x} = \dfrac{56}{x}$

$\tan BPD = \dfrac{BD}{x} = \dfrac{56 + 14}{x} = \dfrac{70}{x}$

$\theta = BPD - BPC$

Hence, $\tan \theta = \tan(BPD - BPC)$

$$= \frac{\tan BPD - \tan BPC}{1 + \tan BPD \, \tan BPC} \quad \text{by the subtraction identity for tangent.}$$

$$\tan \theta = \frac{\dfrac{70}{x} - \dfrac{56}{x}}{1 + \dfrac{70}{x} \cdot \dfrac{56}{x}} = \frac{70x - 56x}{x^2 + 70 \cdot 56} = \frac{14x}{x^2 + 3920}$$

$$\theta = \arctan \frac{14x}{x^2 + 3920}$$

For $x = 150$ feet, $\theta = \arctan \dfrac{14 \cdot 150}{150^2 + 3920} = 4.54°$

Chapter 5 Inverse Trigonometric Functions; Trigonometric Equations

88. From the figure, we see that $y1 = \arctan \dfrac{14x}{x^2 + 3920}$ and

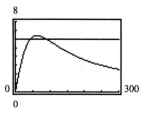

y2 = 6 intersect twice on the interval $0 \leq x \leq 300$.

(Calculator in degree mode)

Using zoom and trace procedures, the solutions to

$\arctan \dfrac{14x}{x^2 + 3920} = 6$ are found to be 43.9 ft and 89.3 ft.

89. (A) The area of segment PABCP = Area of sector OABC + Area of triangle OAC.

Area of sector OABC $= \dfrac{1}{2} r^2 (2\pi - \theta)$

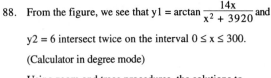

Since $\sin \dfrac{\theta}{2} = \dfrac{AP}{OA} = \dfrac{d}{2} \div r = \dfrac{d}{2r}$,

$\dfrac{\theta}{2} = \sin^{-1} \dfrac{d}{2r}$, $\theta = 2 \sin^{-1} \dfrac{d}{2r}$. Hence,

area of sector OABC $= \dfrac{1}{2} r^2 \left(2\pi - 2 \sin^{-1} \dfrac{d}{2r} \right)$

$= \pi r^2 - r^2 \sin^{-1} \dfrac{d}{2r}$

Area of triangle AOC $= \dfrac{1}{2}$ (base)(altitude) $= \dfrac{1}{2} (AC)(OP)$. AC = d.

From the Pythagorean theorem applied to triangle OAP, $OP^2 + AP^2 = OA^2$, hence

$OP^2 = OA^2 - AP^2 = r^2 - \left(\dfrac{d}{2} \right)^2$. $OP = \sqrt{r^2 - \dfrac{d^2}{4}} = \sqrt{\dfrac{4r^2 - d^2}{4}} = \dfrac{1}{2} \sqrt{4r^2 - d^2}$

Hence, area of triangle AOC $= \dfrac{1}{2} \cdot d \cdot \dfrac{1}{2} \sqrt{4r^2 - d^2} = \dfrac{d}{4} \sqrt{4r^2 - d^2}$.

Thus, area of segment PABCP = area of sector OABC + Area of triangle OAC

$= \pi r^2 - r^2 \sin^{-1} \dfrac{d}{2r} + \dfrac{d}{4} \sqrt{4r^2 - d^2}$

(B) If r = 0.5 in. and d = 0.7 in.

$A = \pi(0.5)^2 - (0.5)^2 \sin^{-1} \dfrac{0.7}{2(0.5)} + \dfrac{0.7}{4} \sqrt{4(0.5)^2 - (0.7)^2}$ (Calculator in radian mode)

$= 0.7165$ in^2

90. We use the formula $A = \pi r^2 - r^2 \sin^{-1} \dfrac{d}{2r} + \dfrac{d}{4} \sqrt{4r^2 - d^2}$ from Problem 89, with A = 0.75 in^2

and $r = \dfrac{1}{2}$ (diameter) $= \dfrac{1}{2}$ (1 inch) = 0.5 inch.

$0.75 = \pi(0.5)^2 - (0.5)^2 \sin^{-1} \dfrac{d}{2(0.5)} + \dfrac{d}{4} \sqrt{4(0.5)^2 - d^2}$

$0.75 = 0.25\pi - 0.25 \sin^{-1} d + \dfrac{d}{4} \sqrt{1 - d^2}$

We graph y1 = 0.75 and

$y2 = 0.25\pi - 0.25 \sin^{-1} d + \dfrac{d}{4} \sqrt{1 - d^2}$ on the

interval from 0 to 1. From the figure, we see that

y1 and y2 intersect once on the given interval.

Using zoom and trace procedures, the solution is

found to be 0.5743 in.

91. Analyzing triangle OAB, we see the following:

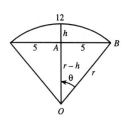

$$OA = r - h$$
$$OA^2 + AB^2 = r^2$$
$$(r - h)^2 + 5^2 = r^2$$
$$r^2 - 2hr + h^2 + 25 = r^2$$
$$h^2 + 25 = 2hr$$
$$r = \frac{h^2 + 25}{2h}$$

$\sin \theta = \dfrac{5}{r}$ $\theta = \sin^{-1} \dfrac{5}{r}$. Using the formula

$s = r\theta$, we can write $12 = 2\theta \cdot r$; $\theta = \dfrac{6}{r}$. Hence,

$\dfrac{6}{r} = \sin^{-1} \dfrac{5}{r}$. $6 \div \dfrac{h^2 + 25}{2h} = \sin^{-1}\left(5 \div \dfrac{h^2 + 25}{2h}\right)$. Thus, $\dfrac{12h}{h^2 + 25} = \sin^{-1} \dfrac{10h}{h^2 + 25}$ is the

equation that must be satisfied by h.

We graph $y1 = \dfrac{12h}{h^2 + 25}$ and $y2 = \sin^{-1} \dfrac{10h}{h^2 + 25}$.

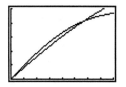

From the figure, we see that y1 and y2 intersect

once for nonzero h on the interval from 0 to 4.

Using zoom and trace procedures, the intersection

is found to be h = 2.82 ft.

92. (A)

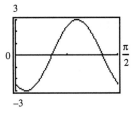

(B) We graph y1 = –1.8 sin 4t – 2.4 cos 4t and y2 = 2.

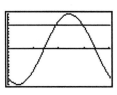

From the figure, we see that y1 and y2 intersect

twice on the indicated interval.

Using zoom and trace procedures, the intersections
are found to be t = 0.74 sec and t = 1.16 sec.

(C) We graph y1 = –1.8 sin 4t – 2.4 cos 4t and y2 = –2.

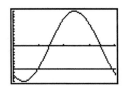

From the figure, we see that y1 and y2 intersect

twice on the indicated interval.

Using zoom and trace procedures, the intersections
are found to be t = 0.37 sec and t = 1.52 sec.

93. (A) $M = -1.8$ and $N = -2.4$

Locate $P(M, N) = P(-1.8, -2.4)$ to determine C:

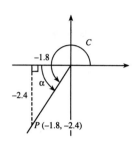

$$R = \sqrt{(-1.8)^2 + (-2.4)^2} = 3$$

$$\sin C = \frac{-2.4}{3} = -0.8 \qquad \tan C = \frac{-2.4}{-1.8} = \frac{4}{3}$$

Find the reference angle α. Then, $C = \pi + \alpha$

$$\tan \alpha = \frac{4}{3} \qquad \alpha = \tan^{-1}\left(\frac{4}{3}\right) = 0.9273$$

Thus, $C \approx 4.07$.

We can now write: $y = -1.8 \sin 4t - 2.4 \cos 4t = 3 \sin(4t + 4.07)$

(B) We are to solve $2 = 3 \sin(4t + 4.07)$ for $0 \le t \le \frac{\pi}{2}$, that is, $0 \le 4t \le 2\pi$.

$$\sin(4t + 4.07) = \frac{2}{3}$$

$$4t + 4.07 = \begin{cases} \sin^{-1}\frac{2}{3} + 2\pi \\ \pi - \sin^{-1}\frac{2}{3} + 2\pi \end{cases}$$

$$4t = \begin{cases} \sin^{-1}\frac{2}{3} - 4.07 \\ 3\pi - \sin^{-1}\frac{2}{3} - 4.07 \end{cases}$$

$$t = \begin{cases} \dfrac{\sin^{-1}\frac{2}{3} - 4.07}{4} \\[3mm] \dfrac{3\pi - \sin^{-1}\frac{2}{3} - 4.07}{4} \end{cases} = \begin{cases} 0.74 \text{ sec} \\ 1.16 \text{ sec} \end{cases}$$

(C) We are to solve $-2 = 3 \sin(4t + 4.07)$ for $0 \le t \le \frac{\pi}{2}$, that is, $0 \le 4t \le 2\pi$.

$$\sin(4t + 4.07) = -\frac{2}{3}$$

$$4t + 4.07 = \begin{cases} 2\pi - \sin^{-1}\frac{2}{3} \\ 3\pi + \sin^{-1}\frac{2}{3} \end{cases}$$

$$4t = \begin{cases} 2\pi - \sin^{-1}\frac{2}{3} - 4.07 \\ 3\pi + \sin^{-1}\frac{2}{3} - 4.07 \end{cases}$$

$$t = \begin{cases} \dfrac{2\pi - \sin^{-1}\frac{2}{3} - 4.07}{4} \\[3mm] \dfrac{3\pi + \sin^{-1}\frac{2}{3} - 4.07}{4} \end{cases} = \begin{cases} 0.37 \text{ sec} \\ 1.52 \text{ sec} \end{cases}$$

Chapter 5   Inverse Trigonometric Functions; Trigonometric Equations

CUMULATIVE REVIEW EXERCISE   Chapters 1–5

1. $\dfrac{\theta_{deg}}{180°} = \dfrac{\theta_{rad}}{\pi\text{ rad}}$ $\theta_{deg} = \dfrac{180°}{\pi\text{ rad}}\,\theta_{rad}$ If $\theta_{rad} = 4.21$, $\theta_{deg} = \dfrac{180}{\pi}(4.21) = 241.22°$

2. $\dfrac{\theta_{deg}}{180°} = \dfrac{\theta_{rad}}{\pi\text{ rad}}$ $\theta_{rad} = \dfrac{\pi\text{ rad}}{180°}\,\theta_{deg}$ If $\theta_{deg} = 505°42'$, or $\left(505 + \dfrac{42}{60}\right)°$,

$$\theta_{rad} = \dfrac{\pi}{180}\left(505 + \dfrac{42}{60}\right) = 8.83\text{ rad}$$

3. Let $c = 7.6$ m, $b = 4.5$ m

    Solve for θ: We will use the sine: $\sin\theta = \dfrac{b}{c} = \dfrac{4.5\text{ m}}{7.6\text{ m}} = 0.5921$

    $\theta = \sin^{-1}(0.5921) = 36°$

    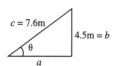

    Solve for the complementary angle: $90° - \theta = 90° - 36° = 54°$

    Solve for a: We choose the cosine to find a. $\cos\theta = \dfrac{a}{c}$

    $a = c\cos\theta = (7.6\text{ m})(\cos 36°) = 6.1$ m

4. $P(a, b) = (-5, -12)$

    $R = \sqrt{a^2 + b^2} = \sqrt{(-5)^2 + (-12)^2} = 13$

    $\cos\theta = \dfrac{a}{R} = \dfrac{-5}{13} = -\dfrac{5}{13}$

    $\cot\theta = \dfrac{a}{b} = \dfrac{-5}{-12} = \dfrac{5}{12}$

    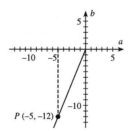

5. (A) Degree mode: $\cos 67°45' = \cos 67.75°$ Convert to decimal degrees

    $= 0.3786$

    (B) Degree mode: $\csc 176.2° = \dfrac{1}{\sin 176.2°} = 15.09$

    (C) Radian mode: $\cot 2.05 = \dfrac{1}{\tan 2.05} = -0.5196$

6. (A) 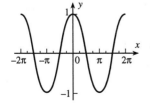   (B)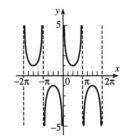

Chapter 5   Inverse Trigonometric Functions; Trigonometric Equations

(C)

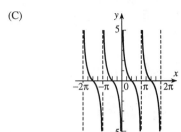

7. $\tan x \csc x = \dfrac{\sin x}{\cos x} \dfrac{1}{\sin x}$ Quotient and reciprocal identities

    $= \dfrac{1}{\cos x}$ Algebra

    $= \sec x$ Reciprocal identity

8. $\csc \theta - \sin \theta = \dfrac{1}{\sin \theta} - \sin \theta$ Reciprocal identity

    $= \dfrac{1}{\sin \theta} - \dfrac{\sin^2 \theta}{\sin \theta}$ Algebra

    $= \dfrac{1 - \sin^2 \theta}{\sin \theta}$ Algebra

    $= \dfrac{\cos^2 \theta}{\sin \theta}$ Pythagorean identity

    $= \cos \theta \cdot \dfrac{\cos \theta}{\sin \theta}$ Algebra

    $= \cos \theta \cot \theta$ Quotient identity

9. $(\sin^2 u)(\tan^2 u + 1) = \sin^2 u \sec^2 u$ Pythagorean identity

    $= \sin^2 u \dfrac{1}{\cos^2 u}$ Reciprocal identity

    $= \dfrac{\sin^2 u}{\cos^2 u}$ Algebra

    $= \tan^2 u$ Quotient identity

    $= \sec^2 u - 1$ Pythagorean identity

250

10. $\dfrac{\sin^2 \alpha - \cos^2 \alpha}{\sin \alpha \cos \alpha} = \dfrac{\dfrac{\sin^2 \alpha}{\sin \alpha \cos \alpha} - \dfrac{\cos^2 \alpha}{\sin \alpha \cos \alpha}}{\dfrac{\sin \alpha \cos \alpha}{\sin \alpha \cos \alpha}}$ Algebra

$= \dfrac{\dfrac{\sin \alpha}{\cos \alpha} - \dfrac{\cos \alpha}{\sin \alpha}}{\dfrac{\sin \alpha}{\cos \alpha} \cdot \dfrac{\cos \alpha}{\sin \alpha}}$ Algebra

$= \dfrac{\tan \alpha - \cot \alpha}{\tan \alpha \cot \alpha}$ Quotient identities

11. Locate the 30° – 60° reference triangle, determine (a, b) and R, then evaluate.

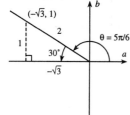

$\sin \dfrac{5\pi}{6} = \dfrac{1}{2}$

12. Locate the 45° reference triangle, determine (a, b) and R, then evaluate.

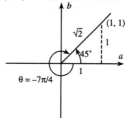

$\cos\left(-\dfrac{7\pi}{4}\right) = \dfrac{1}{\sqrt{2}}$

13. Locate the 30° – 60° reference triangle, determine (a, b) and R, then evaluate.

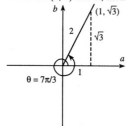

$\tan \dfrac{7\pi}{3} = \dfrac{\sqrt{3}}{1} = \sqrt{3}$

14. (a, b) = (0, –1)   R = 1

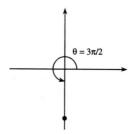

$\sec \dfrac{3\pi}{2} = \dfrac{1}{0}$

Not defined

Chapter 5   Inverse Trigonometric Functions;
Trigonometric Equations

15. $y = \sin^{-1}\left(\dfrac{1}{\sqrt{2}}\right)$ is equivalent to $\sin y = \dfrac{1}{\sqrt{2}}$.

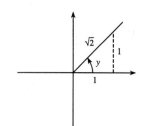

What y between $-\dfrac{\pi}{2}$ and $\dfrac{\pi}{2}$ has sine equal to $\dfrac{1}{\sqrt{2}}$?

y must be associated with a reference triangle in

the first quadrant.

Reference triangle is a special 45° triangle.

$$y = \dfrac{\pi}{4} \qquad \sin^{-1}\left(\dfrac{1}{\sqrt{2}}\right) = \dfrac{\pi}{4}$$

16. $y = \arctan 0$ is equivalent to $\tan y = 0$.

No reference triangle can be drawn, but the

only y between $-\dfrac{\pi}{2}$ and $\dfrac{\pi}{2}$ which has tangent

equal to 0 is $y = 0$. Thus, $\arctan 0 = 0$.

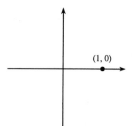

17. $y = \cos^{-1}\left(-\dfrac{\sqrt{3}}{2}\right)$ is equivalent to $\cos y = -\dfrac{\sqrt{3}}{2}$.

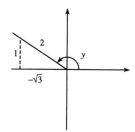

What y between 0 and π has cosine equal to $-\dfrac{\sqrt{3}}{2}$?

y must be associated with a reference triangle in the

second quadrant.

Reference triangle is a special 30° – 60° triangle.

$$y = \dfrac{5\pi}{6} \qquad \cos^{-1}\left(-\dfrac{\sqrt{3}}{2}\right) = \dfrac{5\pi}{6}$$

18. 3 is not in the domain of the inverse sine function. $-1 \le 3 \le 1$ is false. arcsin 3 is not defined.

19. $y = \text{arccot}(-\sqrt{3})$ is equivalent to $\cot y = -\sqrt{3}$ and

$0 < y < \pi$. What number between 0 and π has

cotangent equal to $-\sqrt{3}$? y must be positive and

in the second quadrant.

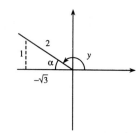

$$\cot y = -\sqrt{3} = \dfrac{-\sqrt{3}}{1} \qquad \alpha = \dfrac{\pi}{6} \qquad y = \dfrac{5\pi}{6}$$

Thus, $\cot^{-1}(-\sqrt{3}) = \dfrac{5\pi}{6}$

20. $y = \sec^{-1} 2$ is equivalent to $\sec y = 2$ and $0 \le y \le \pi$,
$y \ne \dfrac{\pi}{2}$. What number between 0 and π has secant
equal to 2? y must be in the first quadrant.

$\sec y = 2 = \dfrac{2}{1}$ $y = \dfrac{\pi}{3}$

Thus, $\sec^{-1} 2 = \dfrac{\pi}{3}$

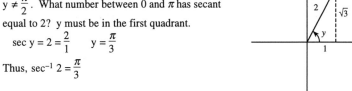

21. Calculator in radian mode: $\sin^{-1}(0.0505) = 0.0505$

22. Calculator in radian mode: $\cos^{-1}(-0.7228) = 2.379$

23. Calculator in radian mode: $\arctan(-9) = -1.460$

24. Calculator in radian mode: $\text{arccot } 3 = \arctan \dfrac{1}{3} = 0.3218$

25. Calculator in radian mode: $\sec^{-1} 2.6 = \cos^{-1} \dfrac{1}{2.6} = 1.176$

26. Calculator in radian mode: $\text{arccsc}(-0.5969) = \arcsin \dfrac{1}{-0.5969} = \text{error}$
$\text{arccsc}(-0.5969)$ is not defined.

-0.5969 is not in the domain of the inverse cosecant function.

27. $2 \sin \theta - 1 = 0$ $\sin \theta = \dfrac{1}{2}$

Sine is positive in the first and second quadrants. Sketch the reference triangles and angles
corresponding to θ. $\alpha = \sin^{-1} \dfrac{1}{2} = 30°$

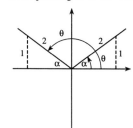

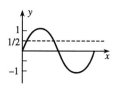

$$\theta = \begin{cases} \alpha \\ 180° - \alpha \end{cases} = \begin{cases} \sin^{-1} \dfrac{1}{2} \\ 180° - \sin^{-1} \dfrac{1}{2} \end{cases} = \begin{cases} 30° \\ 150° \end{cases}$$

Chapter 5   Inverse Trigonometric Functions; Trigonometric Equations

28. $3 \tan x + \sqrt{3} = 0$ $\tan x = -\dfrac{\sqrt{3}}{3}$

Tangent is negative in the fourth quadrant. Sketch the reference triangle and angle corresponding

to x. $\alpha = \tan^{-1}\dfrac{\sqrt{3}}{3} = \dfrac{\pi}{6}$ $x = -\alpha = -\tan^{-1}\dfrac{\sqrt{3}}{3} = -\dfrac{\pi}{6}$.

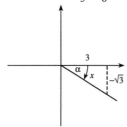 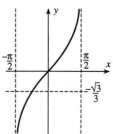

29. $2 \cos x + 2 = 0$

        $\cos x = -1$ The only solution of $\cos x = -1$

                       $-\pi \le x \le \pi$, is $-\pi$

         $x = -\pi$

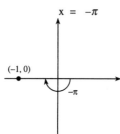

30. $\sin x \cos y = \dfrac{1}{2}[\sin(x + y) + \sin(x - y)]$. Let $x = 7u$ and $y = 3u$

    $\sin 7u \cos 3u = \dfrac{1}{2}[\sin(7u + 3u) + \sin(7u - 3u)] = \dfrac{1}{2}(\sin 10u + \sin 4u)$

31. $\cos x - \cos y = -2 \sin\dfrac{x + y}{2} \sin\dfrac{x - y}{2}$. Let $x = 5w$ and $y = w$

    $\cos 5w - \cos w = -2 \sin\dfrac{5w + w}{2} \sin\dfrac{5w - w}{2} = -2 \sin 3w \sin 2w$

32. $M = \sqrt{3}$ and $N = 1$. Locate $P(M, N) = P(\sqrt{3}, 1)$ to determine C. $R = \sqrt{(\sqrt{3})^2 + 1^2} = 2$

    $\tan C = \dfrac{1}{\sqrt{3}}$ $\sin C = \dfrac{1}{2}$ $C = \dfrac{\pi}{6}$ (Reference triangle is a special $30° - 60°$ triangle.)

    Thus, $y = \sqrt{3} \sin t + \cos t = 2 \sin\left(t + \dfrac{\pi}{6}\right)$

    Amplitude $= |\, 2\, | = 2$

    Period and Phase Shift:

$$t + \frac{\pi}{6} = 0 \qquad t + \frac{\pi}{6} = 2\pi$$

$$t = -\frac{\pi}{6} \qquad t = -\frac{\pi}{6} + 2\pi$$

Period $= 2\pi$ Phase Shift $= -\frac{\pi}{6}$

Frequency $= \dfrac{1}{\text{Period}} = \dfrac{1}{2\pi}$

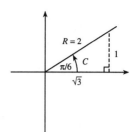

33. $92.462° = 92°(0.462 \times 60)' = 92°27.72' = 92°27'(0.72 \times 60)'' = 92°27'43''$

34. Since $\dfrac{s}{C} = \dfrac{\theta°}{360°}$ then $\dfrac{12 \text{ in.}}{30 \text{ in.}} = \dfrac{\theta}{360°}$ $\theta = \dfrac{12}{30} \cdot 360° = 144°$

35. Since the two right triangles are similar, we can write:

$$\frac{4}{x} = \frac{x}{x + 3}$$

$$x(x + 3)\frac{4}{x} = x(x + 3)\frac{x}{x + 3}$$

$$(x + 3)4 = x^2$$

$$4x + 12 = x^2$$

$$0 = x^2 - 4x - 12$$

$$0 = (x - 6)(x + 2)$$

$x - 6 = 0$ or $x + 2 = 0$

$x = 6$ $x = -2$ We discard the negative answer.

Since $\tan \theta = \dfrac{4}{x}$, we can write $\tan \theta = \dfrac{4}{6}$ $\theta = \tan^{-1}\dfrac{4}{6} = 33.7°$

36. $\dfrac{\theta_{deg}}{180°} = \dfrac{\theta_{rad}}{\pi \text{ rad}}$ $\theta_{rad} = \dfrac{\pi \text{ rad}}{180°}\theta_{deg}$. If $\theta_{deg} = 72°$, $\theta_{rad} = \dfrac{\pi}{180}(72) = \dfrac{2\pi}{5}\text{ rad}$

37. $\alpha = 360° - 45° = 315°$

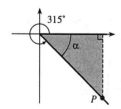

38. We sketch a reference triangle and label what we know. Since $\tan \theta = \dfrac{b}{a} = \dfrac{1}{2} = \dfrac{-1}{-2}$ and $\sin \theta < 0$, the

terminal side of θ must lie in quadrant III. Hence, $b = -1$ and $a = -2$. Use the Pythagorean theorem

to find R: $(-2)^2 + (-1)^2 = R^2$

$$R^2 = 5$$

$R = \sqrt{5}$ (R is never negative)

We can now find the other five functions.

$\sin \theta = \dfrac{b}{R} = \dfrac{-1}{\sqrt{5}}$ $\cot \theta = \dfrac{a}{b} = \dfrac{2}{1} = 2$

$\cos \theta = \dfrac{a}{R} = -\dfrac{2}{\sqrt{5}}$ $\sec \theta = \dfrac{R}{a} = \dfrac{\sqrt{5}}{-2} = -\dfrac{\sqrt{5}}{2}$

$\csc \theta = \dfrac{R}{b} = \dfrac{\sqrt{5}}{-1} = -\sqrt{5}$

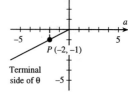

Terminal side of θ

39. $y = 2 - 2 \sin \dfrac{x}{2}$.   Amplitude $= |-2| = 2$.   Period $= 2\pi \div \dfrac{1}{2} = 4\pi$.

This graph is the graph of $y = -2 \sin \dfrac{x}{2}$ moved up 2 units.

We start by drawing a horizontal broken line 2 units above

the x axis, then graph $y = -2 \sin \dfrac{x}{2}$ (an upside down sine

curve with amplitude 2 and period 4π) relative to the

broken line and the original y axis.

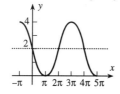

40.   Amplitude $= |A| = |3| = 3$.   Phase Shift and Period: Solve

$\begin{aligned} Bx + C &= 0 \\ 2x - \pi &= 0 \\ 2x &= \pi \\ x &= \dfrac{\pi}{2} \end{aligned}$ and $\begin{aligned} Bx + C &= 2\pi \\ 2x - \pi &= 2\pi \\ 2x &= \pi + 2\pi \\ x &= \dfrac{\pi}{2} + \pi \end{aligned}$

Phase Shift $= \dfrac{\pi}{2}$ Period $= \pi$

Graph one cycle over the interval from $\dfrac{\pi}{2}$ to $\dfrac{\pi}{2} + \pi = \dfrac{3\pi}{2}$.   Then extend the graph from $-\pi$ to 2π.

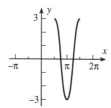 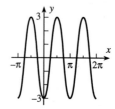

41. We find the period and phase shift by solving

$$\pi x - \frac{\pi}{4} = 0 \qquad \text{and} \qquad \pi x - \frac{\pi}{4} = \pi$$

$$\pi x = \frac{\pi}{4} \qquad\qquad\qquad \pi x = \frac{\pi}{4} + \pi$$

$$x = \frac{1}{4} \qquad\qquad\qquad x = \frac{1}{4} + 1$$

$$\text{Period} = \frac{1}{4} \qquad\qquad \text{Phase shift} = \qquad 1$$

We then sketch one period of the graph starting at $x = \frac{1}{4}$ (the phase shift) and ending at $x = \frac{1}{4} + 1 = \frac{5}{4}$

(the phase shift plus one period). Note that a vertical asymptote is at $x = \frac{3}{4}$. We then extend the

graph from 0 to 3.

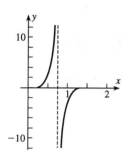 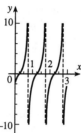

42. Amplitude $= \frac{1}{2}$ (y max – y min) $= \frac{1}{2}(4 - (-2)) = 3$. $k = \frac{1}{2}(y_{max} + y_{min}) = \frac{1}{2}(4 + (-2)) = 1$.

Period $= \frac{2\pi}{B} = 2$. Thus, $B = \frac{2\pi}{2} = \pi$. The form of the graph is that of the basic cosine curve.

Thus, $y = |A| \cos Bx + k = 3 \cos \pi x + 1$

43. Let a = 19.4 cm and b = 41.7 cm.

Solve for c: We will use the Pythagorean theorem

$$c^2 = a^2 + b^2$$

$$c = \sqrt{a^2 + b^2} = \sqrt{(19.4)^2 + (41.7)^2} = 46.0 \text{ cm}$$

Solve for θ: We will use the tangent $\tan \theta = \frac{b}{a} = \frac{41.7 \text{ cm}}{19.4 \text{ cm}} = 2.1495$

$$\theta = \tan^{-1}(2.1495) = 65°0'$$

Solve for the complementary angle: $90° - \theta = 90° - 65°0' = 25°0'$

44. $\dfrac{\cos x}{1 + \sin x} + \tan x = \dfrac{\cos x}{1 + \sin x} + \dfrac{\sin x}{\cos x}$ \qquad Quotient identity

$$= \frac{\cos x \cos x}{\cos x(1 + \sin x)} + \frac{\sin x(1 + \sin x)}{\cos x(1 + \sin x)} \qquad \text{Algebra}$$

$$= \frac{\cos^2 x + \sin x + \sin^2 x}{\cos x(1 + \sin x)} \qquad \text{Algebra}$$

$$= \frac{\cos^2 x + \sin^2 x + \sin x}{\cos x(1 + \sin x)} \qquad \text{Algebra}$$

$$= \frac{1 + \sin x}{\cos x(1 + \sin x)} \qquad \text{Pythagorean identity}$$

$$= \frac{1}{\cos x} \qquad \text{Algebra}$$

$$= \sec x \qquad \text{Reciprocal identity}$$

45. In this problem, it is more straightforward to start with the right-hand side of the identity to be
     verified. The student can confirm that the steps would be valid if reversed.

$$(\sec \theta - \tan \theta)(\sec \theta + 1) = \left(\frac{1}{\cos \theta} - \frac{\sin \theta}{\cos \theta}\right)\left(\frac{1}{\cos \theta} + 1\right) \qquad \begin{array}{l}\text{Quotient and Reciprocal}\\\text{identities}\end{array}$$

$$= \left(\frac{1 - \sin \theta}{\cos \theta}\right)\left(\frac{1}{\cos \theta} + \frac{\cos \theta}{\cos \theta}\right) \qquad \text{Algebra}$$

$$= \frac{1 - \sin \theta}{\cos \theta} \cdot \frac{1 + \cos \theta}{\cos \theta} \qquad \text{Algebra}$$

$$= \frac{(1 - \sin \theta)(1 + \cos \theta)}{\cos^2 \theta} \qquad \text{Algebra}$$

$$= \frac{(1 - \sin \theta)(1 + \cos \theta)}{1 - \sin^2 \theta} \qquad \text{Pythagorean identity}$$

$$= \frac{(1 - \sin \theta)(1 + \cos \theta)}{(1 - \sin \theta)(1 + \sin \theta)} \qquad \text{Algebra}$$

$$= \frac{1 + \cos \theta}{1 + \sin \theta} \qquad \text{Algebra}$$

46. $$\cot \frac{u}{2} = 1 \div \tan \frac{u}{2} \qquad \text{Reciprocal identity}$$

$$= 1 \div \frac{\sin u}{1 + \cos u} \qquad \text{Half-angle identity for tangent}$$

$$= 1 \cdot \frac{1 + \cos u}{\sin u} \qquad \text{Algebra}$$

$$= \frac{1 + \cos u}{\sin u} \qquad \text{Algebra}$$

$$= \frac{1}{\sin u} + \frac{\cos u}{\sin u} \qquad \text{Algebra}$$

$$= \csc u + \cot u \qquad \text{Quotient and Reciprocal identities}$$

47. $\sec^2 \dfrac{x}{2} = 1 \div \cos^2 \dfrac{x}{2}$ Reciprocal identity

$= 1 \div \left(\pm \sqrt{\dfrac{1 + \cos x}{2}} \right)^2$ Half-angle identity

$= 1 \div \dfrac{1 + \cos x}{2}$ Algebra

$= 1 \cdot \dfrac{2}{1 + \cos x}$ Algebra

$= \dfrac{2}{1 + \cos x}$ Algebra

$= \dfrac{2(1 - \cos x)}{(1 + \cos x)(1 - \cos x)}$ Algebra

$= \dfrac{2 - 2 \cos x}{1 - \cos^2 x}$ Algebra

$= \dfrac{2 - 2 \cos x}{\sin^2 x}$ Pythagorean identity

$= \dfrac{2}{\sin^2 x} - \dfrac{2 \cos x}{\sin^2 x}$ Algebra

$= \dfrac{2}{\sin^2 x} - \dfrac{2 \cos x}{\sin x} \cdot \dfrac{1}{\sin x}$ Algebra

$= 2 \csc^2 x - 2 \cot x \csc x$ Reciprocal and Quotient identities

$= 2(1 + \cot^2 x) - 2 \cot x \csc x$ Pythagorean identity

$= 2 + 2 \cot^2 x - 2 \cot x \csc x$ Algebra

$= 2 + 2 \cot x(\cot x - \csc x)$ Algebra

48. $\dfrac{2}{1 + \sec 2\theta} = \dfrac{2}{1 + \dfrac{1}{\cos 2\theta}}$ Reciprocal identity

$= \dfrac{2 \cos 2\theta}{2 \cos 2\theta + 1}$ Algebra

$= \dfrac{2(2 \cos^2 \theta - 1)}{2 \cos^2 \theta - 1 + 1}$ Double-angle identity

$= \dfrac{4 \cos^2 \theta - 2}{2 \cos^2 \theta}$ Algebra

$= \dfrac{4 \cos^2 \theta}{2 \cos^2 \theta} - \dfrac{2}{2 \cos^2 \theta}$ Algebra

$= 2 - \dfrac{1}{\cos^2 \theta}$ Algebra

$$= \ 2 - \sec^2 \theta \qquad\qquad \text{Reciprocal identity}$$

$$= \ 2 - (\tan^2 \theta + 1) \qquad \text{Pythagorean identity}$$

$$= \ 1 - \tan^2 \theta \qquad\qquad \text{Algebra}$$

49. $\dfrac{\sin x - \sin y}{\cos x + \cos y}$

$$= \frac{(\sin x - \sin y)\ \sin(x - y)}{(\cos x + \cos y)\ \sin(x - y)} \qquad\qquad\qquad \text{Algebra}$$

$$= \frac{(\sin x - \sin y)(\sin x \cos y - \cos x \sin y)}{(\cos x + \cos y)\ \sin(x - y)} \qquad\qquad \text{Difference identity for sine}$$

$$= \frac{\sin^2 x \cos y - \sin x \cos x \sin y - \sin x \sin y \cos y + \cos x \sin^2 y}{(\cos x + \cos y)\ \sin(x - y)} \qquad \text{Algebra}$$

$$= \frac{(1 - \cos^2 x) \cos y - \sin x \cos x \sin y - \sin x \sin y \cos y + \cos x(1 - \cos^2 y)}{(\cos x + \cos y)\ \sin(x - y)}$$
$$\text{Pythagorean identity}$$

$$= \frac{\cos y - \cos^2 x \cos y - \sin x \cos x \sin y - \sin x \sin y \cos y + \cos x - \cos x \cos^2 y}{(\cos x + \cos y)\ \sin(x - y)}$$
$$\text{Algebra}$$

$$= \frac{(\cos x - \cos^2 x \cos y - \sin x \cos x \sin y) + (\cos y - \sin x \sin y \cos y - \cos x \cos^2 y)}{(\cos x + \cos y)\ \sin(x - y)}$$
$$\text{Algebra}$$

$$= \frac{\cos x(1 - \cos x \cos y - \sin x \sin y) + \cos y(1 - \sin x \sin y - \cos x \cos y)}{(\cos x + \cos y)\ \sin(x - y)} \qquad \text{Algebra}$$

$$= \frac{(\cos x + \cos y)(1 - \cos x \cos y - \sin x \sin y)}{(\cos x + \cos y)\ \sin(x - y)} \qquad\qquad \text{Algebra}$$

$$= \frac{1 - (\cos x \cos y + \sin x \sin y)}{\sin(x - y)} \qquad\qquad\qquad \text{Algebra}$$

$$= \frac{1 - \cos(x - y)}{\sin(x - y)} \qquad\qquad\qquad\qquad \text{Difference identity for cosine}$$

$$= \tan\frac{x - y}{2} \qquad\qquad\qquad\qquad\qquad \text{Half-angle identity}$$

50. First draw a reference traingle in the third quadrant and
     find cos x. $R = \sqrt{(-15)^2 + (-8)^2} = 17.\ \ \cos x = \dfrac{-15}{17}$.

     We can now find $\sin\dfrac{x}{2}$ from the half-angle identity,
     after determining its sign, as follows:

         If $\pi < x < \dfrac{3\pi}{2}$, then $\dfrac{\pi}{2} < \dfrac{x}{2} < \dfrac{3\pi}{4}$.

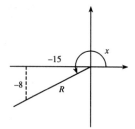

Thus, $\frac{x}{2}$ is in the second quadrant, where sine is positive.

Using half-angle identities, we obtain

$$\sin \frac{x}{2} = \sqrt{\frac{1 - \cos x}{2}} = \sqrt{\frac{1 - \left(-\frac{15}{17}\right)}{2}} = \sqrt{\frac{\frac{32}{17}}{2}} = \sqrt{\frac{16}{17}} \text{ or } \frac{4}{\sqrt{17}}$$

To find cos 2x, we use a double-angle identity.

$$\cos 2x = 2 \cos^2 x - 1 = 2\left(-\frac{15}{17}\right)^2 - 1 = \frac{2}{1} \cdot \frac{225}{289} - 1 = \frac{450 - 289}{289} = \frac{161}{289}$$

51.
$$\cos x + \cos y = 2 \cos \frac{x + y}{2} \cos \frac{x - y}{2}$$
$$\cos 165° + \cos 15° = 2 \cos \frac{165° + 15°}{2} \cos \frac{165° - 15°}{2}$$
$$= 2 \cos 90° \cos 40° = 2 \cdot 0 \cdot \cos 40° = 0$$

52. $M = 1$ and $N = -\sqrt{3}$

Locate $P(M, N) = P(1, -\sqrt{3})$ to determine C.

$R = \sqrt{1^2 + (-\sqrt{3})^2} = 2$ $\sin C = -\frac{\sqrt{3}}{2}$

$\tan C = -\sqrt{3}$ $C = -\frac{\pi}{3}$

| C | is minimum for this choice.

Thus, $y = \sin \pi t - \sqrt{3} \cos \pi t = 2 \sin\left(\pi t - \frac{\pi}{3}\right)$

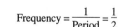

Amplitude $= |2| = 2$

Period and Phase Shift:

$$\pi t - \frac{\pi}{3} = 0 \qquad \pi t - \frac{\pi}{3} = 2\pi$$
$$\pi t = \frac{\pi}{3} \qquad \pi t = \frac{\pi}{3} + 2\pi$$
$$t = \frac{1}{3} \qquad t = \frac{1}{3} + 2$$

Period $= \quad 2$ Phase Shift $= \quad \frac{1}{3}$ Frequency $= \frac{1}{\text{Period}} = \frac{1}{2}$

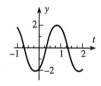

Chapter 5   Inverse Trigonometric Functions; Trigonometric Equations

53 .   (A)   Let $y = \cos^{-1} 0.4$, then $\cos y = 0.4 = \dfrac{2}{5}$, $0 \le y \le \pi$.

Draw the reference triangle associated with y.

Then, $\sin y = \sin(\cos^{-1} 0.4)$ can be determined

directly from the triangle.

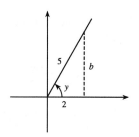

$$b = \sqrt{5^2 - 2^2} = \sqrt{21}$$

$$\sin(\cos^{-1} 0.4) = \sin y = \frac{b}{R} = \frac{\sqrt{21}}{5}$$

(B)   Let $y = \arctan(-\sqrt{5})$, then $\tan y = -\sqrt{5}$, $-\dfrac{\pi}{2} < y < \dfrac{\pi}{2}$.

Draw the reference triangle associated with y.

Then, $\sec y = \sec[\arctan(-\sqrt{5})]$ can be determined

directly from the triangle.

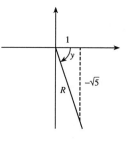

$$R = \sqrt{1^2 + (-\sqrt{5})^2} = \sqrt{6}$$

$$\sec[\arctan(-\sqrt{5})] = \sec y = \frac{R}{a} = \frac{\sqrt{6}}{1} = \sqrt{6}$$

(C) Let $y = \sin^{-1} \dfrac{1}{3}$, then $\sin y = \dfrac{1}{3}$, $-\dfrac{\pi}{2} \le y \le \dfrac{\pi}{2}$. Then, $\csc y = \csc\left(\sin^{-1} \dfrac{1}{3}\right) = \dfrac{1}{\sin y} = \dfrac{1}{\frac{1}{3}} = 3$

(D) Let $y = \sec^{-1} 4$, then $\sec y = 4$, $0 \le y \le \pi$.

Draw the reference triangle associated with y.

Then $\tan y = \tan(\sec^{-1} 4)$ can be determined

directly from the triangle.

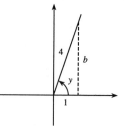

$$b = \sqrt{4^2 - 1^2} = \sqrt{15}$$

$$\tan(\sec^{-1} 4) = \tan y = \frac{b}{a} = \frac{\sqrt{15}}{1} = \sqrt{15}$$

54.   (A) Calculator in radian mode: $\csc[\arctan(-5.624)] = \dfrac{1}{\sin[\arctan(-5.624)]} = -1.016$

(B) Calculator in radian mode: $\cos^{-1}(\sec 2.558) = $ error.

$\cos^{-1}(\sec 2.558) = \cos^{-1}(1.000997)$ is not defined. 1.000997 is not in the domain of the inverse cosine function.

(C) Calculator in radian mode: $\tan^{-1}(\cos 0.1028) = 0.7828$

(D) Calculator in radian mode: $\sin[\text{arcsec}(-4.612)] = \sin\left[\arccos\left(\dfrac{1}{-4.612}\right)\right] = 0.9762$

55. $\theta = \tan^{-1} \sqrt{3}$ is equivalent to $\tan \theta = \sqrt{3}$ $-90° < \theta < 90°$. Thus, $\theta = 60°$

56.   Calculator in degree mode: $\theta = \sin^{-1} 0.8989 = 64.01°$

57. $2 + 3 \sin x = \cos 2x \qquad 0 \le x \le 2\pi$

$$2 + 3 \sin x = 1 - 2 \sin^2 x$$

$$2 \sin^2 x + 3 \sin x + 1 = 0$$

$$(2 \sin x + 1)(\sin x + 1) = 0$$

$2 \sin x + 1 = 0$	$\sin x + 1 = 0$
$\sin x = -\dfrac{1}{2}$	$\sin x = -1$
$x = \dfrac{7\pi}{6}, \dfrac{11\pi}{6}$	$x = \dfrac{3\pi}{2}$

Solutions: $x = \dfrac{7\pi}{6}, \dfrac{3\pi}{2}, \dfrac{11\pi}{6}$

58. First solve for θ over one period, $0 \le \theta < 360°$. Then add integer multiples of $360°$ to find all solutions.

$\sin 2\theta = 2 \cos \theta \qquad 0 \le \theta \le 360°$

$$2 \sin \theta \cos \theta = 2 \cos \theta$$

$$2 \sin \theta \cos \theta - 2 \cos \theta = 0$$

$$2 \cos \theta (\sin \theta - 1) = 0$$

$2 \cos \theta = 0$	$\sin \theta - 1 = 0$
$\cos \theta = 0$	$\sin \theta = 1$
$\theta = 90°, 270°$	$\theta = 90°$

Thus, all solutions over one period, $0° \le \theta < 360°$, are $x = 90°, 270°$. Thus, the solutions, if θ is allowed to range over all degree values, can be written as

$$\theta = \begin{cases} 90° + k360° \\ 270° + k360° \end{cases} \quad k \text{ any integer}$$

More concisely, since the solutions are $90°, 270°, 450°$, and so on, we can write

$\theta = 90° + k180° \qquad k \text{ any integer}$

59. First solve for x over one period of tan x, $0 \le x < \pi$. Then add integer multiples of π to find all solutions.

$$4 \tan^2 x - 3 \sec^2 x = 0$$

$$4 \tan^2 x - 3(\tan^2 x + 1) = 0$$

$$4 \tan^2 x - 3 \tan^2 x - 3 = 0$$

$$\tan^2 x - 3 = 0$$

$$\tan^2 x = 3$$

$$\tan x = \pm\sqrt{3}$$

$$x = \frac{\pi}{3}, \frac{2\pi}{3}$$

Thus, all solutions over one period, $0 \le x < \pi$, are $x = \dfrac{\pi}{3}, \dfrac{2\pi}{3}$.

Thus, the solutions, if x is allowed to range over all real numbers, are

$$x = \begin{cases} \dfrac{\pi}{3} + k\pi \\ \dfrac{2\pi}{3} + k\pi \end{cases} \quad \text{k any integer}$$

60. Sine is negative in the third and fourth quadrants. Sketch the reference triangles and angles

corresponding to x. $\alpha = \sin^{-1}(0.5678) = 0.6038$

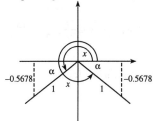 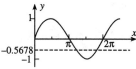

If $0 \le x < 2\pi$ (one period)

$$x = \begin{cases} \pi + \alpha \\ 2\pi - \alpha \end{cases} = \begin{cases} \pi + \sin^{-1}(0.5678) \\ 2\pi - \sin^{-1}(0.5678) \end{cases} = \begin{cases} \pi + 0.6038 \\ 2\pi - 0.6038 \end{cases} = \begin{cases} 3.745 \\ 5.679 \end{cases}$$

We find all solutions by adding integer multples of 2π to these. $x = \begin{cases} 3.745 + 2k\pi \\ 5.679 + 2k\pi \end{cases} \quad \text{k any integer}$

61. Secant is positive in the first and fourth quadrants. Sketch the reference triangles and angles

corresponding to x. $\alpha = \sec^{-1}(2.345) = 1.130$

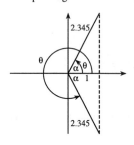 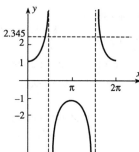

If $0 \le x < 2\pi$ (one period)

$$x = \begin{cases} \alpha \\ 2\pi - \alpha \end{cases} = \begin{cases} \sec^{-1}(2.345) \\ 2\pi - \sec^{-1}(2.345) \end{cases} = \begin{cases} 1.130 \\ 2\pi - 1.130 \end{cases} = \begin{cases} 1.130 \\ 5.153 \end{cases}$$

We find all solutions by adding integer multiples of 2π to these. $x \begin{cases} 1.130 + 2k\pi \\ 5.153 + 2k\pi \end{cases} \quad \text{k any integer}$

Chapter 5   Inverse Trigonometric Functions; Trigonometric Equations

62. First solve for x over one period, $0 \le x < 2\pi$. Then add integer multiples of 2π to find all solutions.

$2 \cos 2x = 7 \cos x$

$$2(2 \cos^2 x - 1) = 7 \cos x$$

$$4 \cos^2 x - 2 = 7 \cos x$$

$$4 \cos^2 x - 7 \cos x - 2 = 0$$

$$(4 \cos x + 1)(\cos x - 2) = 0$$

| $4 \cos x + 1 = 0$ | $0 \le x < 2\pi$ | $\cos x - 2 = 0$ |

$$4 \cos x = -1 \qquad\qquad \cos x = 2$$

$$\cos x = -\frac{1}{4} \qquad\qquad \text{No solution}$$

$$x = \cos^{-1}\left(-\frac{1}{4}\right) \text{ or } \pi + \cos^{-1}\left(\frac{1}{4}\right)$$

$$x = 1.823 \text{ or } 4.460$$

Thus, all solutions over one period, $0 \le x < 2\pi$, are x = 1.823, 4.460.

Thus, the solutions, if x is allowed to range over all real numbers, are $x = \begin{cases} 1.823 + 2k\pi \\ 4.460 + 2k\pi \end{cases}$

63. It should be clear from the figure that s, the length of arc, is given by $s = r|\,\theta\,|$, since θ is an angle that is in standard position, but negative.

Since $\cos \theta = \dfrac{-1.4}{r} = \dfrac{-1.4}{5}$

and $\sin \theta = \dfrac{-4.8}{r} = \dfrac{-4.8}{5}$,

θ is given by $-\cos^{-1}\left(\dfrac{-1.4}{5}\right)$.

Thus, $s = 5\left|-\cos^{-1}\left(\dfrac{-1.4}{5}\right)\right| = 5 \cos^{-1}\left(\dfrac{-1.4}{5}\right) = 9.27$ units.

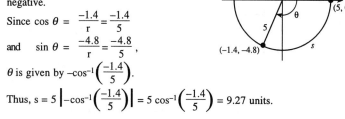

64. Draw a reference triangle and label what we know.

Since $\cos \theta = a = \dfrac{a}{1}$, we can write R = 1.

Use the Pythagorean theorem to find b.

$$a^2 + b^2 = 1$$

$$b^2 = 1 - a^2$$

$$b = -\sqrt{1 - a^2}$$

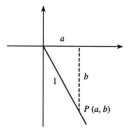

b is negative since P(a, b) is in quadrant IV.

Chapter 5   Inverse Trigonometric Functions; Trigonometric Equations

We can now find the other five functions.

$$\sin \theta = \frac{b}{R} = \frac{-\sqrt{1-a^2}}{1} = -\sqrt{1-a^2} \qquad \cot \theta = \frac{a}{b} = \frac{a}{-\sqrt{1-a^2}} = -\frac{a}{\sqrt{1-a^2}}$$

$$\tan \theta = \frac{b}{a} = \frac{-\sqrt{1-a^2}}{a} \qquad\qquad \sec \theta = \frac{R}{a} = \frac{1}{a}$$

$$\csc \theta = \frac{R}{b} = \frac{1}{-\sqrt{1-a^2}} = -\frac{1}{\sqrt{1-a^2}}$$

65. We first find the period and phase shift by solving

$$4x + \pi = 0 \qquad \text{and} \qquad 4x + \pi = 2\pi$$
$$x = -\frac{\pi}{4} \qquad\qquad x = -\frac{\pi}{4} + \frac{\pi}{2}$$

Period $= \dfrac{\pi}{2}$ Phase shift $= -\dfrac{\pi}{4}$

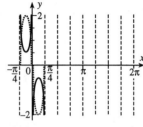

Now, since $0.05 \csc(4x + \pi) = \dfrac{1}{2 \sin(4x + \pi)}$,

we graph $y = 2 \sin(4x + \pi)$ for one cycle from

$-\dfrac{\pi}{4}$ to $-\dfrac{\pi}{4} + \dfrac{\pi}{2} = \dfrac{\pi}{4}$ with a broken line graph,

then take reciprocals. We also place vertical asymptotes through the x intercepts of the sine graph to guide us when we sketch the cosecant function.

We then extend the one cycle over the required interval from 0 to 2π, deleting the portion from

$-\dfrac{\pi}{4}$ to 0, since that was not required.

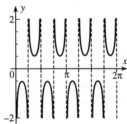

66. $\tan 3x = \tan(x + 2x)$ Algebra

$$= \frac{\tan x + \tan 2x}{1 - \tan x \tan 2x} \qquad\qquad \text{Sum identity for tangent}$$

$$= \frac{\tan x + \dfrac{2 \tan x}{1 - \tan^2 x}}{1 - \tan x \cdot \dfrac{2 \tan x}{1 - \tan^2 x}} \qquad \text{Double-angle identity}$$

$$= \frac{(1 - \tan^2 x)\tan x + 2 \tan x}{(1 - \tan^2 x) - \tan x \cdot 2 \tan x} \qquad \text{Algebra}$$

$$= \frac{\tan x - \tan^3 x + 2 \tan x}{1 - \tan^2 x - 2 \tan^2 x} \qquad \text{Algebra}$$

$$= \frac{3 \tan x - \tan^3 x}{1 - 3 \tan^2 x} \qquad\qquad \text{Algebra}$$

$$= \tan x \frac{3 - \tan^2 x}{1 - 3\tan^2 x} \qquad\qquad \text{Algebra}$$

$$= \tan x \frac{3 - \dfrac{\sin^2 x}{\cos^2 x}}{1 - 3\dfrac{\sin^2 x}{\cos^2 x}} \qquad\qquad \text{Quotient identity}$$

$$= \tan x \frac{3\cos^2 x - \sin^2 x}{\cos^2 x - 3\sin^2 x} \qquad\qquad \text{Algebra}$$

$$= \tan x \frac{3\left(\dfrac{2\cos^2 x - 1 + 1}{2}\right) - \left(\dfrac{1 - (1 - 2\sin^2 x)}{2}\right)}{\dfrac{2\cos^2 x - 1 + 1}{2} - 3\left(\dfrac{1 - (1 - 2\sin^2 x)}{2}\right)} \qquad \text{Algebra}$$

$$= \tan x \frac{3\left(\dfrac{\cos 2x + 1}{2}\right) - \dfrac{1 - \cos 2x}{2}}{\dfrac{\cos 2x + 1}{2} - 3\left(\dfrac{1 - \cos 2x}{2}\right)} \qquad \text{Double-angle identities}$$

$$= \tan x \frac{\dfrac{3}{2}\cos 2x + \dfrac{3}{2} - \dfrac{1}{2} + \dfrac{1}{2}\cos 2x}{\dfrac{1}{2}\cos 2x + \dfrac{1}{2} - \dfrac{3}{2} + \dfrac{3}{2}\cos 2x} \qquad \text{Algebra}$$

$$= \tan x \frac{2\cos 2x + 1}{2\cos 2x - 1} \qquad\qquad \text{Algebra}$$

67. Let $y = \sin^{-1}\dfrac{1}{3}$. Then we are asked to evaluate $\cos(2y)$ which is $1 - 2\sin^2 y$ from the double-angle

identity. Since $\sin y = \sin\left(\sin^{-1}\dfrac{1}{3}\right) = \dfrac{1}{3}$ from the sine-inverse sine identity, we can write

$$\cos\left(2\sin^{-1}\frac{1}{3}\right) = \cos 2y = 1 - 2\sin^2 y = 1 - 2\left(\frac{1}{3}\right)^2 = 1 - 2\cdot\frac{1}{9} = 1 - \frac{2}{9} = \frac{7}{9}$$

68. Let $u = \cos^{-1} x$ and $v = \tan^{-1} x$　　$-1 \le x \le 1$　　　or, equivalently,

$x = \cos u$　　$0 \le u \le \pi$　　and　　$x = \tan v$　　$-\dfrac{\pi}{4} \le v \le \dfrac{\pi}{4}$

Then, $\sin(\cos^{-1} x - \tan^{-1} x) = \sin(u - v) = \sin u \cos v - \cos u \sin v$

For u, geometrically, we have

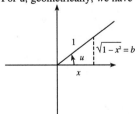 or

In either case, $b = \sqrt{1 - x^2}$ $\sin u = \dfrac{b}{R} = \sqrt{1 - x^2}$ $\cos u = x$

For v, geometrically, we have

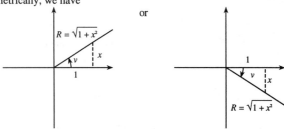

In either case, $R = \sqrt{1 + x^2}$; $\sin v = \dfrac{x}{\sqrt{1 + x^2}}$; $\cos v = \dfrac{1}{\sqrt{1 + x^2}}$

Thus, $\sin(\cos^{-1} x - \tan^{-1} x) = \sin(u - v) = \sin u \cos v - \cos u \sin v$

$$= \sqrt{1 - x^2} \cdot \dfrac{1}{\sqrt{1 + x^2}} - x \cdot \dfrac{x}{\sqrt{1 + x^2}} = \dfrac{\sqrt{1 - x^2} - x^2}{\sqrt{1 + x^2}}$$

69. First solve for x over one period, $0 \le x < 2\pi$. Then add integer multiples of 2π to find all solutions.

$$\sin x = 1 + \cos x$$
$$\pm\sqrt{1 - \cos^2 x} = 1 + \cos x$$
$$1 - \cos^2 x = (1 + \cos x)^2 \qquad \text{Squaring both sides}$$
$$1 - \cos^2 x = 1 + 2 \cos x + \cos^2 x$$
$$0 = 2 \cos x + 2 \cos^2 x$$
$$0 = 2 \cos x(1 + \cos x)$$

$2 \cos x = 0$	$1 + \cos x = 0$
$\cos x = 0$	$\cos x = -1$
$x = \dfrac{\pi}{2}, \dfrac{3\pi}{2}$	$x = \pi$

In squaring both sides, we may have introduced extraneous solutions; hence, it is necessary to check solutions of these equations in the original equation.

Check: $x = \dfrac{\pi}{2}$ $x = \dfrac{3\pi}{2}$ $x = \pi$

$\sin x = 1 + \cos x$ $\sin x = 1 + \cos x$ $\sin \pi = 1 + \cos x$

$\sin \dfrac{\pi}{2} = 1 + \cos \dfrac{\pi}{2}$ $\sin \dfrac{3\pi}{2} = 1 + \cos \dfrac{3\pi}{2}$ $\sin \pi = 1 + \cos \pi$

$\quad 1 = 1 + 0$ $-1 \ne 1 + 0$ $0 = 1 + (-1)$

$\quad$ A solution   Not a solution   A solution

Thus, all solutions over one period, $0 \le x < 2\pi$, are $x = \dfrac{\pi}{2}$, π.

Thus, the solutions, if x is allowed to range over all real numbers, are

$$x = \begin{cases} \dfrac{\pi}{2} + 2k\pi \\ \pi + 2k\pi \end{cases} \qquad \text{k any integer}$$

70. $\sin x = \cos^2 x \qquad 0 \le x \le 2\pi$

$$\sin x = 1 - \sin^2 x$$

$$\sin^2 x + \sin x - 1 = 0 \qquad\qquad \text{Quadratic in sin x}$$

$$\sin x = \frac{-1 \pm \sqrt{1^2 - 4(1)(-1)}}{2 \cdot 1} = \frac{-1 \pm \sqrt{5}}{2}$$

$\sin x = -1.618 \qquad \sin x = 0.618$

$$x = \begin{cases} \sin^{-1} 0.618 \\ \pi - \sin^{-1} 0.618 \end{cases} = \begin{cases} 0.6662 \\ 2.475 \end{cases}$$

71. $M = -2.4$ and $N = 3.2$

Locate $P(M, N) = P(-2.4, 3.2)$ to determine C.

$R = \sqrt{(-2.4)^2 + (3.2)^2} = 4$

$\sin C = \dfrac{3.2}{4} = 0.8 \qquad \cos C = \dfrac{-2.4}{4} = -0.6$

Find the reference angle α. Then, $C = \pi - \alpha$

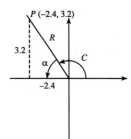

$\sin \alpha = 0.8$

$\alpha = \sin^{-1} 0.8 \approx 0.927$

Thus, $C = \pi - 0.927 \approx 2.21$. $|C|$ is minimum for this choice of C. We can now write

$$y = -2.4 \sin \frac{t}{2} + 3.2 \sin \frac{t}{2} = 4 \sin\left(\frac{t}{2} + 2.21\right).$$

Amplitude $= |4| = 4$

Period and Phase Shift

$\dfrac{t}{2} + 2.21 = 0 \qquad\qquad \dfrac{t}{2} + 2.21 = 2\pi$

$\qquad \dfrac{t}{2} = -2.21 \qquad\qquad\qquad \dfrac{t}{2} = -2.21 + 2\pi$

$\qquad t = -4.42 \qquad\qquad\qquad\quad t = -4.42 + 4\pi$

Period $= 4\pi \qquad$ Phase Shift $= -4.42 \qquad$ Frequency $= \dfrac{1}{\text{Period}} = \dfrac{1}{4\pi}$

Chapter 5   Inverse Trigonometric Functions; Trigonometric Equations

72.   The graph of f(x) is shown in the figure.

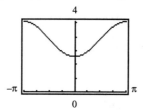

The graph appears to be an upside down cosine curve with

period 2π, amplitude $= \dfrac{1}{2}$ (y max − y min) $= \dfrac{1}{2}(4 - 2) = 1$,

displaced upward by k = 3 units.

It appears that g(x) = 3 − cos x would be an appropriate choice.

We verify f(x) = g(x) as follows:

$$f(x) = \frac{\sin^2 x}{1 - \cos x} + \frac{2 \tan^2 x \cos^2 x}{1 + \cos x}$$

$$= \frac{1 - \cos^2 x}{1 - \cos x} + \frac{2 \dfrac{\sin^2 x}{\cos^2 x} \cos^2 x}{1 + \cos x} \qquad \text{Pythagorean and quotient identities}$$

$$= \frac{1 - \cos^2 x}{1 - \cos x} + \frac{2 \sin^2 x}{1 + \cos x} \qquad \text{Algebra}$$

$$= \frac{1 - \cos^2 x}{1 - \cos x} + \frac{2(1 - \cos^2 x)}{1 + \cos x} \qquad \text{Pythagorean identity}$$

$$= \frac{(1 - \cos x)(1 + \cos x)}{1 - \cos x} + \frac{2(1 - \cos x)(1 + \cos x)}{1 + \cos x} \qquad \text{Algebra}$$

$$= 1 + \cos x + 2(1 - \cos x) \qquad \text{Algebra}$$

$$= 1 + \cos x + 2 - 2 \cos x \qquad \text{Algebra}$$

$$= 3 - \cos x = g(x) \qquad \text{Algebra}$$

73.   The graph of f(x) is shown in the figure.

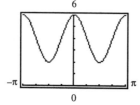

The graph appears to be a basic cosine curve with period π,

amplitude $= \dfrac{1}{2}$ (y max − y min) $= \dfrac{1}{2}(6 - 2) = 2$, displaced upward by k = 4 units.

It appears that $g(x) = 4 + 2 \cos 2x$ would be an appropriate choice. We verify $f(x) = g(x)$ as follows:

$$
\begin{aligned}
f(x) &= 2 \sin^2 x + 6 \cos^2 x \\
&= 2(1 - \cos^2 x) + 6 \cos^2 x \qquad && \text{Pythagorean identity} \\
&= 2 - 2 \cos^2 x + 6 \cos^2 x \qquad && \text{Algebra} \\
&= 2 + 4 \cos^2 x \qquad && \text{Algebra} \\
&= 2 + 2(2 \cos^2 x - 1) + 2 \qquad && \text{Algebra} \\
&= 4 + 2(2 \cos^2 x - 1) \qquad && \text{Algebra} \\
&= 4 + 2 \cos 2x = g(x) \qquad && \text{Double-angle identity}
\end{aligned}
$$

74. The graph of $f(x)$ is shown in the figure.

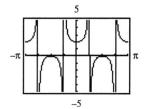

The graph appears to have vertical asymptotes $x = -\dfrac{3\pi}{4}$, $-\dfrac{\pi}{4}$, $\dfrac{\pi}{4}$, and $\dfrac{3\pi}{4}$ and period π. It appears to have high and low points with y coordinates 0 and 2, respectively.

It appears that $g(x) = 1 + \sec 2x$ would be an appropriate choice. We verify $f(x) = g(x)$ as follows:

$$
f(x) = \frac{2 - 2 \sin^2 x}{2 \cos^2 x - 1}
$$

$$
= \frac{1 + 1 - 2 \sin^2 x}{2 \cos^2 x - 1} \qquad \text{Algebra}
$$

$$
= \frac{1 + \cos 2x}{\cos 2x} \qquad \text{Double-angle identity}
$$

$$
= \frac{1}{\cos 2x} + \frac{\cos 2x}{\cos 2x} \qquad \text{Algebra}
$$

$$
= \sec 2x + 1 = g(x) \qquad \text{Quotient identity, algebra}
$$

75. The graph $f(x)$ is shown in the figure.

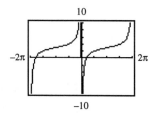

The graph appears to have vertical asymptotes $x = -2\pi$, $x = 0$, and $x = 2\pi$, and period 2π. Its x intercepts are difficult to determine, but since there appears to be symmetry with respect to the points where the curve crosses the line $y = 3$, it appears to be an upside down cotangent curve displaced upward by $k = 3$ units.

It appears that $g(x) = 3 - \cot \dfrac{x}{2}$ would be an appropriate choice. We verify $f(x) = g(x)$ as follows:

$$f(x) = \frac{3 \cos x + \sin x - 3}{\cos x - 1}$$

$$= \frac{3(\cos x - 1) + \sin x}{\cos x - 1} \qquad \text{Algebra}$$

$$= \frac{3(\cos x - 1)}{\cos x - 1} - \frac{\sin x}{1 - \cos x} \qquad \text{Algebra}$$

$$= 3 - \frac{\sin x}{1 - \cos x} \qquad \text{Algebra}$$

$$= 3 - 1 \div \frac{1 - \cos x}{\sin x} \qquad \text{Algebra}$$

$$= 3 - 1 \div \tan \frac{x}{2} \qquad \text{Half-angle identity}$$

$$= 3 - \cot \frac{x}{2} = g(x) \qquad \text{Reciprocal identity}$$

76.

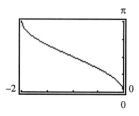

77.

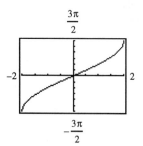

78.

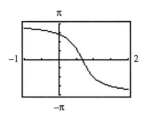

79. The solutions of the equation tan x = 3 are the x coordinates of the points of intersection of the graphs of y1 = tan x and y2 = 3 in the following graphing calculator figure.

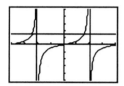

Examining the graph, we see that there are two points of intersection on the indicated interval. Using the zoom and trace procedures of the calculator, we obtain the following approximations to the x coordinates of these intersection points: $x = \begin{cases} -1.893 \\ 1.249 \end{cases}$

80. From the figure, we see that y1 = cos x and y2 = $\sqrt{x}$ intersect only once on the interval $0 \le x \le 2\pi$.

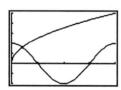

Using zoom and trace procedures, the solution is found to be 0.6417. Since | cos x | ≤ 1, while $\sqrt{x} > 1$ for real x not shown, there can be no other solutions.

81. From the figure, we see that y1 = cos $\frac{x}{2}$ − 2 sin x and y2 = 1 intersect five times on the indicated interval.

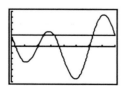

Using zoom and trace procedures, the solutions are found to be 0, 3.895, 5.121, 9.834, and 12.566.

82. The graph of $y = -2.4 \sin \dfrac{t}{2} + 3.2 \cos \dfrac{t}{2}$ is shown in the figure.

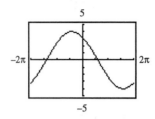

We use the zoom feature or the built-in approximation

routine to locate the t intercepts in this interval at

$t = -4.43$ and $t = 1.85$. The phase shift for

$y = -2.4 \sin \dfrac{t}{2} + 3.2 \cos \dfrac{t}{2}$ is -4.43 (compare to -4.42

from Problem 71).

83. We use the figure and reason as follows: Since the

cities have the same longitude, θ is given by their

difference in latitude.

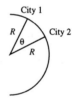

$$\theta = 40°26' - 37°32' = 2°54' = \left(2 + \dfrac{54}{60}\right)^{\circ}$$

Since $\dfrac{s}{C} = \dfrac{\theta}{360°}$ and $C = 2\pi R$, then $\dfrac{s}{2\pi R} = \dfrac{\theta}{360°}$

$s = 2\pi R \cdot \dfrac{\theta}{360°}$

$$s \approx 2(3.14)(3960)\dfrac{2 + \dfrac{54}{60}}{360} \approx 200 \text{ miles}$$

84. Labelling the diagram, we note: In triangle ACD, $\cot 32° = \dfrac{1000 + x}{h}$

In triangle BCD, $\cot 48° = \dfrac{x}{h}$

We solve the system of equations

$\cot 32° = \dfrac{1000 + x}{h} \qquad \cot 48° = \dfrac{x}{h}$

by clearing of fractions, then eliminating x.

$h \cot 32° = 1000 + x \qquad h \cot 48° = x$

$h \cot 32° = 1000 + h \cot 48°$

$h \cot 32° - h \cot 48° = 1000$

$h(\cot 32° - \cot 48°) = 1000$

$h = \dfrac{1000}{\cot 32° - \cot 48°} = 1429 \text{ meters}$

85. $\sqrt{u^2 - a^2} = \sqrt{(a\csc x)^2 - a^2}$ Using the given substitution

       $= \sqrt{a^2\csc^2 x - a^2}$ Algebra

       $= \sqrt{a^2(\csc^2 x - 1)}$ Algebra

       $= \sqrt{a^2\cot^2 x}$ Pythagorean identity

       $= |a|\,|\cot x|$ Algebra

       $= a\cot x$ Since $a > 0$ and x is in quadrant I

                              $\left(\text{given } 0 < x < \dfrac{\pi}{2}\right)$, thus, $\cot x > 0$.

86.   We note that $\tan\theta = \dfrac{2}{x}$ and $\tan 2\theta = \dfrac{2+4}{x} = \dfrac{6}{x}$ (see figure).

     Then, $\tan 2\theta = \dfrac{2\tan\theta}{1 - \tan^2\theta}$

$$\frac{6}{x} = \frac{2\left(\frac{2}{x}\right)}{1 - \left(\frac{2}{x}\right)^2} = \frac{\frac{4}{x}}{1 - \left(\frac{2}{x}\right)^2} = \frac{x^2 \cdot \frac{4}{x}}{x^2 \cdot 1 - x^2 \cdot \frac{4}{x^2}} = \frac{4x}{x^2 - 4}$$

$$x(x^2 - 4)\cdot\frac{6}{x} = x(x^2 - 4)\cdot\frac{4x}{x^2 - 4}$$

$$6(x^2 - 4) = x\cdot 4x$$

$$6x^2 - 24 = 4x^2$$

$$2x^2 - 24 = 0$$

$$x^2 = 12$$

$$x = \sqrt{12} \text{ (we discard the negative solution)}$$

$$= 2\sqrt{3}$$

Since $\tan\theta = \dfrac{2}{x} = \dfrac{2}{2\sqrt{3}} = \dfrac{1}{\sqrt{3}}$, $\theta = 30°$

87.   We are to find the smallest positive solution to $0 = -7.2\sin 5t - 9.6\cos 5t$

     $7.2\sin 5t = -9.6\cos 5t$

$$\frac{7.2\sin 5t}{7.2\cos 5t} = \frac{-9.6\cos 5t}{7.2\cos 5t}$$

$$\tan 5t = -\frac{9.6}{7.2}$$

$$5t = \tan^{-1}\left(-\frac{9.6}{7.2}\right) + \pi$$

$$t = \frac{1}{5}\left[\tan^{-1}\left(-\frac{9.6}{7.2} + \pi\right)\right] = 0.443 \text{ sec}$$

88.   We graph y1 = –7.2 sin 5t – 9.6 cos 5t and y2 = 5. From the figure, we see that y1 and y2
      intersect twice   on the indicated interval.

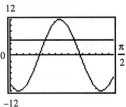

Using zoom and trace procedures, the intersections are found to be t = 0.529 sec and t = 0.985 sec.

89.   (A) M = –7.2 and N = –9.6

Locate P(M, N) = P(–7.2, –9.6) to determine C.

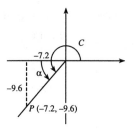

$$R = \sqrt{(-7.2)^2 + (-9.6)^2} = 12$$

$$\sin C = \frac{-9.6}{12} = -0.8 \qquad \tan C = \frac{-9.6}{-7.2} = \frac{4}{3}$$

Find the reference angle α. Then, C = π + α.

$$\tan \alpha = \frac{4}{3}, \; \alpha = \tan^{-1}\left(\frac{4}{3}\right) = 0.9273$$

Thus, C ≈ 4.069. We can now write y = –7.2 sin 5t – 9.6 cos 5t = 12 sin(5t + 4.069)

(B) We are to solve –2 = 12 sin(5t + 4.069) for $0 \le t \le \frac{\pi}{2}$, that is $0 \le 5t \le \frac{5\pi}{2}$

$$\sin(5t + 4.069) = -\frac{1}{6}$$

$$5t + 4.069 = \begin{cases} \pi + \sin^{-1}\left(\frac{1}{6}\right) + 2\pi \\[2mm] 2\pi - \sin^{-1}\left(\frac{1}{6}\right) \end{cases}$$

$$5t = \begin{cases} 3\pi + \sin^{-1}\frac{1}{6} - 4.069 \\[2mm] 2\pi - \sin^{-1}\frac{1}{6} - 4.069 \end{cases}$$

$$t = \begin{cases} \dfrac{3\pi + \sin^{-1}\frac{1}{6} - 4.069}{5} \\[5mm] \dfrac{2\pi - \sin^{-1}\frac{1}{6} - 4.069}{5} \end{cases} = \begin{cases} 1.105 \text{ sec} \\[2mm] 0.409 \text{ sec} \end{cases}$$

90.   Amplitude $| A | = | 60 | = 60$

Phase Shift and Period: Solve

$$Bx + C = 0 \quad \text{and} \quad Bx + C = 2\pi$$

$$90\pi t - \frac{\pi}{2} = 0 \qquad\qquad 90\pi t - \frac{\pi}{2} = 2\pi$$

$$90\pi t = \frac{\pi}{2} \qquad\qquad 90\pi t = 2\pi + \frac{\pi}{2}$$

$$t = \frac{1}{180} \qquad\qquad t = \frac{1}{45} + \frac{1}{180}$$

$$\text{Period} = \frac{1}{45} \quad \text{Phase Shift} = \frac{1}{180} \qquad \text{Frequency} = \frac{1}{\text{Period}} = \frac{1}{\frac{1}{45}} = 45 \text{ Hz}$$

Graph one cycle over the interval from $\frac{1}{180}$ Extend the graph from 0 to $\frac{4}{45}$.

to $\left(\frac{1}{180} + \frac{1}{45} \right) = \frac{1}{36}$.

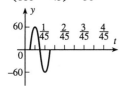   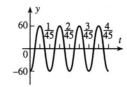

91.   We are to solve $35 = 60 \sin\left(90\pi t - \frac{\pi}{2} \right)$

$$\sin\left(90\pi t - \frac{\pi}{2} \right) = \frac{35}{60}$$

$$90\pi t - \frac{\pi}{2} = \sin^{-1}\frac{35}{60}$$

$$90\pi t = \frac{\pi}{2} + \sin^{-1}\frac{35}{60}$$

$$t = \frac{1}{90\pi}\left(\frac{\pi}{2} + \sin^{-1}\frac{35}{60} \right) \approx 0.007758 \text{ sec}$$

92.(A)   From the figure, it should be clear that $\cot \theta = \frac{d}{200}$.

Thus, $d = 200 \cot \theta$

(B)   Period $= \pi$

One period of the graph would therefore extend from

0 to π, with vertical asymptotes at $t = 0$ and $t = \pi$.

We sketch half of one period, since the required

interval is from 0 to $\frac{\pi}{2}$ only. Ordinates can be

determined from a calculator, thus:

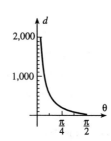

θ	$\dfrac{\pi}{20}$	$\dfrac{\pi}{10}$	$\dfrac{3\pi}{20}$	$\dfrac{\pi}{5}$	$\dfrac{\pi}{4}$	$\dfrac{3\pi}{10}$	$\dfrac{7\pi}{20}$	$\dfrac{2\pi}{5}$	$\dfrac{9\pi}{20}$	$\dfrac{\pi}{2}$
$200\cot\theta$	1263	616	393	275	200	145	101	65	32	0

93.(A)   Labelling the diagram as shown, we note

In triangle ABC, $\tan\alpha = \dfrac{100}{x}$

In triangle ABD, $\tan(\theta + \alpha) = \dfrac{200}{x}$

Hence,

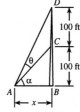

$$\tan\theta = \tan(\theta + \alpha - \alpha) = \frac{\tan(\theta + \alpha) - \tan\alpha}{1 + \tan(\theta + \alpha)\tan\alpha} = \frac{\dfrac{200}{x} - \dfrac{100}{x}}{1 + \dfrac{200}{x}\cdot\dfrac{100}{x}}$$

$$= \frac{x^2\cdot\dfrac{200}{x} - x^2\cdot\dfrac{100}{x}}{x^2 + x^2\cdot\dfrac{200}{x}\cdot\dfrac{100}{x}} = \frac{200x - 100x}{x^2 + 20,000} = \frac{100x}{x^2 + 20,000}$$

$$\theta = \arctan\frac{100x}{x^2 + 20,000}$$

(B)   We are given x = 50 feet. Thus, $\theta = \arctan\dfrac{100\cdot 50}{50^2 + 20,000} = 12.5°$

94.   We graph $y1 = \tan^{-1}\dfrac{100x}{x^2 + 20,000}$ and y2 = 15 on the interval from 0 to 400. We see that the

curves intersect twice on the interval.

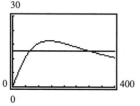

Using zoom and trace procedures, the solutions are found to be x = 64.9 ft and x = 308.3 ft.

95.   Since the sprockets are connected by the bicycle chain, the distance (arc length) that the rear sprocket turns is equal to the distance that the pedal sprocket turns. Let R_1 = radius of rear sprocket and R_2 = radius of pedal sprocket. $s = R_1\theta_1$ $s = R_2\theta_2$ $R_1\theta_1 = R_2\theta_2$ $\theta_1 = \dfrac{R_2}{R_1}\theta_2$

Note that the angle through which the rear wheel turns is equal to the angle through which the rear sprocket turns. Thus, $\theta_1 = \dfrac{11.0}{4.0}\,18\pi = 49.5\pi$ rad

96. We use the formula $\theta_1 = \dfrac{R_2}{R_1} \theta_2$ from the previous problem, and note:

$\omega_1 = \dfrac{\theta_1}{t}$ angular velocity of rear wheel and sprocket

$\omega_2 = \dfrac{\theta_2}{t}$ angular velocity of pedal sprocket.

Hence, $\dfrac{\theta_1}{t} = \dfrac{R_2}{R_1}\dfrac{\theta_2}{t}$, $\omega_1 = \dfrac{R_2}{R_1}\omega_2 = \dfrac{11.0}{4.0}$ 60.0 rpm = 165 rpm or $165 \cdot 2\pi$ rad/min

Then, the linear velocity of the rear wheel (i.e., that of the bicycle), v, is given by

$$v = R_{wheel}\omega_1 = \frac{70.0 \text{ cm}}{2} \cdot 165 \cdot 2\pi \text{ rad/min} = 11{,}550\pi \text{ cm/min} \approx 36{,}300 \text{ cm/min}$$

97. In Exercise 5.1, Problem 93, a precisely analogous situation was analyzed. We omit a second derivation for reasons of space. The student can confirm that L, the length of the bicycle chain, is given by the formula derived there, with D replaced by $2R_1$, d replaced by $2R_2$, and C replaced by d. Thus, $L = \pi 2R_1 + (2R_2 - 2R_1)\theta + 2d \sin \theta$, with $\theta = \cos^{-1}\dfrac{2R_1 - 2R_2}{2d}$. Hence,

$$\theta = \cos^{-1}\frac{2 \cdot 11.0 - 2 \cdot 4.0}{2 \cdot 50.0} = 1.430 \text{ rad}$$

$$L = \pi \cdot 2 \cdot 11.0 + (2 \cdot 4.0 - 2 \cdot 11.0)\theta + 2 \cdot 50.0 \sin \theta$$

$$= 22.0\pi - 14.0\theta + 100.0 \sin \theta = 148 \text{ cm}$$

98. We use the formula from the previous problem,

with L = 130 cm, R_1 = 11.0 cm, R_2 = 4.0 cm.

Then,

$130 = \pi \cdot 2 \cdot 11.0 - (2 \cdot 4.0 - 2 \cdot 11.0)\theta + 2d \sin \theta$,

with $\theta = \cos^{-1}\dfrac{2 \cdot 11.0 - 2 \cdot 4.0}{2d}$.

Hence,

$130 = 22.0\pi - 14.0 \cos^{-1}\dfrac{14.0}{2d} + 2d \sin\left(\cos^{-1}\dfrac{14.0}{2d}\right)$

We graph $y1 = 22.0\pi - 14.0 \cos^{-1}\dfrac{7.0}{d} + 2d \sin\left(\cos^{-1}\dfrac{7.0}{d}\right)$ and $y2 = 130$ on the interval from 20 to 60. We see that the curves intersect once on the interval.

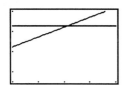

Using zoom and trace procedures, the solution is found to be d = 40.8 cm.

99. (A)

x (months)	1, 13	2, 14	3, 15	4, 16	5, 17	6, 18	7, 19	8, 20	9, 21
y $\left(\dfrac{\text{daylight}}{\text{duration}}\right)$	6.52	9.17	11.83	14.60	17.55	19.27	18.45	15.85	12.95

x (months)	10, 22	11, 23	12, 24
y $\left(\dfrac{\text{daylight}}{\text{duration}}\right)$	10.15	7.32	5.60

20

1 ⋯⋯⋯ 24

6

(B) From the table, Max y = 19.27 and Min y = 5.60.

Then,

$$A = \frac{\text{Max } y - \text{Min } y}{2} = \frac{19.27 - 5.60}{2} = 6.835$$

$$B = \frac{2\pi}{\text{Period}} = \frac{2\pi}{12} = \frac{\pi}{6}$$

$$k = \text{Min } y + A = 5.60 + 6.835 = 12.435$$

From the plot in (A) or the table, we estimate the smallest value of x for which y = k = 12.435 to be approximately 3.1. Then, this is the phase-shift for the graph. Substitute $B = \dfrac{\pi}{6}$ and x = 3.1 into the phase-shift equation $x = -\dfrac{C}{B}$, $3.1 = \dfrac{-C}{\dfrac{\pi}{6}}$, $C = \dfrac{-3.1\pi}{6} \approx -1.6$.

Thus, the equation required is $y = 12.435 + 6.835 \sin\left(\dfrac{\pi x}{6} - 1.6\right)$.

(C)

20

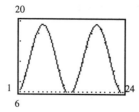

1 ⋯⋯⋯ 24

6

CHAPTER 6   Additional Triangle Topics;   Vectors

EXERCISE 6.1   Law of Sines

1. *Solve for γ:* $\alpha + \beta + \gamma = 180°$

   $\gamma = 180° - (41° + 33°) = 106°$

   Solve for a: $\dfrac{\sin \alpha}{a} = \dfrac{\sin \gamma}{c}$

   $a = \sin \alpha \dfrac{c}{\sin \gamma} = (\sin 41°)\dfrac{21 \text{ mi}}{\sin 106°}$

   $= 14 \text{ mi}$

   Solve for b: $\dfrac{\sin \beta}{b} = \dfrac{\sin \gamma}{c}$

   $b = \sin \beta \dfrac{c}{\sin \gamma} = (\sin 33°)\dfrac{21 \text{ mi}}{\sin 106°}$

   $= 12 \text{ mi}$

3. *Solve for α:* $\alpha + \beta + \gamma = 180°$

   $\alpha = 180° - (36° + 43°) = 101°$

   Solve for b: $\dfrac{\sin \alpha}{a} = \dfrac{\sin \beta}{b}$

   $b = \sin \beta \dfrac{a}{\sin \alpha} = (\sin 43°)\dfrac{92 \text{ cm}}{\sin 101°}$

   $= 64 \text{ cm}$

   Solve for c: $\dfrac{\sin \alpha}{a} = \dfrac{\sin \gamma}{c}$

   $c = \sin \gamma \dfrac{a}{\sin \alpha} = (\sin 36°)\dfrac{92 \text{ cm}}{\sin 101°}$

   $= 55 \text{ cm}$

5. *Solve for α:* $\alpha + \beta + \gamma = 180°$

   $\alpha = 180° - (27.5° + 54.5°) = 98.0°$

   Solve for b: $\dfrac{\sin \alpha}{a} = \dfrac{\sin \beta}{b}$

   $b = \sin \beta \, \backslash F(a,\sin \alpha) = (\sin 27.5°)$

   $\dfrac{9.27 \text{ mm}}{\sin 98.0°}$

   $= 4.32 \text{ mm}$

   Solve for c: $\dfrac{\sin \alpha}{a} = \dfrac{\sin \gamma}{c}$

   $c = \sin \gamma \dfrac{a}{\sin \alpha} = (\sin 54.5°)\dfrac{9.27 \text{ mm}}{\sin 98.0°}$

   $= 7.62 \text{ mm}$

Chapter 6   Additional Triangle Topics;   Vectors

7. *Solve for γ:* $\alpha + \beta + \gamma = 180°$

$\gamma = 180° - (122.7° + 34.4°)$

$\quad = 22.9°$

Solve for a: $\dfrac{\sin \alpha}{a} = \dfrac{\sin \beta}{b}$

$a = \sin \alpha \dfrac{b}{\sin \beta} = (\sin 122.7°) \dfrac{18.3 \text{ cm}}{\sin 34.4°}$

$\quad = 27.3 \text{ cm}$

Solve for c: $\dfrac{\sin \beta}{b} = \dfrac{\sin \gamma}{c}$

$c = \sin \gamma \dfrac{b}{\sin \beta} = (\sin 22.9°) \dfrac{18.3 \text{ cm}}{\sin 34.4°}$

$\quad = 12.6 \text{ cm}$

9. *Solve for α:* $\alpha + \beta + \gamma = 180°$
$\alpha = 180° - (100°0' + 12°40')$

$\quad = 67°20'$

Solve for a: $\dfrac{\sin \alpha}{a} = \dfrac{\sin \beta}{b}$

$a = \sin \alpha \dfrac{b}{\sin \beta} = (\sin 67°20') \dfrac{13.1 \text{ km}}{\sin 12°40'}$

$\quad = 55.1 \text{ km}$

Solve for c: $\dfrac{\sin \beta}{b} = \dfrac{\sin \gamma}{c}$

$c = \sin \gamma \dfrac{b}{\sin \beta} = (\sin 100°0') \dfrac{13.1 \text{ km}}{\sin 12°40'}$

$\quad = 58.8 \text{ km}$

11. *Solve for α:* $\dfrac{\sin \alpha}{a} = \dfrac{\sin \beta}{b}$

$\sin \alpha = \dfrac{a \sin \beta}{b} = \dfrac{(8.00 \text{ cm})(\sin 52.0°)}{12.0 \text{ cm}}$

$\quad\quad = 0.5253$

$\alpha = \sin^{-1} 0.5253 = 31.7°$

(There is another solution of $\sin \alpha = 0.5253$ that deserves brief consideration:

$\alpha^l = 180° - \sin^{-1} 0.5253 = 148.3°$. However, there is not enough room in a triangle for an angle of

148.3° and an angle of 52.0°, since their sum is greater than 180°.)

Solve for γ: $\alpha + \beta + \gamma = 180°$

$\gamma = 180° - (31.7° + 52.0°) = 96.3°$

Solve for c: $\dfrac{\sin \beta}{b} = \dfrac{\sin \gamma}{c}$

$c = \dfrac{b \sin \gamma}{\sin \beta} = \dfrac{(12.0 \text{ cm})(\sin 96.3°)}{\sin 52.0°} = 15.1 \text{ cm}$

13. *Solve for β*:
$$\frac{\sin \beta}{b} = \frac{\sin \alpha}{a}$$
$$\sin \beta = \frac{b \sin \alpha}{a} = \frac{(36.4 \text{ mm})(\sin 25.5°)}{15.0 \text{ mm}}$$
$$= 1.045$$

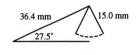

Since sin β = 1.045 has no solution, no triangle exists with

the given measurements. No solution.

15. *Solve for β*:
$$\frac{\sin \alpha}{a} = \frac{\sin \beta}{b}$$
$$\sin \beta = \frac{b \sin \alpha}{a} = \frac{(105.0 \text{ yd})(\sin 18.92°)}{48.35 \text{ yd}}$$
$$= 0.7042$$

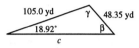

Since β is specified acute, we choose the acute solution of

this equation, $β = \sin^{-1} 0.7042 = 44.76°$

Solve for γ. $\alpha + \beta + \gamma = 180°$

$\gamma = 180° - (18.92° + 44.76°) = 116.3°$

Solve for c:
$$\frac{\sin \alpha}{a} = \frac{\sin \gamma}{c}$$
$$c = \frac{a \sin \gamma}{\sin \alpha} = \frac{(48.35 \text{ yd})(\sin 116.3°)}{\sin 18.92°} = 133.7 \text{ yd}$$

17. Here we are given the same data as in Problem 15, except that β

is specified obtuse.

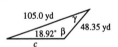

Solve for β:
$$\frac{\sin \alpha}{a} = \frac{\sin \beta}{b}$$
$$\sin \beta = \frac{b \sin \alpha}{a} = \frac{(105.0 \text{ yd})(\sin 18.92°)}{48.35 \text{ yd}}$$
$$= 0.7042$$

Since β is specified obtuse, we choose the obtuse solution of this equation:

$β = 180° - \sin^{-1} 0.7042 = 180° - 44.76° = 135.24°$

Solve for γ. $\alpha + \beta + \gamma = 180°$

$\gamma = 180° - (18.92° + 135.24°) = 25.84°$

Solve for c:
$$\frac{\sin \alpha}{a} = \frac{\sin \gamma}{c}$$
$$c = \frac{a \sin \gamma}{\sin \alpha} = \frac{(48.35 \text{ yd})(\sin 25.84°)}{\sin 18.92°} = 64.99 \text{ yd}$$

19. *Solve for β*:
$$\frac{\sin \beta}{b} = \frac{\sin \alpha}{a}$$
$$\sin \beta = \frac{b \sin \alpha}{a} = \frac{(18.3 \text{ m})(\sin 135°20')}{(14.6 \text{ m})}$$

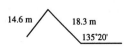

$$= 0.8811$$

$$\beta = \sin^{-1}(0.8811) = 61°50'$$

Since there is not enough room in a triangle for an angle of 135°20′ and an angle of 61°50′ (their

sum is more than 180°), no triangle exists with the given measurements. No solution.

21. β is acute   b > a   Only one triangle is possible.

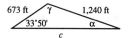

Solve for α: $\dfrac{\sin \alpha}{a} = \dfrac{\sin \beta}{b}$

$$\sin \alpha = \frac{a \sin \beta}{b} = \frac{(673 \text{ ft})(\sin 33°50')}{1,240 \text{ ft}}$$

$$= 0.3022$$

$$\alpha = \sin^{-1} 0.3022 = 17°40'$$

(There is another solution of sin α = 0.3022 that deserves brief consideration:

$\alpha' = 180° - \sin^{-1} 0.3022 = 162°20'$. However, there is not enough room in a triangle for an angle

of 162°20′ and an angle of 33°50′, since their sum is greater than 180°.)

Solve for γ: $\alpha + \beta + \gamma = 180°$

$$\gamma = 180° - (17°40' + 33°50') = 128°30'$$

Solve for c: $\dfrac{\sin \beta}{b} = \dfrac{\sin \gamma}{c}$

$$c = \frac{b \sin \gamma}{\sin \beta} = \frac{(1,240 \text{ ft})(\sin 128°30')}{\sin 33°50'} = 1,740 \text{ ft}$$

23. *Solve for* α: $\dfrac{\sin \alpha}{a} = \dfrac{\sin \beta}{b}$

$$\sin \alpha = \frac{a \sin \beta}{b} = \frac{(244 \text{ ft})(\sin 27.3°)}{135 \text{ ft}} = 0.829$$

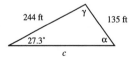   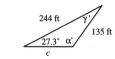

Angle α can be either acute or obtuse.

$$\alpha = \sin^{-1} 0.829 = 56.0° \qquad\qquad\qquad \alpha' = 180° - \sin^{-1} 0.829 = 124°$$

Solve for γ *and* γ':

$$\gamma = 180° - (\alpha + \beta) \qquad\qquad\qquad \gamma' = 180° - (\alpha' + \beta)$$

$$= 180° - (56.0° + 27.3°) = 96.7° \qquad = 180° - (124° + 27.3°) = 28.7°$$

Solve for c and c′:

$$\frac{\sin \alpha}{a} = \frac{\sin \gamma}{c} \qquad\qquad\qquad\qquad \frac{\sin \alpha'}{a} = \frac{\sin \gamma'}{c'}$$

$$c = \frac{a \sin \gamma}{\sin \alpha} = \frac{(244 \text{ ft})(\sin 96.7°)}{\sin 56.0°} \qquad c' = \frac{a \sin \gamma'}{\sin \alpha'} = \frac{(244 \text{ ft})(\sin 28.7°)}{\sin 124°}$$

$$= 292 \text{ ft} \qquad\qquad\qquad\qquad\qquad = 141 \text{ ft}$$

25. Using the given and the calculated data, we have

$$(a - b)\cos\frac{\gamma}{2} = c \sin\frac{\alpha - \beta}{2}$$

$$(14 - 12)\cos\frac{106°}{2} = 21 \sin\frac{41° - 33°}{2}$$

$$1.20 \approx 1.46$$

27. Using the law of sines in the form $\dfrac{a}{\sin\alpha} = \dfrac{b}{\sin\beta}$, we have $b = \dfrac{a \sin\beta}{\sin\alpha}$, which we use as follows:

$$\frac{a - b}{a + b} = \frac{a - \dfrac{a\sin\beta}{\sin\alpha}}{a + \dfrac{a\sin\beta}{\sin\alpha}} \qquad \text{Law of sines as above}$$

$$= \frac{a\sin\alpha - a\sin\beta}{a\sin\alpha + a\sin\beta} \qquad \text{Algebra}$$

$$= \frac{a(\sin\alpha - \sin\beta)}{a(\sin\alpha + \sin\beta)} \qquad \text{Algebra}$$

$$= \frac{\sin\alpha - \sin\beta}{\sin\alpha + \sin\beta} \qquad \text{Algebra}$$

$$= \frac{2\cos\dfrac{\alpha + \beta}{2}\sin\dfrac{\alpha - \beta}{2}}{2\sin\dfrac{\alpha + \beta}{2}\cos\dfrac{\alpha - \beta}{2}} \qquad \text{Sum-product identities}$$

$$= \frac{\dfrac{\sin\dfrac{\alpha - \beta}{2}}{\cos\dfrac{\alpha - \beta}{2}}}{\dfrac{\sin\dfrac{\alpha + \beta}{2}}{\cos\dfrac{\alpha + \beta}{2}}} \qquad \text{Algebra}$$

$$= \frac{\tan\dfrac{\alpha - \beta}{2}}{\tan\dfrac{\alpha + \beta}{2}} \qquad \text{Quotient identities}$$

29. From the diagram, we can see that

k = altitude of any possible triangle.

Thus, $\sin\beta = \dfrac{k}{a}$, $k = a \sin\beta$.

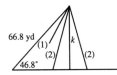

k = (66.8 yd)sin 46.8° = 48.7 yd
If 0 < b < k, there is no solution: (1) in the diagram.

If b = k, there is one solution.

If k < b < a, there are two solutions: (2) in the diagram.

Chapter 6   Additional Triangle Topics; Vectors

31. $\angle BAC + \angle ABC + \angle ACB = 180°$

$\angle ACB = 180° - (\angle BAC + \angle ABC) = 180° - (118.1° + 58.1°) = 3.8°$

Now apply the law of sines to find AC

$$\frac{\sin ABC}{AC} = \frac{\sin ACB}{AB}$$

$$AC = \frac{AB \sin ABC}{\sin ACB} = \frac{1.00 \sin 58.1°}{\sin 3.8°} = 12.8 \text{ mi}$$

33. First draw a figure and label known information.

$\angle ABF + \angle BAF + \angle BFA = 180°$

$\angle BFA = 180° - (\angle ABF + \angle BAF)$

$= 180° - (52.6° + 25.3°) = 102.1°$

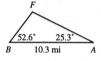

Now apply the law of sines to find AF and BF:

$$\frac{\sin ABF}{AF} = \frac{\sin BFA}{AB} \qquad\qquad \frac{\sin BAF}{BF} = \frac{\sin BFA}{AB}$$

$$AF = \frac{AB \sin ABF}{\sin BFA} \qquad\qquad BF = \frac{AB \sin BAF}{\sin BFA}$$

$$= \frac{(10.3 \text{ mi})(\sin 52.6°)}{\sin 102.1°} = 8.37 \text{ mi} \qquad = \frac{(10.3 \text{ mi})(\sin 25.3°)}{\sin 102.1°} = 4.50 \text{ mi}$$

The fire is 8.37 mi from A, 4.50 mi from B.

35. Label known information in the figure.

$\angle ABC + \angle CAB + \angle ACB = 180°$

$\angle ACB = 180° - (\angle ABC + \angle CAB)$

$= 180° - (19.2° + 118.4°) = 42.4°$

Now apply th

e law of sines to find AB.

$$\frac{\sin ACB}{AB} = \frac{\sin ABC}{AC}$$

$$AB = \frac{AC \sin ACB}{\sin ABC} = \frac{(112 \text{ m})\sin 42.4°}{\sin 19.2°} = 230 \text{ m}$$

37

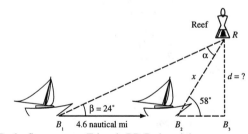

In the figure, note: Triangle RB_2B_3 is a right

286

triangle, hence, $\dfrac{d}{x} = \sin 58°$. Triangle RB_1B_2 is not a right triangle; but, from the law of

sines, $\dfrac{\sin \alpha}{B_1B_2} = \dfrac{\sin \beta}{x}$.

Hence, $x = \dfrac{B_1B_2 \sin \beta}{\sin \alpha}$

$d = x \sin 58° = \dfrac{B_1B_2 \sin \beta \sin 58°}{\sin \alpha}$

Given $B_1B_2 = 4.6$ nautical miles, $\beta = 24°$, we can find α since the exterior angle of a triangle has measure equal to the sum of the two nonadjacent interior angles. Hence, $\alpha + \beta = 58°$, $\alpha = 58° - \beta = 58° - 24° = 34°$.

Hence, $d = \dfrac{(4.6 \text{ naut. mi})(\sin 24°)(\sin 58°)}{\sin 34°} = 2.8$ nautical miles

39. In the figure, note: Triangle ADC is a right triangle, hence $\angle ACB = 90° - 42° = 48°$. Triangle ABC is not a right triangle; but, from the law of sines,

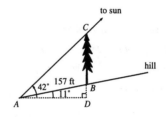

$\dfrac{\sin CAB}{BC} = \dfrac{\sin ACB}{AB}$.

$\angle CAB = \angle CAD - \angle BAD = 42° - 11° = 31°$

$BC = \dfrac{AB \sin CAB}{\sin ACB} = \dfrac{(157 \text{ ft})\sin 31°}{\sin 48°} = 109 \text{ ft}$

41. Labelling the figure as shown, we have, from the law of sines,

$\dfrac{\sin \beta}{b} = \dfrac{\sin \alpha}{a}$

$\sin \beta = \dfrac{b \sin \alpha}{a} = \dfrac{(18.2 \text{ m})(\sin 33.7°)}{11.0 \text{ m}}$

$= 0.9180$

$\beta = 180° - \sin^{-1}(0.9180)$ (since β is obtuse)

$= 113.4°$

Since $\gamma = 180° - (\alpha + \beta) = 180° - (33.7° + 113.4°) = 32.9°$, we have, from the law of sines

$\dfrac{\sin \gamma}{c} = \dfrac{\sin \alpha}{a}$ $c = \dfrac{a \sin \gamma}{\sin \alpha} = \dfrac{(11.0 \text{ m})(\sin 32.9°)}{\sin 33.7°} = 10.8 \text{ m}$

Chapter 6   Additional Triangle Topics;   Vectors

43. Labelling the diagram as shown, we note:

    triangle OAB is isosceles, hence $\alpha = \beta$.

    Thus, $\alpha + \alpha + 63.2° = 180°$

    $$2\alpha = 116.8°$$

    $\alpha = 58.4°$

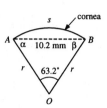

Now apply the law of sines to find r.

$$\frac{\sin \alpha}{r} = \frac{\sin AOB}{AB}$$

$$r = \frac{AB \sin \alpha}{\sin AOB} = \frac{(10.2 \text{ mm})(\sin 58.4°)}{\sin 63.2°} = 9.73 \text{ mm}$$

To find s, we use the formula $s = \frac{\pi}{180°} \theta r$ from Chapter 2, thus

$$s = \frac{\pi}{180°} (63.2°)(9.73 \text{ mm}) = 10.7 \text{ mm}$$

45. Following the hint, we find all angles for triangle ACS first.

    $$\angle SAC = \theta + 90° = 24.9° + 90° = 114.9°$$

    For $\angle ACS$, we use the formula $s = \frac{\pi}{180°} \theta R$ from Chapter 2, with s = 504 miles, $\theta = \angle ACS$,

    R = 3,964 miles. Then,

    $$\angle ACS = \frac{180°s}{\pi R} = \frac{180°(504 \text{ mi})}{\pi(3,964 \text{ mi})} = 7.28°$$

    Since $\angle SAC + \angle ACS + \angle ASC = 180°$, we have $\angle ASC = 180° - (\angle ACS + \angle SAC)$

    $$= 180° - (114.9° + 7.28°) = 57.8°$$

    Now apply the law of sines to find side CS:

    $$\frac{\sin SAC}{CS} = \frac{\sin ASC}{AC}$$

    $$CS = \frac{AC \sin SAC}{\sin ASC} = \frac{(3,964 \text{ mi})\sin 114.9°}{\sin 57.8°} = 4,248 \text{ miles}$$

    Hence, the height of the satellite above B = BS = CS – BC.

    $$BC = 4,248 \text{ mi} - 3,964 \text{ mi} = 284 \text{ miles}$$

47. First, draw a figure. We are given:

    $SE = \frac{1}{2}$ (diameter of earth's orbit) $= \frac{1}{2}$ (2.99 × 10⁸ km)

    $= 1.495 \times 10^8$ km

    $SV = SV' = \frac{1}{2}$ (diameter of Venus' orbit)

    $= \frac{1}{2}$ (2.17 × 10⁸ km) = 1.085 × 10⁸ km

    $\alpha = \angle SEV = 18°40'$

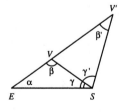

Chapter 6   Additional Triangle Topics;   Vectors

There are two possible triangles; hence, two possible values of the required distance. Call them EV and EV'. The law of sines gives two possible values for angle VSE or V'SE; we denote them by γ and γ' in the figure. We start by calculating angle EVS, or β, and angle EV'S, or β', from the law of sines.

$$\frac{\sin \beta}{SE} = \frac{\sin \alpha}{SV} \qquad \sin \beta = \frac{SE \sin \alpha}{SV} = \frac{(1.495 \times 10^8 \text{ km})(\sin 18°40')}{1.085 \times 10^8 \text{ km}} = 0.4410$$

Hence, the two possibilities are:

β obtuse β' acute

$\beta = 180° - \sin^{-1} 0.4410 = 153°50'$ $\beta' = \sin^{-1} 0.4410 = 26.2° = 26°10'$

Hence, the two possibilities for angle VSE, or γ, become:

$\gamma = 180° - (\beta + \alpha)$ $\gamma' = 180° - (\beta' + \alpha')$

$\quad = 180° - (153°50' + 18°40') = 7°30'$ $\quad = 180° - (26°10' + 18°40') = 135°10'$

Applying the law of sines again to calculate EV and EV' from these two values of γ, we have

$$\frac{\sin \gamma}{EV} = \frac{\sin \alpha}{SV} \qquad\qquad\qquad \frac{\sin \gamma'}{EV'} = \frac{\sin \alpha'}{SV'}$$

$$EV = \frac{SV \sin \gamma}{\sin \alpha} \qquad\qquad\qquad EV' = \frac{SV' \sin \gamma'}{\sin \alpha'}$$

$$= \frac{(1.085 \times 10^8 \text{ km})(\sin 7°30')}{\sin 18°40'} \qquad\qquad = \frac{(1.085 \times 10^8 \text{ km})(\sin 135°10')}{\sin 18°40'}$$

$$= 4.42 \times 10^7 \text{ km} \qquad\qquad\qquad\qquad = 2.39 \times 10^8 \text{ km}$$

49. In the above diagram, we are to calculate c based on the given information. There are two possible triangles, but the requirement that the distance c be as long as possible leads to the choice of α acute.

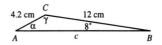

Solve for α: $\dfrac{\sin \alpha}{a} = \dfrac{\sin \beta}{b}$

$\qquad\qquad\qquad \sin \alpha = \dfrac{a \sin \beta}{b} = \dfrac{(12 \text{ cm})(\sin 8°)}{4.2 \text{ cm}} = 0.397 \ldots$

$\qquad\qquad\qquad\quad \alpha = \sin^{-1} 0.397 = 23°$

Solve for γ: $\alpha + \beta + \gamma = 180°$ $\gamma = 180° - (8° + 23°) = 149°$

Solve for c: $\dfrac{\sin \beta}{b} = \dfrac{\sin \gamma}{c}$ $c = \dfrac{b \sin \gamma}{\sin \beta} = \dfrac{(4.2 \text{ cm})(\sin 149°)}{\sin 8°} = 16 \text{ cm}$

Chapter 6   Additional Triangle Topics;   Vectors

51. In the figure, note: ABC is a right triangle;

hence, $\dfrac{h}{x} = \tan \gamma$

(1) $h = x \tan \gamma$

BCD is not a right triangle, but, from the law of sines,

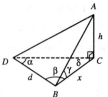

$\dfrac{\sin \delta}{d} = \dfrac{\sin \alpha}{x}$

(2) $x = \dfrac{d \sin \alpha}{\sin \delta}$

Since $\alpha + \beta + \delta = 180°$, $\delta = 180° - (\alpha + \beta)$

$\sin \delta = \sin[180° - (\alpha + \beta)] = \sin 180° \cos(\alpha + \beta) - \cos 180° \sin(\alpha + \beta)$

$= 0 \cos(\alpha + \beta) - (-1)\sin(\alpha + \beta) = \sin(\alpha + \beta)$

Hence, $\dfrac{1}{\sin \delta} = \dfrac{1}{\sin(\alpha + \beta)} = \csc(\alpha + \beta)$. Thus,

$h = x \tan \gamma = \dfrac{d \sin \alpha}{\sin \delta} \tan \gamma = d \sin \alpha \dfrac{1}{\sin \delta} \tan \gamma = d \sin \alpha \csc(\alpha + \beta)\tan \gamma$

EXERCISE 6.2 Law of Cosines

1. *Solve for a*: We use the law of cosines.

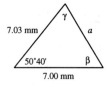

$a^2 = b^2 + c^2 - 2bc \cos \alpha$

$= (7.03)^2 + (7.00)^2 - 2(7.03)(7.00)\cos 50°40'$

$= 36.039253 \ldots$

$a = 6.00$ mm

Solve for β: We use the law of sines.

$\dfrac{\sin \alpha}{a} = \dfrac{\sin \beta}{b}$

$\sin \beta = \dfrac{b \sin \alpha}{a} = \dfrac{(7.03 \text{ mm})(\sin 50°40')}{6.00} = 0.9063$

$\beta = \sin^{-1} 0.9063 = 65°0'$ since the other possibility, $180° - \sin^{-1}(0.9063)$, would lead to a contradiction (since c is almost exactly equal to b, γ must be almost equal to β, but $180° - 65°0'$ is obtuse, and there cannot be two obtuse angles in a triangle).

Solve for γ: $\gamma = 180° - (\alpha + \beta) = 180° - (50°40' + 65°0') = 64°20'$

3. *Solve for c*: We use the law of cosines.

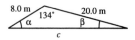

$c^2 = a^2 + b^2 - 2ab \cos \gamma$

$= (20.0)^2 + (8.00)^2 - 2(20.0)(8.00)\cos 134°$

$$= 686.29068 \ldots$$

$$c = 26.2 \text{ m}$$

Solve for α: We use the law of sines.

$$\frac{\sin \alpha}{a} = \frac{\sin \gamma}{c}$$

$$\sin \alpha = \frac{a \sin \gamma}{c} = \frac{(20.0 \text{ m})(\sin 134°)}{26.2 \text{ m}} = 0.5492$$

$\alpha = \sin^{-1} 0.5492 = 33.3°$ since the solution, $180° - \sin^{-1} 0.5492$, would lead to a second obtuse

angle in this triangle, an impossibility.

Solve for β: $\beta = 180° - (\alpha + \gamma) = 180° - (134° + 33.3°) = 12.7°$

5.   *Solve for* α: We use the law of cosines.

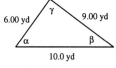

$$a^2 = b^2 + c^2 - 2bc \cos \alpha$$

$$\cos \alpha = \frac{b^2 + c^2 - a^2}{2bc}$$

$$\alpha = \cos^{-1}\frac{b^2 + c^2 - a^2}{2bc}$$

$$= \cos^{-1}\frac{(6.00)^2 + (10.0)^2 - (9.00)^2}{2(6.00)(10.0)} = \cos^{-1} 0.4583 = 62.7°$$

Solve for β: We can use either the law of cosines or the law of sines. We choose the latter because

it involves simpler calculations.

$$\frac{\sin \alpha}{a} = \frac{\sin \beta}{b} \qquad\qquad \sin \beta = \frac{b \sin \alpha}{a}$$

$$\beta = \sin^{-1}\frac{b \sin \alpha}{a} \qquad \beta \text{ is acute (see figure)}$$

$$= \sin^{-1}\frac{(6.00 \text{ yd})(\sin 62.7°)}{9.00 \text{ yd}} = 36.3°$$

Solve for γ: $\gamma = 180° - (\alpha + \beta) = 180° - (62.7° + 36.3°) = 81.0°$

7.   *Solve for* α: We use the law of cosines.

$$a^2 = b^2 + c^2 - 2bc \cos \alpha$$

$$\cos \alpha = \frac{b^2 + c^2 - a^2}{2bc}$$

$$\alpha = \cos^{-1}\frac{b^2 + c^2 - a^2}{2bc}$$

$$= \cos^{-1}\frac{(770.0)^2 + (860.0)^2 - (420.0)^2}{2(770.0)(860.0)} = \cos^{-1} 0.8729 = 29.20° \text{ or } 29°12'$$

Chapter 6   Additional Triangle Topics;   Vectors

Solve for β: We can use either the law of cosines or the law of sines. We choose the latter because it involves simpler calculations.

$$\frac{\sin \alpha}{a} = \frac{\sin \beta}{b} \qquad\qquad \sin \beta = \frac{b \sin \alpha}{a}$$

$$\beta = \sin^{-1}\frac{b \sin \alpha}{a} \qquad\qquad \beta \text{ is acute (see figure)}$$

$$= \sin^{-1}\frac{(770.0 \text{ km})(\sin 29°12')}{420 \text{ km}} = 63.43° \text{ or } 63°26'$$

Solve for γ: $\gamma = 180° - (\alpha + \beta) = 180° - (29°12' + 63°26') = 87°22'$

9. We are given two angles and a non-included side (AAS).

    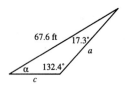

    Solve for α: $\alpha + \beta + \gamma = 180°$

    $$\alpha = 180° - (17.3° + 132.4°) = 30.3°$$

    Now use the law of sines to find the remaining sides.

    Solve for a: $\dfrac{\sin \alpha}{a} = \dfrac{\sin \beta}{b}$

    $$a = \frac{b \sin \alpha}{\sin \beta} = \frac{(67.6 \text{ ft})(\sin 30.3°)}{\sin 132.4°} = 46.2 \text{ ft}$$

    Solve for c: $\dfrac{\sin \beta}{b} = \dfrac{\sin \gamma}{c}$ $c = \dfrac{b \sin \gamma}{\sin \beta} = \dfrac{(67.6 \text{ ft})(\sin 17.3°)}{\sin 132.4°} = 27.2 \text{ ft}$

11. We are given two sides and the included angle (SAS). We use the law of cosines to find the third side, then the law of sines to find a second angle.

    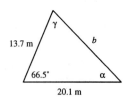

    Solve for b:

    $$b^2 = a^2 + c^2 - 2ac \cos \beta$$

    $$= (13.7)^2 + (20.1)^2 - 2(13.7)(20.1)\cos 66.5°$$

    $$= 372.09294 \ldots$$

    $$b = 19.3 \text{ m}$$

    Solve for α: $\dfrac{\sin \alpha}{a} = \dfrac{\sin \beta}{b}$

    $$\sin \alpha = \frac{a \sin \beta}{b} = \frac{(13.7 \text{ m})(\sin 66.5°)}{19.3 \text{ m}} = 0.6513$$

    $$\alpha = \sin^{-1} 0.6513 = 40.6°$$

Chapter 6   Additional Triangle Topics;   Vectors

(There is another solution of sin $\alpha = 0.6513$ that deserves brief consideration:

$\alpha^1 = 180° - \sin^{-1} 0.6513 = 139°$. However, there is not enough room in a triangle for an angle of 66.5° and an angle of 139°, since their sum is greater than 180°.)

Solve for γ: $\alpha + \beta + \gamma = 180°$

$\qquad\qquad \gamma = 180° - (66.5° + 40.6°) = 72.9°$

13. It is impossible to draw or form a triangle with this data, since angles $\beta + \gamma$ together add up to more than 180°. No solution.

15. We are given two sides and a non-included angle (SSA). Only one triangle can be formed from the given data, and this triangle is in fact isosceles, with two equal (congruent) sides. Therefore, the angles opposite the equal sides must also be equal; hence, $\beta = 66.4°$.

Solve for α: $\alpha + \beta + \gamma = 180°$

$\qquad\qquad \alpha = 180° - (66.4° + 66.4°) = 47.2°$

Solve for a: We use the law of sines.

$\dfrac{\sin \alpha}{a} = \dfrac{\sin \beta}{b}$

$\qquad a = \dfrac{b \sin \alpha}{\sin \beta} = \dfrac{(25.5 \text{ yd})(\sin 47.2°)}{\sin 66.4°} = 20.4 \text{ yd}$

17. We are given three sides (SSS). We solve for the largest

angle, α (largest because it is opposite the

largest side, a) using the law of cosines. We then

solve for a second angle using the law of sines,

because it involves simpler calculations.

Solve for α: $a^2 = b^2 + c^2 - 2bc \cos \alpha$

$\qquad\qquad \cos \alpha = \dfrac{b^2 + c^2 - a^2}{2bc}$

$\qquad\qquad \alpha = \cos^{-1} \dfrac{b^2 + c^2 - a^2}{2bc} = \cos^{-1} \dfrac{(5.23)^2 + (9.66)^2 - (10.5)^2}{2(5.23)(9.66)}$

$\qquad\qquad\quad = \cos^{-1} 0.1031 = 84.1°$

Solve for β: $\dfrac{\sin \alpha}{a} = \dfrac{\sin \beta}{b}$

$\qquad\qquad \sin \beta = \dfrac{b \sin \alpha}{a} = \dfrac{(5.23 \text{ in.})(\sin 84.1°)}{10.5 \text{ in.}} = 0.4954$

$\qquad\qquad \beta = \sin^{-1} 0.4954 = 29.7°$

293

(There is another solution of sin $\beta = 0.4954$ that deserves brief consideration:

$\beta' = 180° - \sin^{-1} 0.4954 = 150°$. However, there is not enough room in a triangle for an angle of 84.1° and an angle of 150°, since their sum is greater than 180°.)

Solve for γ. $\gamma = 180° - (\alpha + \beta) = 180° - (84.1° + 29.7°) = 66.2°$

19.   We are given two sides and the included angle (SAS). We use the law of cosines to find the third side, then the law of sines to find a second angle.

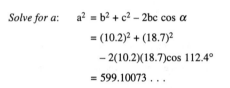

Solve for a: $a^2 = b^2 + c^2 - 2bc \cos \alpha$

$= (10.2)^2 + (18.7)^2$

$- 2(10.2)(18.7)\cos 112.4°$

$= 599.10073 \ldots$

a $= 24.5$ cm

Solve for β: $\dfrac{\sin \alpha}{a} = \dfrac{\sin \beta}{b}$ $\sin \beta = \dfrac{b \sin \alpha}{a}$

$\beta = \sin^{-1}\dfrac{b \sin \alpha}{a}$ β is acute because there is not enough room for two obtuse angles in a triangle.

$= \sin^{-1}\dfrac{(10.2 \text{ cm})(\sin 112.4°)}{24.5 \text{ cm}} = 22.6°$

Solve for γ. $\alpha + \beta + \gamma = 180°$

$\gamma = 180° - (22.6° + 112.4°) = 45.0°$

21.   We are given two sides and a non-included angle (SSA).

Solve for α: $\dfrac{\sin \alpha}{a} = \dfrac{\sin \gamma}{c}$

$\sin \alpha = \dfrac{a \sin \gamma}{c}$

$= \dfrac{(14.5 \text{ mm})(\sin 80.3°)}{10.0 \text{ mm}} = 1.429$

Since sin $\alpha = 1.429$ has no solution, no triangle exists with the given measurements. No solution.

23.   We are given two angles and the included side (ASA).

We use the law of sines.

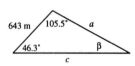

Solve for β: $\alpha + \beta + \gamma = 180°$
$\beta = 180° - (46.3° + 105.5°) = 28.2°$

Chapter 6   Additional Triangle Topics;   Vectors

Solve for a: $\dfrac{\sin\,\alpha}{a}=\dfrac{\sin\,\beta}{b}$

$$a=\dfrac{b\,\sin\,\alpha}{\sin\,\beta}$$

$$=\dfrac{(643\text{ m})(\sin\,46.3°)}{\sin\,28.2°}$$

$$=984\text{ m}$$

Solve for c: $\dfrac{\sin\,\beta}{b}=\dfrac{\sin\,\gamma}{c}$

$$c=\dfrac{b\,\sin\,\gamma}{\sin\,\beta}$$

$$=\dfrac{(643\text{ m})(\sin\,105.5°)}{\sin\,28.2°}$$

$$=1310\text{ m}$$

25. It is impossible to form a triangle with this data, since the triangle inequality (a + b > c) is not satisfied. No solution.

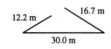

27. We are given two sides and a non-included angle (SSA). Two triangles are possible. There are two possible values for β. We use the law of sines.

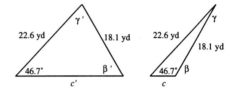

Solve for β: $\dfrac{\sin\,\alpha}{a}=\dfrac{\sin\,\beta}{b}$

$\sin\,\beta=\dfrac{b\,\sin\,\alpha}{a}=\dfrac{(22.6\text{ yd})(\sin\,46.7°)}{18.1\text{ yd}}=0.9087$

Angle β can be either obtuse or acute.

$\beta=180°-\sin^{-1}0.9087=114.7°$ $\beta'=\sin^{-1}0.9087=65.3°$

Solve for γ and γ':

$\gamma=180°-(\alpha+\beta)$ $\gamma'=180°-(\alpha'+\beta')$

$=180°-(46.7°+114.7°)=18.6°$ $=180°-(46.7°+65.3°)=68.0°$

Solve for c and c':

$\dfrac{\sin\,\alpha}{a}=\dfrac{\sin\,\gamma}{c}$ $\dfrac{\sin\,\alpha'}{a'}=\dfrac{\sin\,\gamma'}{c'}$

$c=\dfrac{a\,\sin\,\gamma}{\sin\,\alpha}$ $c'=\dfrac{a'\,\sin\,\gamma'}{\sin\,\alpha'}$

$=\dfrac{(18.1\text{ yd})(\sin\,18.6°)}{\sin\,46.7°}$ $=\dfrac{(18.1\text{ yd})(\sin\,68.0°)}{\sin\,46.7°}$

$=7.93\text{ yd}$ $=23.1\text{ yd}$

29.

  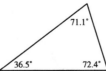

We are given three angles (AAA). An infinite number of triangles, all similar, can be drawn from the given values, but no one triangle is determined. No solution.

Chapter 6   Additional Triangle Topics;   Vectors

31. The law of cosines states that $b^2 = c^2 + a^2 - 2ac \cos \beta$ for any triangle.

If $\beta = 90°$, $\cos \beta = 0$; hence, $b^2 = c^2 + a^2 - 2ac \cos(90°) = c^2 + a^2 - 0 = c^2 + a^2$

33. Using the given and the calculated data, we have:

$$(a - b)\cos\frac{\gamma}{2} = c \sin\frac{(\alpha - \beta)}{2}$$

$$(6.00 - 7.03)\cos\frac{64°20'}{2} = 7.00 \sin\frac{(50°40' - 65°0')}{2}$$

$$-0.872 \approx -0.873$$

35. We can write the law of cosines two different ways as follows:

(1) $a^2 = b^2 + c^2 - 2bc \cos \alpha$

(2) $b^2 = a^2 + c^2 - 2ac \cos \beta$

Adding (1) and (2), we have

$$a^2 + b^2 = a^2 + b^2 + 2c^2 - 2bc \cos \alpha - 2ac \cos \beta$$

$$0 = 2c^2 - 2bc \cos \alpha - 2a \cos \beta$$

$$-2c^2 = -2bc \cos \alpha - 2ac \cos \beta$$

Dividing both sides by $-2c$ (which is never 0), we obtain $c = b \cos \alpha + a \cos \beta$

37. We are given two sides and an included angle. We use the law of cosines to find side BC.

$$BC^2 = AC^2 + AB^2 - 2(AC)(AB) \cos CAB = 425^2 + 384^2 - 2(425)(384) \cos 98.3°$$

$$= 375, 198.864 \ldots$$

$$BC = 613 \text{ m}$$

39. In triangle OAB, we are given OA = OB = 8.26 cm and AB = 13.8 cm. From the law of cosines, we can determine the central angle AOB.

$$\cos AOB = \frac{OA^2 + OB^2 - AB^2}{2(OA)(OB)}$$

$$\angle AOB = \cos^{-1}\frac{OA^2 + OB^2 - AB^2}{2(OA)(OB)}$$

$$= \cos^{-1}\frac{(8.26)^2 + (8.26)^2 - (13.8)^2}{2(8.26)(8.26)} = \cos^{-1}(-0.3956) = 113.3°$$

41. First, complete and label the figure.

From the given information, we know:

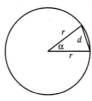

γ = angle between east and northeast = 45°

d_a = (rate of plane A)(2 hours)

   = (400 km/hr)(2 hr) = 800 km

d_b = (rate of plane B)(2 hours)

   = (500 km/hr)(2 hr) = 1,000 km

Hence, from the law of cosines,

c^2 = $d_a^2 + d_b^2 - 2d_a d_b \cos \gamma = (800)^2 + (1,000)^2 - 2(800)(1,000)(\cos 45°) = 508629.15 \ldots$

c = 710 km

43. In the figure, we are given d = 58.3 cm,
$\alpha = 27.8°$. From the law of cosines, we
know $d^2 = r^2 + r^2 - 2r \cdot r \cos \alpha$

$d^2 = r^2(2 - 2 \cos \alpha)$

$r^2 = \dfrac{d^2}{2 - 2 \cos \alpha}$

$r = \dfrac{d}{\sqrt{2 - 2 \cos \alpha}}$

Thus, r = $\dfrac{58.3 \text{ cm}}{\sqrt{2 - 2 \cos 27.8°}}$ = 121 cm

45. From the figure, it should be clear that the sides of the triangle are:

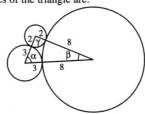

a = 2 cm + 8 cm = 10 cm

b = 2 cm + 3 cm = 5 cm

c = 3 cm + 8 cm = 11 cm

Solve for α:

We use the law of cosines.

$a^2 = b^2 + c^2 - 2bc \cos \alpha$ $\cos \alpha = \dfrac{b^2 + c^2 - a^2}{2bc}$

$\alpha = \cos^{-1} \dfrac{b^2 + c^2 - a^2}{2bc} = \cos^{-1} \dfrac{5^2 + 11^2 - 10^2}{2 \cdot 5 \cdot 11} = \cos^{-1} 0.4182 = 65°20'$ to nearest 10'

Solve for β: We can use either the law of cosines or the law of sines. We choose the latter
because it involves simpler calculations.

$\dfrac{\sin \alpha}{a} = \dfrac{\sin \beta}{b}$ $\sin \beta = \dfrac{b \sin \alpha}{a}$

$$\beta = \sin^{-1}\frac{b\sin\alpha}{a} \qquad\qquad \beta \text{ is acute (see figure)}$$

$$= \sin^{-1}\frac{(5\text{ cm})(\sin 65°20')}{10\text{ cm}} = 27°0' \text{ to nearest } 10'$$

Solve for γ. $\gamma = 180° - (\alpha + \beta) = 180° - (65°20' + 27°0') = 87°40'$

47. In triangle ACD, angle ADC = 8° + 90° = 98° (Why?) DC = 12.0 ft and AD = 18.0 ft.

We know two sides and the included angle; hence we can apply the law of cosines to find side AC.

$$AC^2 = AD^2 + DC^2 - 2(AD)(DC)\cos ADC$$
$$= (18.0)^2 + (12.0)^2 - 2(18.0)(12.0)\cos 98° = 528.1227 \ldots$$

AC = 23.0 ft

To find side AB, we need to find angle ADB, then apply the law of cosines again, in triangle ADB.
Since angle ADB + angle BDC = 98°, angle ADB = 98° – angle BDC. By symmetry,
angle BDC = angle ACD of triangle ACD, which we can find from the law of sines. Thus,

$$\frac{\sin ACD}{AD} = \frac{\sin ADC}{AC} \qquad\qquad \sin ACD = \frac{(AD)(\sin ADC)}{AC}$$

$$\text{angle } ACD = \sin^{-1}\frac{(AD)(\sin ADC)}{AC} = \sin^{-1}\frac{18.0\sin 98°}{23.0} = 50.9°$$

Hence, angle ADB = 98° – 50.9° = 47.1°. Then, applying the law of cosines to triangle ADB, we
have (AC = BD by symmetry.)

$$AB^2 = AD^2 + BD^2 - 2(AD)(BD)\cos ADB$$
$$= (18.0)^2 + (23.0)^2 - 2(18.0)(23.0)\cos 47.1° = 290.3765 \ldots$$

AB = 17.0 ft

49. In triangle CST, we are given TS = 1,034 miles
and TC = 3,964 miles.

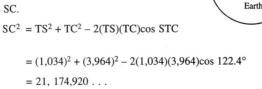

$\angle STC = \theta + 90° = 32.4° + 90° = 122.4°$

We are given two sides and the included angle;

hence, we can apply the law of cosines to find

side SC.

$$SC^2 = TS^2 + TC^2 - 2(TS)(TC)\cos STC$$

$$= (1,034)^2 + (3,964)^2 - 2(1,034)(3,964)\cos 122.4°$$

$$= 21,174,920 \ldots$$

SC = 4,602 mi

Hence, the height of the satellite = SC – radius of earth

$$= 4,602 \text{ mi} - 3,964 \text{ mi} = 638 \text{ mi}$$

51. The three sides of triangle ABC are each in turn the hypotenuse of a right triangle formed with two edges of the solid. Hence, by the Pythagorean theorem.

$$AB^2 = 6.0^2 + 3.0^2 = 45.00 \qquad AB = 6.7 \text{ cm}$$

$$AC^2 = 6.0^2 + 4.0^2 = 52.00 \qquad AC = 7.2 \text{ cm}$$

$$BC^2 = 3.0^2 + 4.0^2 = 25.00 \qquad BC = 5.0 \text{ cm}$$

To find $\angle ABC$, we apply the law of cosines.

$$AC^2 = BC^2 + AB^2 - 2(BC)(AB)\cos ABC \qquad \cos ABC = \frac{AB^2 + BC^2 - AC^2}{2(BC)(AB)}$$

$$\angle ABC = \cos^{-1}\frac{AB^2 + BC^2 - AC^2}{2(BC)(AB)} = \cos^{-1}\frac{6.7^2 + 5.0^2 - 7.2^2}{2(5.0)(6.7)} = 74.4°$$

53. We will first find AD by applying the law of sines to triangle ABD, in which we know all angles and a side. We then will find AC by applying the law of sines to triangle ABC, in which we also know all angles and a side. We can then apply the law of cosines to triangle ADC to find DC. In triangle ABD:

$$\frac{\sin ABD}{AD} = \frac{\sin ADB}{AB} \qquad AD = \frac{AB \sin ABD}{\sin ADB} = \frac{(120 \text{ m})\sin 79°}{\sin 29°} = 243.0 \text{ m}$$

In triangle ABC:

$$\frac{\sin ABC}{AC} = \frac{\sin ACB}{AB} \qquad AC = \frac{AB \sin ABC}{\sin ACB} = \frac{(120 \text{ m})\sin (79° + 44°)}{\sin 26°} = 229.6 \text{ m}$$

In triangle ACD:

$$CD^2 = AC^2 + AD^2 - 2(AC)(AD)\cos CAD$$
$$= (243.0)^2 + (229.6)^2 - 2(243.0)(229.6)\cos 41° = 27544.55 \ldots$$
$$CD = 166 \text{ m}$$

EXERCISE 6.3 Areas of Triangles

1. The base and the height of the triangle are given; hence, we use the formula

$$A = \frac{1}{2}bh = \frac{1}{2}(17.0 \text{ m})(12.0 \text{ m}) = 102 \text{ m}^2$$

3. The given information consists of two sides and the included angle; hence, we use the formula

$$A = \frac{ab}{2}\sin \theta \text{ in the form } A = \frac{1}{2}bc \sin \alpha = \frac{1}{2}(6.0 \text{ cm})(8.0 \text{ cm})\sin 30° = 12 \text{ cm}^2$$

Chapter 6   Additional Triangle Topics;   Vectors

5. The given information consists of three sides; hence, we use Heron's formula. First, find the semiperimeter s:

$$s = \frac{a + b + c}{2} = \frac{4.00 + 6.00 + 8.00}{2} = 9.00 \text{ in.}$$

Then, $s - a = 9.00 - 4.00 = 5.00$

$s - b = 9.00 - 6.00 = 3.00$

$s - c = 9.00 - 8.00 = 1.00$

Thus, $A = \sqrt{s(s - a)(s - b)(s - c)} = \sqrt{(9.00)(5.00)(3.00)(1.00)} = \sqrt{135} = 11.6 \text{ in}^2$

7. The given information consists of two sides and the included angle; hence, we use the formula

$$A = \frac{ab}{2} \sin \theta \text{ in the form } A = \frac{1}{2} bc \sin \alpha = \frac{1}{2}(403)(512)\sin 23°20' = 40,900 \text{ ft}^2$$

9. The given information consists of two sides and the included angle; hence, we use the formula

$$A = \frac{ab}{2} \sin \theta \text{ in the form } A = \frac{1}{2} bc \sin \alpha \qquad \alpha \text{ is obtuse; this does not alter the use of the formula.}$$

$$= \frac{1}{2}(12.1)(10.2)\sin 132.67° = 45.4 \text{ cm}^2$$

11. The given information consists of three sides; hence, we use Heron's formula. First, find the semiperimeter s:

$$s = \frac{a + b + c}{2} = \frac{12.7 + 20.3 + 24.4}{2} = 28.7 \text{ m}$$

Then, $s - a = 28.7 - 12.7 = 16.0$

$s - b = 28.7 - 20.3 = 8.4$

$s - c = 28.7 - 24.4 = 4.3$

Thus, $A = \sqrt{s(s - a)(s - b)(s - c)} = \sqrt{(28.7)(16.0)(8.4)(4.3)} = 129 \text{ m}^2$

13. Starting with a half-angle identity for cosine in the form $\cos^2 \dfrac{\alpha}{2} = \dfrac{1 + \cos \alpha}{2}$ and substituting (4) of the text, $\cos \alpha = \dfrac{b^2 + c^2 - a^2}{2bc}$, we can write

$$\cos^2 \frac{\alpha}{2} = \frac{1 + \dfrac{b^2 + c^2 - a^2}{2bc}}{2}$$

$$= \frac{2bc\left(1 + \dfrac{b^2 + c^2 - a^2}{2bc}\right)}{2bc(2)} \qquad \text{Conversion to simple fraction}$$

$$= \frac{2bc + b^2 + c^2 - a^2}{4bc}$$

$$= \frac{(b^2 + 2bc + c^2) - a^2}{4bc}$$ Group terms

$$= \frac{(b + c + a)(b + c - a)}{4bc}$$ Factor (difference of squares)

$$= \frac{(a + b + c)(a + b + c - 2a)}{4bc}$$

$$= \frac{2\left(\dfrac{a + b + c}{2}\right) 2\left(\dfrac{a + b + c}{2} - a\right)}{4bc}$$

$$= \frac{2(s)2(s - a)}{4bc} = \frac{s(s - a)}{bc}$$

Thus, $\cos \dfrac{\alpha}{2} = \sqrt{\dfrac{s(s - a)}{bc}}$.

15. All four triangles have two sides a and b given. For triangles with areas A_1 and A_3, the included angle is $180° - \theta$; hence, $A_1 = A_3 = \dfrac{1}{2} ab \sin(180° - \theta)$.

For triangles with areas A_2 and A_4, the included angle is θ ; hence, $A_2 = A_4 = \dfrac{1}{2} ab \sin \theta$

But, $\sin(180° - \theta) = \sin 180° \cos \theta - \cos 180° \sin \theta = 0 \cos \theta - (-1)\sin \theta = \sin \theta$

Hence, $A_1 = A_3 = \dfrac{1}{2} ab \sin(180° - \theta) = \dfrac{1}{2} ab \sin \theta = A_2 = A_4$.

EXERCISE 6.4 Vectors: Geometrically Defined

Figure for Problems 1, 3, and 5:

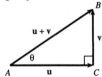

1. To find $|u + v|$: Apply the Pythagorean theorem to triangle ABC.

$$|u + v|^2 = AB^2 = AC^2 + BC^2 = |u|^2 + |v|^2 = 62^2 + 34^2 = 5,000$$

$$|u + v| = \sqrt{5,000} = 71 \text{ km/hr}$$

Solve triangle ABC for θ: $\tan \theta = \dfrac{BC}{AC} = \dfrac{|v|}{|u|}$ $\theta = \tan^{-1} \dfrac{|v|}{|u|}$ θ is acute

$$= \tan^{-1} \dfrac{34}{62} = 29°$$

3. To find $|u + v|$: Apply the Pythagorean theorem to triangle ABC.

$$|u + v|^2 = AB^2 = AC^2 + BC^2 = |u|^2 + |v|^2 = 48^2 + 31^2 = 3,265$$

$$|u + v| = 57 \text{ lb}$$

Solve triangle ABC for θ: $\tan \theta = \dfrac{BC}{AC} = \dfrac{|v|}{|u|}$ $\theta = \tan^{-1} \dfrac{|v|}{|u|}$ θ is acute

$$= \tan^{-1} \dfrac{31}{48} = 33°$$

Chapter 6   Additional Triangle Topics;   Vectors

5. To find |u + v|: Apply the Pythagorean theorem to triangle ABC.

$$|u + v|^2 = AB^2 = AC^2 + BC^2 = |u|^2 + |v|^2 = 143^2 + 57.4^2 = 23,743.76$$

$$|u + v| = 154 \text{ knots}$$

Solve triangle ABC for θ: $\tan \theta = \dfrac{BC}{AC} = \dfrac{|v|}{|u|}$

$\theta = \tan^{-1}\dfrac{|v|}{|u|}$ θ is acute

$= \tan^{-1}\dfrac{57.4}{143} = 21.9°$

Figure for Problems 7, 9, and 11:

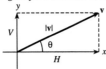

7. $|v| = 42 \text{ lb}$ $\theta = 34°$

Horizontal component H: $\cos 34° = \dfrac{H}{42}$ $H = 42 \cos 34° = 35 \text{ lb}$

Vertical component V: $\sin 34° = \dfrac{V}{42}$ $V = 42 \sin 34° = 23 \text{ lb.}$

9. $|v| = 23 \text{ knots}$ $\theta = 62°$

Horizontal component H: $\cos 62° = \dfrac{H}{23}$ $H = 23 \cos 62° = 11 \text{ knots}$

Vertical component V: $\sin 62° = \dfrac{V}{23}$ $V = 23 \sin 62° = 20 \text{ knots}$

11. $|v| = 244 \text{ km/hr}$ $\theta = 43.2°$

Horizontal component H: $\cos 43.2° = \dfrac{H}{244}$ $H = 244 \cos 43.2° = 178 \text{ km/hr}$

Vertical component V: $\sin 43.2° = \dfrac{V}{244}$ $V = 244 \sin 43.2° = 167 \text{ km/hr}$

Figure for Problems 13, 15, and 17:

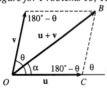

13. $\angle\theta = 44°$. Hence, $\angle OCB = 180° - \theta = 180° - 44° = 136°$.

We can find |u + v| using the law of cosines:

$$|u + v|^2 = |u|^2 + |v|^2 - 2 |u| |v| \cos(OCB)$$

$$= 125^2 + 84^2 - 2(125)(84)\cos 136° = 37,787.136 \ldots$$

$$|u + v| = \sqrt{37,787.136 \ldots} = 190 \text{ lb}$$

To find α, we use the law of sines: $\dfrac{\sin \alpha}{|v|} = \dfrac{\sin OCB}{|u + v|}$

$$\frac{\sin \alpha}{84} = \frac{\sin 136°}{190}$$

$$\sin \alpha = \frac{84}{190} \sin 136°$$

$$\alpha = \sin^{-1}\left(\frac{84}{190} \sin 136°\right) = 18°$$

15. $\angle\theta = 64°$. Hence, $\angle OCB = 180° - \theta = 180° - 64° = 116°$

We can find $|\mathbf{u} + \mathbf{v}|$ using the law of cosines:

$$|\mathbf{u} + \mathbf{v}|^2 = |\mathbf{u}|^2 + |\mathbf{v}|^2 - 2|\mathbf{u}||\mathbf{v}|\cos(OCB)$$
$$= (8.0)^2 + (2.0)^2 - 2(8.0)(2.0)\cos 116° = 82.027877 \ldots$$
$$|\mathbf{u} + \mathbf{v}| = \sqrt{82.027877 \ldots} = 91 \text{ knots}$$

To find α, we use the law of sines: $\dfrac{\sin \alpha}{|\mathbf{v}|} = \dfrac{\sin OCB}{|\mathbf{u} + \mathbf{v}|}$

$$\frac{\sin \alpha}{2.0} = \frac{\sin 116°}{9.1}$$

$$\sin \alpha = \frac{2.0}{9.1} \sin 116°$$

$$\alpha = \sin^{-1}\left(\frac{2.0}{9.1} \sin 116°\right) = 11°$$

17. $\angle\theta = 66.8°$. Hence, $\angle OCB = 180° - 66.8° = 113.2°$

We can find $|\mathbf{u} + \mathbf{v}|$ using the law of cosines:

$$|\mathbf{u} + \mathbf{v}|^2 = |\mathbf{u}|^2 + |\mathbf{v}|^2 - 2|\mathbf{u}||\mathbf{v}|\cos(OCB)$$
$$= 655^2 + (97.3)^2 - 2(655)(97.3)\cos 113.2° = 488,705.31 \ldots$$
$$|\mathbf{u} + \mathbf{v}| = \sqrt{488,705.31 \ldots} = 699 \text{ km/hr}$$

To find α, we use the law of sines. $\dfrac{\sin \alpha}{|\mathbf{v}|} = \dfrac{\sin OCB}{|\mathbf{u} + \mathbf{v}|}$

$$\frac{\sin \alpha}{97.3} = \frac{\sin 113.2°}{699}$$

$$\sin \alpha = \frac{97.3}{699} \sin 113.2°$$

$$\alpha = \sin^{-1}\left(\frac{97.3}{699} \sin 113.2°\right) = 7.4°$$

19. The actual velocity **v** of the boat is the vector sum of the apparent velocity **B** of the boat and the velocity **R** of the river.

$$|B| = 4.0 \text{ km/hr} \qquad\qquad |R| = 3.0 \text{ km/hr}$$

Using the Pythagorean theorem, we find the magnitude of the resultant vector to be

$$|v| = \sqrt{4.0^2 + 3.0^2} = 5.0 \text{ km/hr}$$

To find θ, we see that

$$\tan \alpha = \frac{|B|}{|R|} = \frac{4.0}{3.0} \qquad \alpha = \tan^{-1}\frac{4.0}{3.0} = 53°$$

$$\theta = \text{actual heading} = 90° + \alpha = 90° + 53° = 143.$$

21. We require θ such that the actual velocity **R** will be the resultant of the apparent velocity **v** and the wind velocity **w**. The heading α will then be $360° - \theta$. From the diagram it should be clear that

$$\sin \theta = \frac{|w|}{|v|} = \frac{46}{255}$$

$$\theta = \sin^{-1}\frac{46}{255} = 10° \qquad \alpha = 360° - 10° = 350°$$

The ground speed for this course will be the magnitude $|R|$ of the actual velocity. In the right triangle ABC, we have

$$\cos \theta = \frac{|R|}{|v|} \qquad\qquad |R| = |v| \cos \theta = 255 \cos\left(\sin^{-1}\frac{46}{255}\right)$$

$$= 255\sqrt{1 - \left(\frac{46}{255}\right)^2} = 250 \text{ mi/hr}$$

23. In triangle ABC, $\beta = 180° - 32° = 148°$. We can find the magnitude M of the resulting force using the law of cosines.

$$M^2 = |F_1|^2 + |F_2|^2 - 2|F_1||F_2| \cos \beta = 1,500^2 + 1,100^2 - 2(1,500)(1,100)\cos 148°$$

$$= 6,258,558.7 \ldots$$

$$M = \sqrt{6,258,558.7 \ldots} = 2,500 \text{ lb}$$

To find α, we use the law of sines.

$$\frac{\sin \alpha}{|F_2|} = \frac{\sin \beta}{M}$$

$$\frac{\sin \alpha}{1,100} = \frac{\sin 148°}{2,500}$$

$$\sin \alpha = \frac{1,100}{2,500} \sin 148° \qquad \alpha = \sin^{-1}\left(\frac{1,100}{2,500} \sin 148°\right) = 13° \text{ (relative to } F_1)$$

25. From the analysis in the text, we have $\theta = \tan^{-1} \dfrac{v^2}{gr}$, v in m/sec, r in m, g = 9.81 m/sec^2

    Hence, $\theta = \tan^{-1} \dfrac{(29)^2}{(9.81)(138)} = 32°$

27. The force parallel to the hill is the component of

    W parallel to the hill, that is, the magnitude of **CD**.

    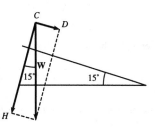

    $\dfrac{|CD|}{|W|} = \sin 15°$

    $|CD| = |W| \sin 15° = 2500 \sin 15° = 650$ lb

    The force perpendicular to the hill is the component of **W** perpendicular to the hill, that is, the

    magnitude of **CH**.

    $\dfrac{|CH|}{|W|} = \cos 15°$ $|CH| = |W| \cos 15° = 2500 \cos 15° = 2400$ lb

29. The weights will slide to the left if the horizontal component H_1 pointing left is greater than the

    horizontal component H_2 pointing right. They will slide to the right if H_2 is greater than H_1.

    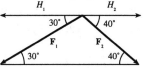

    H_1 = $|F_1| \cos 30° = (40\ g)\cos 30° \approx 34\ g$

    H_2 = $|F_2| \cos 40° = (30\ g)\cos 40° \approx 23\ g$

    Since H_1 is greater than H_2, they will slide to the left.

EXERCISE 6.5 Vectors: Algebraically Defined

1. The coordinates of P(x, y) are given by

   $x = x_b - x_a = 5 - 2 = 3$

   $y = y_b - y_a = 1 - (-3) = 4$

   Thus, P(x, y) = P(3, 4)

   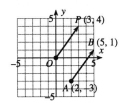

3. The coordinates of P(x, y) are given by

   $x = x_b - x_a = (-3) - (-1) = -2$

   $y = y_b - y_a = (-1) - 3 = -4$

   Thus, P(x, y) = P(-2, -4)

   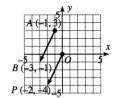

Chapter 6   Additional Triangle Topics; Vectors

5. The algebraic vector $\langle a, b \rangle$ has coordinates given by

$a = x_b - x_a = 3 - (-1) = 4$ $\qquad\qquad$ $b = y_b - y_a = 0 - (-2) = 2$

Hence, $\langle a, b \rangle = \langle 4, 2 \rangle$

7. The algebraic vector $\langle a, b \rangle$ has coordinates given by

$a = x_b - x_a = 4 - 0 = 4$ $\qquad\qquad$ $b = y_b - y_a = (-2) - 2 = -4$

Hence, $\langle a, b \rangle = \langle 4, -4 \rangle$

9. $|\langle a, b \rangle| = \sqrt{a^2 + b^2} = \sqrt{(-3)^2 + 4^2} = 5$

11. $|\langle a, b \rangle| = \sqrt{a^2 + b^2} = \sqrt{(-5)^2 + (-2)^2} = \sqrt{29}$

13. (A) $\mathbf{u} + \mathbf{v} = \langle 1, 4 \rangle + \langle -3, 2 \rangle = \langle -2, 6 \rangle$

   (B) $\mathbf{u} - \mathbf{v} = \langle 1, 4 \rangle - \langle -3, 2 \rangle = \langle 4, 2 \rangle$

   (C) $2\mathbf{u} - 3\mathbf{v} = 2\langle 1, 4 \rangle - 3\langle -3, 2 \rangle = \langle 2, 8 \rangle + \langle 9, -6 \rangle = \langle 11, 2 \rangle$

   (D) $3\mathbf{u} - \mathbf{v} + 2\mathbf{w} = 3\langle 1, 4 \rangle - \langle -3, 2 \rangle + 2\langle 0, 4 \rangle = \langle 3, 12 \rangle + \langle 3, -2 \rangle + \langle 0, 8 \rangle = \langle 6, 18 \rangle$

15. (A) $\mathbf{u} + \mathbf{v} = \langle 2, -3 \rangle + \langle -1, -3 \rangle = \langle 1, -6 \rangle$

   (B) $\mathbf{u} - \mathbf{v} = \langle 2, -3 \rangle - \langle -1, -3 \rangle = \langle 3, 0 \rangle$

   (C) $2\mathbf{u} - 3\mathbf{v} = 2\langle 2, -3 \rangle - 3\langle -1, -3 \rangle = \langle 4, -6 \rangle + \langle 3, 9 \rangle = \langle 7, 3 \rangle$

   (D) $3\mathbf{u} - \mathbf{v} - 2\mathbf{w} = 3\langle 2, -3 \rangle - \langle -1, -3 \rangle + 2\langle -2, 0 \rangle = \langle 6, -9 \rangle + \langle 1, 3 \rangle + \langle -4, 0 \rangle = \langle 3, -6 \rangle$

17. $|\mathbf{v}| = \sqrt{4^2 + (-3)^2} = 5$ $\qquad$ $\mathbf{u} = \dfrac{1}{|\mathbf{v}|}\mathbf{v} = \dfrac{1}{5}\langle 4, -3 \rangle = \left\langle \dfrac{4}{5}, -\dfrac{3}{5} \right\rangle$

19. $|\mathbf{v}| = \sqrt{2^2 + (-3)^2} = \sqrt{13}$ $\qquad$ $\mathbf{u} = \dfrac{1}{|\mathbf{v}|}\mathbf{v} = \dfrac{1}{\sqrt{13}}\langle 2, -3 \rangle = \left\langle \dfrac{2}{\sqrt{13}}, -\dfrac{3}{\sqrt{13}} \right\rangle$

21. $\mathbf{v} = \langle 3, -2 \rangle = \langle 3, 0 \rangle + \langle 0, -2 \rangle = 3\langle 1, 0 \rangle - 2\langle 0, 1 \rangle = 3\mathbf{i} - 2\mathbf{j}$

23. $\mathbf{v} = \langle 0, 4 \rangle = 4\langle 0, 1 \rangle = 4\mathbf{j}$

25. $\mathbf{v} = \overline{AB}$ $\ = \langle 0 - (-2), 2 - (-1) \rangle = \langle 2, 3 \rangle = \langle 2, 0 \rangle + \langle 0, 3 \rangle$

$= 2\langle 1, 0 \rangle + 3\langle 0, 1 \rangle = 2\mathbf{i} + 3\mathbf{j}$

27. $\mathbf{u} - \mathbf{v}\ = (2\mathbf{i} - 3\mathbf{j}) - (3\mathbf{i} + 4\mathbf{j}) = 2\mathbf{i} - 3\mathbf{j} - 3\mathbf{i} - 4\mathbf{j} = -\mathbf{i} - 7\mathbf{j}$

29. $3\mathbf{u} - 2\mathbf{v}\ = 3(2\mathbf{i} - 3\mathbf{j}) - 2(3\mathbf{i} + 4\mathbf{j}) = 6\mathbf{i} - 9\mathbf{j} - 6\mathbf{i} - 8\mathbf{j} = -17\mathbf{j}$

31. $\mathbf{u} - 2\mathbf{v} + 2\mathbf{w} = (2\mathbf{i} - 3\mathbf{j}) - 2(3\mathbf{i} + 4\mathbf{j}) + 2(5\mathbf{j}) = 2\mathbf{i} - 3\mathbf{j} - 6\mathbf{i} - 8\mathbf{j} + 10\mathbf{j} = -4\mathbf{i} - \mathbf{j}$

33. $\mathbf{u} + \mathbf{v} = \langle a, b \rangle + \langle c, d \rangle$

$= \langle a + c, b + d \rangle$	Definition of vector addition
$= \langle c + a, d + b \rangle$	Commutative property for addition of real numbers[*]
$= \langle c, d \rangle + \langle a, b \rangle$	Definition of vector addition
$= \mathbf{v} + \mathbf{u}$	

35. $\mathbf{v} + (-\mathbf{v}) = \langle c, d \rangle + (-\langle c, d \rangle)$

$= \langle c, d \rangle + \langle -c, -d \rangle$	Definition of scalar multiplication
$= \langle c + (-c), d + (-d) \rangle$	Definition of vector addition
$= \langle 0, 0 \rangle$	Additive inverse property for real numbers*
$= \mathbf{0}$	Definition of zero vector

37. $m(\mathbf{u} + \mathbf{v}) = m(\langle a, b \rangle + \langle c, d \rangle)$

$= m\langle a + c, b + d \rangle$	Definition of vector addition
$= \langle m(a + c), m(b + d) \rangle$	Definition of scalar multiplication
$= \langle ma + mc, mb + md \rangle$	Distributive property for real numbers*
$= \langle ma, mb \rangle + \langle mc, md \rangle$	Definition of vector addition
$= m\langle a, b \rangle + m\langle c, d \rangle$	Definition of scalar multiplication
$= m\mathbf{u} + m\mathbf{v}$	

39. $1\mathbf{u} = 1\langle a, b \rangle$

$= \langle 1a, 1b \rangle$	Definition of scalar multiplication
$= \langle a, b \rangle$	Multiplicative identity property for real numbers[*]
$= \mathbf{u}$	

41. First, form a force diagram with all force vectors in standard position at the origin.

Let $\mathbf{F}_1$ = the tension in one rope

$\mathbf{F}_2$ = the tension in the other rope

Write each force vector in terms of $\mathbf{i}$ and $\mathbf{j}$ unit vectors.

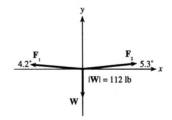

$$\mathbf{F}_1 = |\mathbf{F}_1|(-\cos 4.2°)\mathbf{i} + |\mathbf{F}_1|(\sin 4.2°)\mathbf{j}$$
$$\mathbf{F}_2 = |\mathbf{F}_2|(\cos 5.3°)\mathbf{i} + |\mathbf{F}_2|(\sin 5.3°)\mathbf{j}$$

---

[*] The basic properties of the set of real numbers are listed in the text, Appendix A.1.

Chapter 6   Additional Triangle Topics;   Vectors

$\mathbf{W} = -112\mathbf{j}$

For the system to be in static equilibrium, we must have $\mathbf{F}_1 + \mathbf{F}_2 + \mathbf{W} = 0$ which becomes, on addition,

$$[-|\mathbf{F}_1|(\cos 4.2°) + |\mathbf{F}_2|(\cos 5.3°)]\ \mathbf{i} + [|\mathbf{F}_1|(\sin 4.2°) + |\mathbf{F}_2|(\sin 5.3°) - 112]\ \mathbf{j} = 0\mathbf{i} + 0\mathbf{j}$$

Since two vectors are equal if and only if their corresponding components are equal, we are led to the following system of equations in $|\mathbf{F}_1|$ and $|\mathbf{F}_2|$:

$-|\mathbf{F}_1|\cos 4.2° + |\mathbf{F}_2|\cos 5.3° = 0$

$|\mathbf{F}_1|\sin 4.2° + |\mathbf{F}_1|\sin 5.3° - 112 = 0$

Solving, $|\mathbf{F}_2| = |\mathbf{F}_1|\dfrac{\cos 4.2°}{\cos 5.3°}$

$|\mathbf{F}_1|\sin 4.2° + |\mathbf{F}_1|\dfrac{\cos 4.2°}{\cos 5.3°}\sin 5.3° = 112$

$|\mathbf{F}_1|[\sin 4.2° + \cos 4.2° \tan 5.3°] = 112$

$|\mathbf{F}_1| = \dfrac{112}{\sin 4.2° + \cos 4.2° \tan 5.3°} = 676\ \text{lb}$

$|\mathbf{F}_2| = 676\dfrac{\cos 4.2°}{\cos 5.3°} = 677\ \text{lb}$

43. First, form a force diagram with all force vectors in standard position at the origin.

Let $\mathbf{F}_1$ = the force on the horizontal member BC

$\mathbf{F}_2$ = the force on the supporting member AB

$\mathbf{W}$ = the downward force (5,000 lb)

We note: $\cos\theta = \dfrac{5.0}{6.0}$ $\theta = \cos^{-1}\dfrac{5.0}{6.0} = 33.6°$

Then write each force vector in terms of $\mathbf{i}$ and $\mathbf{j}$ unit vectors.

$\mathbf{F}_1 = -|\mathbf{F}_1|\ \mathbf{i}$

$\mathbf{F}_2 = |\mathbf{F}_2|(\cos 33.6°)\ \mathbf{i} + |\mathbf{F}_2|(\sin 33.6°)\ \mathbf{j}$

$\mathbf{W} = -5,000\mathbf{j}$

For the system to be in static equilibrium, we must have $\mathbf{F}_1 + \mathbf{F}_2 + \mathbf{W} = 0$ which becomes, on addition,

$$[-|\mathbf{F}_1| + |\mathbf{F}_2|(\cos 33.6°)]\ \mathbf{i} + [|\mathbf{F}_2|(\sin 33.6°) - 5,000]\ \mathbf{j} = 0\mathbf{i} + 0\mathbf{j}$$

Since two vectors are equal if and only if their corresponding components are equal, we are led to the following system of equations in $|\mathbf{F}_1|$ and $|\mathbf{F}_2|$:

$-|\mathbf{F}_1| + |\mathbf{F}_2|(\cos 33.6°) = 0$

$|\mathbf{F}_1|(\sin 33.6°) - 5,000 = 0$

Solving, $|\mathbf{F}_2| = \dfrac{5,000}{\sin 33.6°} = 9,050\ \text{lb}$

$|\mathbf{F}_1| = |\mathbf{F}_2|\cos 33.6° = 7,540\ \text{lb}$

The force in the member AB is directed oppositely to the diagram — a compression of 9,050 lb.

The force in the member BC is also directed oppositely to the diagram — a tension of 7,540 lb.

EXERCISE 6.6 The Dot Product

1. $\langle\, 5, 3\,\rangle \cdot \langle\, -2, 3\,\rangle = 5 \cdot (-2) + 3 \cdot 3 = -1$

3. $(5\mathbf{i} - 3\mathbf{j}) \cdot (-2\mathbf{i} + 3\mathbf{j}) = 5 \cdot (-2) - 3 \cdot 3 = -19$

5. $\langle\, 2, 8\,\rangle \cdot \langle\, 12, -3\,\rangle = 2 \cdot 12 + 8 \cdot (-3) = 0$

7. $3\mathbf{i} \cdot 4\mathbf{j} = (3\mathbf{i} + 0\mathbf{j}) \cdot (0\mathbf{i} + 4\mathbf{j}) = 3 \cdot 0 + 0 \cdot 4 = 0$

9. $|\mathbf{u}| = \sqrt{(-3)^2 + 2^2} = \sqrt{13}$ $|\mathbf{v}| = \sqrt{0^2 + 4^2} = 4$

$$\cos\theta = \frac{\mathbf{u} \cdot \mathbf{v}}{|\mathbf{u}|\,|\mathbf{v}|} = \frac{\langle\, -3, 2\,\rangle \cdot \langle\, 0, 4\,\rangle}{(\sqrt{13})(4)} = \frac{0 + 8}{4\sqrt{13}} = \frac{2}{\sqrt{13}} \qquad \theta = \cos^{-1}\frac{2}{\sqrt{13}} = 56.3°$$

11. $|\mathbf{u}| = \sqrt{3^2 + 3^2} = \sqrt{18}$ $|\mathbf{v}| = \sqrt{2^2 + (-5)^2} = \sqrt{29}$

$$\cos\theta = \frac{\mathbf{u} \cdot \mathbf{v}}{|\mathbf{u}|\,|\mathbf{v}|} = \frac{\langle\, 3, 3\,\rangle \cdot \langle\, 2, -5\,\rangle}{(\sqrt{18})(\sqrt{29})} = \frac{6 + (-15)}{\sqrt{18}\,\sqrt{29}} = \frac{-9}{\sqrt{18}\,\sqrt{29}}$$

$$\theta = \cos^{-1}\frac{-9}{\sqrt{18}\,\sqrt{29}} = 113.2°$$

13. $|\mathbf{u}| = \sqrt{2^2 + (-3)^2} = \sqrt{13}$ $|\mathbf{v}| = \sqrt{6^2 + 4^2} = \sqrt{52}$

$$\cos\theta = \frac{\mathbf{u} \cdot \mathbf{v}}{|\mathbf{u}|\,|\mathbf{v}|} = \frac{\langle\, 2, -3\,\rangle \cdot \langle\, 6, 4\,\rangle}{(\sqrt{13})(\sqrt{52})} = \frac{12 - 12}{\sqrt{13}\,\sqrt{52}} = 0 \qquad \theta = \cos^{-1} 0 = 90°$$

15. $\mathbf{u} \cdot \mathbf{v} = (2\mathbf{i} + \mathbf{j}) \cdot (\mathbf{i} - 2\mathbf{j}) = 2 - 2 = 0$ Thus, $\mathbf{u}$ and $\mathbf{v}$ are orthogonal.

17. $\mathbf{u} \cdot \mathbf{v} = \langle\, 1, 3\,\rangle \cdot \langle\, -3, -1\,\rangle = -3 - 3 = -6 \neq 0$ Thus, $\mathbf{u}$ and $\mathbf{v}$ are not orthogonal.

19. $\text{comp}_{\mathbf{v}}\,\mathbf{u} = \dfrac{\mathbf{u} \cdot \mathbf{v}}{|\mathbf{v}|} = \dfrac{\langle\, -2, 4\,\rangle \cdot \langle\, -3, -1\,\rangle}{|\langle\, -3, -1\,\rangle|} = \dfrac{6 - 4}{\sqrt{(-3)^2 + (-1)^2}} = \dfrac{2}{\sqrt{10}} \approx 0.632$

21. $\text{comp}_{\mathbf{v}}\,\mathbf{u} = \dfrac{\mathbf{u} \cdot \mathbf{v}}{|\mathbf{v}|} = \dfrac{\langle\, -2, -4\,\rangle \cdot \langle\, 6, -3\,\rangle}{|\langle\, 6, -3\,\rangle|} = \dfrac{-12 + 12}{\sqrt{6^2 + (-3)^2}} = 0$

23. $\text{comp}_{\mathbf{v}}\,\mathbf{u} = \dfrac{\mathbf{u} \cdot \mathbf{v}}{|\mathbf{v}|} = \dfrac{(7\mathbf{i} - 2\mathbf{j}) \cdot (\mathbf{i} + \mathbf{j})}{|\mathbf{i} + \mathbf{j}|} = \dfrac{7 - 2}{\sqrt{1^2 + 1^2}} = \dfrac{5}{\sqrt{2}} \approx 3.54$

25. $\mathbf{u} \cdot \mathbf{u} = \langle\, a, b\,\rangle \cdot \langle\, a, b\,\rangle = a^2 + b^2 = \left(\sqrt{a^2 + b^2}\right)^2 = |\mathbf{u}|^2$

27. $\mathbf{u} \cdot (\mathbf{v} + \mathbf{w}) = \langle\, a, b\,\rangle \cdot (\langle\, c, d\,\rangle + \langle\, e, f\,\rangle)$

                 $= \langle\, a, b\,\rangle \cdot \langle\, c + e, d + f\,\rangle$ Definition of vector addition

                 $= a(c + e) + b(d + f)$ Definition of dot product

                 $= ac + ae + bd + bf$ Distributive property of real numbers

                 $= (ac + bd) + (ae + bf)$ Commutative and associative properties of real numbers

$$= \langle a, b \rangle \cdot \langle c, d \rangle + \langle a, b \rangle \cdot \langle e, f \rangle \qquad \text{Definition of dot product}$$

$$= \mathbf{u} \cdot \mathbf{v} + \mathbf{u} \cdot \mathbf{w}$$

29. $k(\mathbf{u} \cdot \mathbf{v}) = k(\langle a, b \rangle \cdot \langle c, d \rangle)$

$\quad = k(ac + bd)$ Definition of dot product

$\quad = k(ac) + k(bd)$ Distributive property of real numbers

$\quad = (ka)c + (kb)d$ Associative property for multiplication of real numbers

$\quad = \langle ka, kb \rangle \cdot \langle c, d \rangle$ Definition of dot product

$\quad = (k\langle a, b \rangle) \cdot \langle c, d \rangle$ Definition of scalar multiplication

$\quad = (k\mathbf{u}) \cdot \mathbf{v}$

Also, $k(\mathbf{u} \cdot \mathbf{v}) = k(\langle a, b \rangle \cdot \langle c, d \rangle)$

$\quad = k(ac + bd)$ Definition of dot product

$\quad = k(ac) + k(bd)$ Distributive property of real numbers

$\quad = a(kc) + b(kd)$ Commutative and associative properties of real numbers

$\quad = \langle a, b \rangle \cdot \langle kc, kd \rangle$ Definition of dot product

$\quad = \langle a, b \rangle \cdot (k\langle c, d \rangle)$ Definition of scalar multiplication

$\quad = \mathbf{u} \cdot (k\mathbf{v})$

31. $\text{Proj}_{\mathbf{v}}\, \mathbf{u} = \dfrac{\mathbf{u} \cdot \mathbf{v}}{\mathbf{v} \cdot \mathbf{v}}\, \mathbf{v} = \dfrac{\langle 3, 4 \rangle \cdot \langle 4, 0 \rangle}{\langle 4, 0 \rangle \cdot \langle 4, 0 \rangle}\langle 4, 0 \rangle = \dfrac{12 + 0}{16 + 0}\langle 4, 0 \rangle = \dfrac{3}{4}\langle 4, 0 \rangle = \langle 3, 0 \rangle$

33. $\text{Proj}_{\mathbf{v}}\, \mathbf{u} = \dfrac{\mathbf{u} \cdot \mathbf{v}}{\mathbf{v} \cdot \mathbf{v}}\, \mathbf{v} = \dfrac{(-6\mathbf{i} + 3\mathbf{j}) \cdot (-3\mathbf{i} - 2\mathbf{j})}{(-3\mathbf{i} - 2\mathbf{j}) \cdot (-3\mathbf{i} - 2\mathbf{j})}(-3\mathbf{i} - 2\mathbf{j}) = \dfrac{18 - 6}{9 + 4}(-3\mathbf{i} - 2\mathbf{j})$

$\qquad = \dfrac{12}{13}(-3\mathbf{i} - 2\mathbf{j}) = -\dfrac{36}{13}\mathbf{i} - \dfrac{24}{13}\mathbf{j}$

35. $\text{Proj}_{\mathbf{v}}\, \mathbf{u} = \dfrac{\mathbf{u} \cdot \mathbf{v}}{\mathbf{v} \cdot \mathbf{v}}\, \mathbf{v} = \dfrac{(-2\mathbf{i} + 2\mathbf{j}) \cdot (-\mathbf{i} - 4\mathbf{j})}{(-\mathbf{i} - 4\mathbf{j}) \cdot (-\mathbf{i} - 4\mathbf{j})}(-\mathbf{i} - 4\mathbf{j})$

$\qquad = \dfrac{2 - 8}{1 + 16}(-\mathbf{i} - 4\mathbf{j}) = -\dfrac{6}{17}(-\mathbf{i} - 4\mathbf{j}) = \dfrac{6}{17}\mathbf{i} + \dfrac{24}{17}\mathbf{j}$

37. $W = \left(\begin{array}{c}\text{component of force in}\\ \text{the direction of motion}\end{array}\right)(\text{displacement}) = [(15\ \text{lb})\cos 42°](440\ \text{ft}) = 4{,}900\ \text{ft-lb}$

39. $\mathbf{d} = \langle 8, 1 \rangle \qquad W = \mathbf{F} \cdot \mathbf{d} = \langle 10, 5 \rangle \cdot \langle 8, 1 \rangle = 85\ \text{ft-lb}$

41. $\mathbf{d} = \langle -3, 1 \rangle = -3\mathbf{i} + \mathbf{j} \qquad W = \mathbf{F} \cdot \mathbf{d} = (-2\mathbf{i} + 3\mathbf{j}) \cdot (-3\mathbf{i} + \mathbf{j}) = 9\ \text{ft-lb}$

43. $\mathbf{d} = \langle 1, 1 \rangle = \mathbf{i} + \mathbf{j} \qquad W = \mathbf{F} \cdot \mathbf{d} = (10\mathbf{i} + 10\mathbf{j}) \cdot (\mathbf{i} + \mathbf{j}) = 20\ \text{ft-lb}$

45. To prove that $\angle ACB$, an angle inscribed in a semicircle, is a right angle, we need only show that $\mathbf{c} - \mathbf{a}$ and $\mathbf{c} + \mathbf{a}$ are orthogonal.

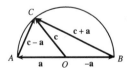

But $(\mathbf{c} - \mathbf{a}) \cdot (\mathbf{c} + \mathbf{a}) = (\mathbf{c} - \mathbf{a}) \cdot \mathbf{c} + (\mathbf{c} - \mathbf{a}) \cdot \mathbf{a}$ Distributive property of dot product

$\qquad\qquad\qquad\qquad = \mathbf{c} \cdot \mathbf{c} - \mathbf{a} \cdot \mathbf{c} + \mathbf{c} \cdot \mathbf{a} - \mathbf{a} \cdot \mathbf{a}$ Distributive property of dot product

$\qquad\qquad\qquad\qquad = \mathbf{c} \cdot \mathbf{c} - \mathbf{a} \cdot \mathbf{a} - \mathbf{a} \cdot \mathbf{c} + \mathbf{c} \cdot \mathbf{a}$ Commutative and associative properties of real numbers

$\qquad\qquad\qquad\qquad = \mathbf{c} \cdot \mathbf{c} - \mathbf{a} \cdot \mathbf{a} - \mathbf{a} \cdot \mathbf{c} + \mathbf{a} \cdot \mathbf{c}$ Commutative property of dot product

$\qquad\qquad\qquad\qquad = \mathbf{c} \cdot \mathbf{c} - \mathbf{a} \cdot \mathbf{a}$ Additive inverse and additive identity properties of real numbers

$\qquad\qquad\qquad\qquad = |\mathbf{c}|\,|\mathbf{c}| \cos 0° - |\mathbf{a}|\,|\mathbf{a}| \cos 0°$ Definition of dot product

$\qquad\qquad\qquad\qquad = (\text{Radius})^2 \cdot 1 - (\text{Radius})^2 \cdot 1$

$\qquad\qquad\qquad\qquad = 0$

Therefore, $\mathbf{c} - \mathbf{a}$ and $\mathbf{c} + \mathbf{a}$ are orthogonal, and an arbitary angle ACB, inscribed in a semicircle, is a right angle.

CHAPTER 6 REVIEW EXERCISE

1.

We are given three angles (AAA). An infinite number of triangles, all similar, can be drawn from the given values.

2.

We are given two angles and the included side (ASA). One triangle can be constructed.

3. It is impossible to draw or form a triangle with this data, since angle α and β together add up to more than 180°. No triangle can be constructed.

4.

We are given two angles and a non-included side (AAS). One triangle can be constructed.

5.

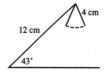

We are given two sides and a non-included angle (SSA). β is acute

Chapter 6   Additional Triangle Topics; Vectors

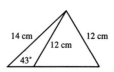

$h = a \sin \beta = 12 \sin 43° = 8.2$

$a = 4 < 8.2 = h$

No triangle can be constructed.

6.

7.

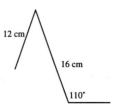

We are given two sides and a non-included angle (SSA). α is acute.

$h = b \sin \alpha = 9 \sin 43° = 6.1$

$a > b.$

One triangle can be constructed.

We are given two sides and a non-included angle (SSA). α is acute.

$h = b \sin \alpha = 14 \sin 43° = 9.5$

$h < a < b$

Two triangles can be constructed.

8.

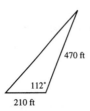

9.

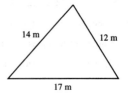

We are given two sides and a non-included angle (SSA). α is obtuse

$12 = a > b = 11.$

One triangle can be constructed.

We are given two sides and a non-included angle (SSA). α is obtuse.

$12 = a < b = 16.$

No triangle can be constructed.

10.

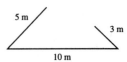

11.

We are given two sides and the included angle (SAS). One triangle can be constructed.

We are given three sides (SSS). One triangle can be constructed.

12. It is impossible to form a triangle with this data, since the triangle inequality $(a + b > c)$ is not satisfied. No triangle can be constructed.

Chapter 6   Additional Triangle Topics; Vectors

13. We are given two angles and the included side (ASA). We use the law of sines.

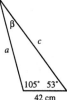

Solve for β: $\alpha + \beta + \gamma = 180°$

$\beta = 180° - (105° + 53°) = 22°$

Solve for β: $\dfrac{\sin \alpha}{a} = \dfrac{\sin \beta}{b}$

$a = \dfrac{b \sin \alpha}{\sin \beta} = \dfrac{(42 \text{ cm})(\sin 53°)}{\sin 22°} = 90 \text{ cm}$

Solve for c: $\dfrac{\sin \beta}{b} = \dfrac{\sin \gamma}{c}$

$c = \dfrac{b \sin \gamma}{\sin \beta} = \dfrac{(42 \text{ cm})(\sin 105°)}{\sin 22°} = 110 \text{ cm}$

14. We are given two angles and a non-included side (AAS). We use the law of sines.

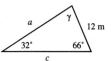

Solve for γ: $\alpha + \beta + \gamma = 180°$

$\gamma = 180° - (66° + 32°) = 82°$

Solve for a: $\dfrac{\sin \alpha}{a} = \dfrac{\sin \beta}{b}$ $a = \dfrac{b \sin \alpha}{\sin \beta} = \dfrac{(12 \text{ m})(\sin 66°)}{\sin 32°} = 21 \text{ m}$

Solve for c: $\dfrac{\sin \gamma}{c} = \dfrac{\sin \beta}{b}$ $c = \dfrac{b \sin \gamma}{\sin \beta} = \dfrac{(12 \text{ m})(\sin 82°)}{\sin 32°} = 22\text{m}$

15. We are given two sides and the included angle (SAS). We use the law of cosines to find the third side, then the law of sines to find a second angle.

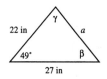

Solve for a: $a^2 = b^2 + c^2 - 2bc \cos \alpha$

$= 22^2 + 27^2 - 2(22)(27)\cos 49° = 433.60187 \ldots$

$a = \sqrt{433.60187 \ldots} = 21 \text{ in.}$

Solve for β: We use the law of sines.

$\dfrac{\sin \beta}{b} = \dfrac{\sin \alpha}{a}$ $\sin \beta = \dfrac{b \sin \alpha}{a} = \dfrac{(22 \text{ in.})(\sin 49°)}{21 \text{ in.}} = 0.7974$

$\beta = \sin^{-1} 0.7974 = 53°$ since the other solution, $180° - \sin^{-1} 0.7974$, would lead to a contradiction (since c is greater than b, γ must be greater than β, yielding an impossible second obtuse angle in the triangle).

Solve for γ: $\alpha + \beta + \gamma = 180°$ $\gamma = 180° - (49° + 53°) = 78°$

Chapter 6   Additional Triangle Topics;   Vectors

16. We are given two sides and a non-included angle (SSA).

α is acute. a > b. One triangle is possible. We use

the law of sines.

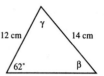

Solve for β: $\dfrac{\sin \alpha}{a} = \dfrac{\sin \beta}{b}$

$$\sin \beta = \frac{b \sin \alpha}{a} = \frac{(12 \text{ cm})(\sin 62°)}{14 \text{ cm}} = 0.7568$$

$$\beta = \sin^{-1} 0.7568 = 49°$$

(There is another solution of sin β = 0.7568 that deserves brief consideration:

β' = 180° – sin^{-1} 0.7568 = 131°. However, there is not enough room in a triangle for an angle of

62° and an angle of 131°, since their sum is greater than 180°.)

Solve for γ. $\alpha + \beta + \gamma = 180°$ $\gamma = 180° - (62° + 49°) = 69°$

Solve for c: $\dfrac{\sin \beta}{b} = \dfrac{\sin \gamma}{c}$ $c = \dfrac{b \sin \gamma}{\sin \beta} = \dfrac{(12 \text{ cm})(\sin 69°)}{\sin 49°} = 15 \text{ cm}$

17. The given information consists of two sides and the included angle; hence, we use the formula

$$A = \frac{ab}{2} \sin \theta \text{ in the form } A = \frac{1}{2} bc \sin \alpha = \frac{1}{2} (22 \text{ in.})(27 \text{ in.})\sin 49° = 224 \text{ in}^2$$

18. The given information consists of two sides and a non-included angle. Hence, we use the

information computed in Problem 16 in the formula

$$A = \frac{ab}{2} \sin \theta \text{ in the form } A = \frac{1}{2} ab \sin \gamma = \frac{1}{2} (14 \text{ cm})(12 \text{ cm})\sin 69° = 79 \text{ cm}^2$$

19. To find $|\mathbf{u} + \mathbf{v}|$: Apply the Pythagorean theorem to

triangle OCB.

$$|\mathbf{u} + \mathbf{v}|^2 = OB^2 = OC^2 + BC^2 = 8.0^2 + 5.0^2 = 89.00$$

$|\mathbf{u} + \mathbf{v}| = \sqrt{89.00} = 9.4$
Solve triangle OCB for θ:

$\tan \theta = \dfrac{BC}{OC} = \dfrac{|\mathbf{v}|}{|\mathbf{u}|}$ $\theta = \tan^{-1}\dfrac{|\mathbf{v}|}{|\mathbf{u}|}$ θ is acute $\theta = \tan^{-1}\dfrac{5.0}{8.0} = 32°$

20. $|\mathbf{v}| = 12$ $\theta = 35°$

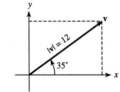

Horizontal component H: $\cos 35° = \dfrac{H}{12}$

H = 12 cos 35° = 9.8

Vertical component V: $\sin 35° = \dfrac{V}{12}$

V = 12 sin 35° = 6.9

Chapter 6   Additional Triangle Topics;   Vectors

21. The algebraic vector $\langle a, b \rangle$ has coordinates given by

$a = x_b - x_a = (-1) - (-3) = 2$ $b = y_b - y_a = (-3) - 2 = -5$

Hence, $\langle a, b \rangle = \langle 2, -5 \rangle$

22. Magnitude of $\langle -5, 12 \rangle = |\langle -5, 12 \rangle| = \sqrt{a^2 + b^2} = \sqrt{(-5)^2 + 12^2} = 13$

23. $\langle 2, -1 \rangle \cdot \langle -3, 2 \rangle = 2 \cdot (-3) + (-1) \cdot 2 = -8$

24. $(2\mathbf{i} + \mathbf{j}) \cdot (3\mathbf{i} - 2\mathbf{j}) = 2 \cdot 3 + 1 \cdot (-2) = 4$

25. $|\mathbf{u}| = \sqrt{4^2 + 3^2} = 5$ $|\mathbf{v}| = \sqrt{3^2 + 0^2} = 3$

$\cos \theta = \dfrac{\mathbf{u} \cdot \mathbf{v}}{|\mathbf{u}|\,|\mathbf{v}|} = \dfrac{\langle 4, 3 \rangle \cdot \langle 3, 0 \rangle}{(5)(3)} = \dfrac{12 + 0}{15} = \dfrac{12}{15}$ $\theta = \cos^{-1} \dfrac{12}{15} = 36.9°$

26. $|\mathbf{u}| = \sqrt{5^2 + 1^2} = \sqrt{26}$ $|\mathbf{v}| = \sqrt{(-2)^2 + 2^2} = \sqrt{8}$

$\cos \theta = \dfrac{\mathbf{u} \cdot \mathbf{v}}{|\mathbf{u}|\,|\mathbf{v}|} = \dfrac{(5\mathbf{i} + \mathbf{j}) \cdot (-2\mathbf{i} + 2\mathbf{j})}{\sqrt{26}\,\sqrt{8}} = \dfrac{5(-2) + 1 \cdot 2}{\sqrt{26}\,\sqrt{8}} = \dfrac{-8}{\sqrt{26}\,\sqrt{8}}$

$\theta = \cos^{-1} \dfrac{-8}{\sqrt{26}\,\sqrt{8}} = 123.7°$

27. We are given two sides and the included angle (SAS).
We use the law of cosines to find the third side, then
the law of sines to find a second angle.

Solve for a: $a^2 = b^2 + c^2 - 2bc \cos \alpha = (103)^2 + (72.4)^2 - 2(103)(72.4)\cos 65.0°$

$= 9{,}547.6622$

$a = \sqrt{9{,}547.6622 \ldots} = 97.7 \text{ m}$

Solve for γ: $\dfrac{\sin \gamma}{c} = \dfrac{\sin \alpha}{a}$ $\sin \gamma = \dfrac{c \sin \alpha}{a} = \dfrac{(72.4 \text{ m})(\sin 65.0°)}{97.7 \text{ m}} = 0.6715$

$\gamma = \sin^{-1} 0.6715 = 42.2°$ since the other solution, $180° - \sin^{-1} 0.6715$, would lead to a contradiction
(since b is greater than c, β must be greater than γ, yielding an impossible second obtuse angle in
the triangle).

Solve for β: $\alpha + \beta + \gamma = 180°$ $\beta = 180° - (\alpha + \gamma) = 180° - (65.0° + 42.2°) = 72.8°$

28. We are given two sides and a non-included
angle (SSA). α is acute.

$h = b \sin \alpha = 15.7 \sin 35°20' = 9.08$

$h < a < b$. Two triangles are possible,

but β is specified acute.

We use the law of sines.

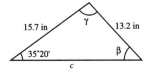

Solve for β: $\dfrac{\sin \alpha}{a} = \dfrac{\sin \beta}{b}$ $\sin \beta = \dfrac{b \sin \alpha}{a} = \dfrac{(15.7 \text{ in.})(\sin 35°20')}{13.2 \text{ in.}} = 0.6879$

Since β is specified acute, we choose the acute angle solution to this equation,

$\beta = \sin^{-1} 0.6879 = 43°30'$

Solve for γ. $\alpha + \beta + \gamma = 180°$ $\gamma = 180° - (\alpha + \beta) = 180° - (35°20' + 43°30') = 101°10'$

Solve for c: $\dfrac{\sin \gamma}{c} = \dfrac{\sin \alpha}{a}$ $c = \dfrac{a \sin \gamma}{\sin \alpha} = \dfrac{(13.2 \text{ in.})(\sin 101°10')}{\sin 35°20'} = 22.4 \text{ in.}$

29. We are given the same information as in Problem 28, except that β is specified obtuse.

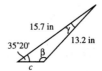

Solve for β: $\dfrac{\sin \alpha}{a} = \dfrac{\sin \beta}{b}$

$\sin \beta = \dfrac{b \sin \alpha}{a}$

$= \dfrac{(15.7 \text{ in.})(\sin 35°20')}{13.2 \text{ in.}} = 0.6879$

Since β is specified obtuse, we choose the obtuse angle solution to this equation,

$\beta = 180° - \sin^{-1} 0.6879 = 136°30'$

Solve for γ. $\alpha + \beta + \gamma = 180°$ $\gamma = 180° - (\alpha + \beta) = 180° - (35°20' + 136°30') = 8°10'$

Solve for c: $\dfrac{\sin \gamma}{c} = \dfrac{\sin \alpha}{a}$ $c = \dfrac{a \sin \gamma}{\sin \alpha} = \dfrac{(13.2 \text{ in.})(\sin 8°10')}{\sin 35°20'} = 3.24 \text{ in.}$

30.

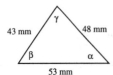

We are given three sides (SSS). We solve for the largest angle, γ (largest because it is opposite the largest side, c), using the law of cosines. We then solve for a second angle using the law of sines, because it involves simpler calculations.

Solve for γ. $c^2 = a^2 + b^2 - 2ab \cos \gamma$

$\cos \gamma = \dfrac{a^2 + b^2 - c^2}{2ab}$

$\gamma = \cos^{-1}\dfrac{a^2 + b^2 - c^2}{2ab} = \cos^{-1}\dfrac{43^2 + 48^2 - 53^2}{2 \cdot 43 \cdot 48} = 71°$

Solve for β: $\dfrac{\sin \beta}{b} = \dfrac{\sin \gamma}{c}$ $\sin \beta = \dfrac{b \sin \gamma}{c}$

$$\beta = \sin^{-1}\frac{b \sin \gamma}{c}$$ $\begin{cases} \beta \text{ is acute, because if it were obtuse, it would be} \\ \text{larger than } \gamma, \text{ which must be the largest angle in} \\ \text{the triangle, as noted above.} \end{cases}$

$$= \sin^{-1}\frac{(48 \text{ mm})\sin 71°}{53 \text{ mm}} = 59°$$

Solve for α: $\alpha + \beta + \gamma = 180°$ $\alpha = 180° - (\beta + \gamma) = 180° - (59° + 71°) = 50°$

31. The given information consists of two sides and the included angle; hence we use the formula

$$A = \frac{ab}{2} \sin \theta \text{ in the form } A = \frac{1}{2} bc \sin \alpha = \frac{1}{2}(103 \text{ m})(72.4 \text{ m})\sin 65.0° = 3{,}380 \text{ m}^2$$

32. The given information consists of three sides; hence, we use Heron's formula. First, find the semiperimeter s:

$$s = \frac{a + b + c}{2} = \frac{43 + 48 + 53}{2} = 72 \text{ mm}$$

Then, $s - a = 72 - 43 = 29$

$s - b = 72 - 48 = 24$

$s - c = 72 - 53 = 19$

Thus, $A = \sqrt{s(s-a)(s-b)(s-c)} = \sqrt{72(29)(24)(19)} = 980 \text{ mm}^2$

33.

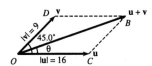 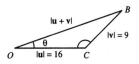

$\angle DOC = 45.0°$. Hence, $\angle OCB = 180° - 45.0° = 135.0°$. We can find the magnitude of **u** + **v** using the law of cosines.

$$|\mathbf{u} + \mathbf{v}|^2 = |\mathbf{u}|^2 + |\mathbf{v}|^2 - 2 |\mathbf{u}| |\mathbf{v}| \cos OCB = 16^2 + 9^2 - 2(16)(9)\cos 135.0° = 540.64675 \ldots$$

$$|\mathbf{u} + \mathbf{v}| = \sqrt{540.64675 \ldots} \approx 23.3$$

To find θ, we use the law of sines.

$$\frac{\sin \theta}{|\mathbf{v}|} = \frac{\sin OCB}{|\mathbf{u} + \mathbf{v}|} \qquad \frac{\sin \theta}{9} = \frac{135.0°}{23.3} \qquad \sin \theta = \frac{9 \sin 135.0°}{23.3}$$

$$\theta = \sin^{-1}\left(\frac{9 \sin 135.0°}{23.3}\right) \approx 15.9°$$

34. (A) $\mathbf{u} + \mathbf{v} = \langle 4, 0 \rangle + \langle -2, -3 \rangle = \langle 2, -3 \rangle$

(B) $\mathbf{u} - \mathbf{v} = \langle 4, 0 \rangle - \langle -2, -3 \rangle = \langle 6, 3 \rangle$

(C) $3\mathbf{u} - 2\mathbf{v} = 3\langle 4, 0 \rangle - 2\langle -2, -3 \rangle = \langle 12, 0 \rangle + \langle 4, 6 \rangle = \langle 16, 6 \rangle$

(D) $2\mathbf{u} - 3\mathbf{v} + \mathbf{w} = 2\langle 4, 0 \rangle - 3\langle -2, -3 \rangle + \langle 1, -1 \rangle = \langle 8, 0 \rangle + \langle 6, 9 \rangle + \langle 1, -1 \rangle = \langle 15, 8 \rangle$

Chapter 6   Additional Triangle Topics;   Vectors

35. (A) $\mathbf{u} + \mathbf{v} = (3\mathbf{i} - \mathbf{j}) + (2\mathbf{i} - 3\mathbf{j}) = 3\mathbf{i} + 2\mathbf{i} - \mathbf{j} - 3\mathbf{j} = 5\mathbf{i} - 4\mathbf{j}$

(B) $\mathbf{u} - \mathbf{v} = (3\mathbf{i} - \mathbf{j}) - (2\mathbf{i} - 3\mathbf{j}) = 3\mathbf{i} - \mathbf{j} - 2\mathbf{i} + 3\mathbf{j} = \mathbf{i} + 2\mathbf{j}$

(C) $3\mathbf{u} - 2\mathbf{v} = 3(3\mathbf{i} - \mathbf{j}) - 2(2\mathbf{i} - 3\mathbf{j}) = 9\mathbf{i} - 3\mathbf{j} - 4\mathbf{i} + 6\mathbf{j} = 5\mathbf{i} + 3\mathbf{j}$

(D) $2\mathbf{u} - 3\mathbf{v} + \mathbf{w} = 2(3\mathbf{i} - \mathbf{j}) - 3(2\mathbf{i} - 3\mathbf{j}) + (-2\mathbf{j}) = 6\mathbf{i} - 2\mathbf{j} - 6\mathbf{i} + 9\mathbf{j} - 2\mathbf{j} = 0\mathbf{i} + 5\mathbf{j}$ or $5\mathbf{j}$

36. $|\mathbf{v}| = \sqrt{(-8)^2 + 15^2} = 17$ $\mathbf{u} = \dfrac{1}{|\mathbf{v}|}\mathbf{v} = \dfrac{1}{17}\langle -8, 15 \rangle = \left\langle -\dfrac{8}{17}, \dfrac{15}{17} \right\rangle$

37. (A) $\mathbf{v} = \langle -5, 7 \rangle = \langle -5, 0 \rangle + \langle 0, 7 \rangle = -5\langle 1, 0 \rangle + 7\langle 0, 1 \rangle = -5\mathbf{i} + 7\mathbf{j}$

(B) $\mathbf{v} = \langle 0, -3 \rangle = -3\langle 0, 1 \rangle = -3\mathbf{j}$

(C) $\mathbf{v} = \overline{AB} = \langle 0 - 4, (-3) - (-2) \rangle = \langle -4, -1 \rangle = \langle -4, 0 \rangle + \langle 0, -1 \rangle$

$= -4\langle 1, 0 \rangle - \langle 0, 1 \rangle = -4\mathbf{i} - \mathbf{j}$

38. (A) $\mathbf{u} \cdot \mathbf{v} = \langle -12, 3 \rangle \cdot \langle 2, 8 \rangle = -24 + 24 = 0$. Thus, $\mathbf{u}$ and $\mathbf{v}$ are orthogonal.

(B) $\mathbf{u} \cdot \mathbf{v} = (-4\mathbf{i} + \mathbf{j}) \cdot (-\mathbf{i} + 4\mathbf{j}) = 4 + 4 = 8 \neq 0$. Thus, $\mathbf{u}$ and $\mathbf{v}$ are not orthogonal.

39. (A) $\text{comp}_\mathbf{v}|\mathbf{u}| = \dfrac{\mathbf{u} \cdot \mathbf{v}}{|\mathbf{v}|} = \dfrac{\langle 4, 5 \rangle \cdot \langle 3, 1 \rangle}{|\langle 3, 1 \rangle|} = \dfrac{12 + 5}{\sqrt{3^2 + 1^2}} = \dfrac{17}{\sqrt{10}} \approx 5.38$

(B) $\text{comp}_\mathbf{v}|\mathbf{u}| = \dfrac{\mathbf{u} \cdot \mathbf{v}}{|\mathbf{v}|} = \dfrac{(-\mathbf{i} + 4\mathbf{j}) \cdot (3\mathbf{i} - \mathbf{j})}{|3\mathbf{i} - \mathbf{j}|} = \dfrac{-3 - 4}{\sqrt{3^2 + (-1)^2}} = -\dfrac{7}{\sqrt{10}} \approx -2.21$

40. $(a - b)\cos\dfrac{\gamma}{2} = c \sin\dfrac{\alpha - \beta}{2}$

$(8.42 - 11.5)\cos\dfrac{59.1°}{2} = 10.2 \sin\dfrac{45.1° - 75.8°}{2}$

$-2.68 \approx -2.70$

The results agree to two significant digits; the results check.

41. From the diagram, we can see that

k = altitude of any possible triangle. Thus,

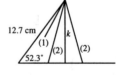

$\sin \beta = \dfrac{k}{a}$ $k = a \sin \beta$

$k = (12.7 \text{ cm})\sin 52.3° = 10.0 \text{ cm}$

If $0 < a < k$, there is no solution: (1) in the diagram.

If $a = k$, there is one solution.

If $k < a < b$, there are two solutions: (2) in the diagram.

42. $\mathbf{u} + \mathbf{v}$ $= \langle a, b \rangle + \langle c, d \rangle$

  $= \langle a + c, b + d \rangle$ Definition of vector addition

  $= \langle c + a, d + b \rangle$ Commutative property for addition of real numbers

  $= \langle c, d \rangle + \langle a, b \rangle$ Definition of vector addition

  $= \mathbf{v} + \mathbf{u}$

43. $\mathbf{u} \cdot \mathbf{v}$ $= \langle a, b \rangle \cdot \langle c, d \rangle$

  $= ac + bd$ Definition of dot product

  $= ca + db$ Commutative property for multiplication of real numbers

  $= \langle c, d \rangle \cdot \langle a, b \rangle$ Definition of dot product

  $= \mathbf{v} \cdot \mathbf{u}$

44. $(mn)\mathbf{v}$ $= (mn)\langle c, d \rangle$

  $= \langle (mn)c, (mn)d \rangle$ Definition of scalar multiplication

  $= \langle m(nc), m(nd) \rangle$ Associative property for multiplication of real numbers

  $= m\langle nc, nd \rangle$ Definition of scalar multiplication

  $= m(n\langle c, d \rangle)$ Definition of scalar multiplication

  $= m(n\mathbf{v})$

45. $m(\mathbf{u} \cdot \mathbf{v}) = m(\langle a, b \rangle \cdot \langle c, d \rangle)$

  $= m(ac + bd)$ Definition of dot product

  $= m(ac) + m(bd)$ Distributive property of real numbers

  $= (ma)c + (mb)d$ Associative property for multiplication of real numbers

  $= \langle ma, mb \rangle \cdot \langle c, d \rangle$ Definition of dot product

  $= (m\langle a, b \rangle) \cdot \langle c, d \rangle$ Definition of scalar multiplication

  $= (m\mathbf{u}) \cdot \mathbf{v}$

Also, $m(\mathbf{u} \cdot \mathbf{v})$ $= m(\langle a, b \rangle \cdot \langle c, d \rangle)$

  $= m(ac + bd)$ Definition of dot product

  $= m(ac) + m(bd)$ Distributive property of real numbers

  $= a(mc) + b(md)$ Commutative and associative properties of real numbers

  $= \langle a, b \rangle \cdot \langle mc, md \rangle$ Definition of dot product

  $= \langle a, b \rangle \cdot (m\langle c, d \rangle)$ Definition of scalar multiplication

  $= \mathbf{u} \cdot (m\mathbf{v})$

Chapter 6   Additional Triangle Topics;   Vectors

46. $\mathbf{u} \cdot (\mathbf{v} + \mathbf{w}) = \langle a, b \rangle \cdot (\langle c, d \rangle + \langle e, f \rangle)$

$\qquad = \langle a, b \rangle \cdot \langle c + e, d + f \rangle$ Definition of vector addition

$\qquad = a(c + e) + b(d + f)$ Definition of dot product

$\qquad = ac + ae + bd + bf$ Distributive property of real numbers

$\qquad = (ac + bd) + (ae + bf)$ Commutative and associative properties of real numbers

$\qquad = \langle a, b \rangle \cdot \langle c, d \rangle + \langle a, b \rangle \cdot \langle e, f \rangle$ Definition of dot product

$\qquad = \mathbf{u} \cdot \mathbf{v} + \mathbf{u} \cdot \mathbf{w}$

47. $\mathbf{u} \cdot \mathbf{u} = \langle a, b \rangle \cdot \langle a, b \rangle$

$\qquad = a^2 + b^2$ Definition of dot product

$\qquad = \left(\sqrt{a^2 + b^2} \right)^2$ Definition of the square root of a real number

$\qquad = |\mathbf{u}|^2$ Definition of magnitude

48. Since the diagonals of a parallelogram bisect each other,

we see that

$a = \dfrac{1}{2} (16.0 \text{ cm}) = 8.0 \text{ cm} \qquad b = \dfrac{1}{2} (20.0 \text{ cm}) = 10.0 \text{ cm}$

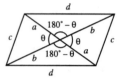

$\theta = 36.4° \qquad\qquad 180° - \theta = 143.6°$

To find c and d, we use the law of cosines.

$c^2 = a^2 + b^2 - 2ab \cos \theta$

$\qquad = (8.0)^2 + (10.0)^2 - 2(8.0)(10.0)\cos 36.4$

$\qquad = 35.216992 \ldots$

$c = \sqrt{35.216992 \ldots} = 5.9 \text{ cm}$

$d^2 = a^2 + b^2 - 2ab \cos(180° - \theta) = (8.0)^2 + (10.0)^2 - 2(8.0)(10.0)\cos(143.6°) = 292.78301 \ldots$

$d \; = \sqrt{292.78301 \ldots} = 17 \text{ cm}$

49. In triangle ACD, we note $\dfrac{AD}{h} = \cot 31°20'.$

In triangle BCD, we note $\dfrac{BD}{h} = \cot 49°40'.$

Hence, $AD = h \cot 31°20'$, $BD = h \cot 49°40'$.

Since $AD - BD = AB = 200$ m, we have

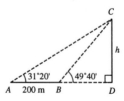

$200 \text{ m} = h \cot 31°20' - h \cot 49°40'$ or

$h = \dfrac{200}{\cot 31°20' - \cot 49°40'} = 252 \text{ m}$

Alternatively, we can note $ABC = 180° - 49°40' = 130°20'$, and use the law of sines to

determine BC.

$\dfrac{\sin CAB}{BC} = \dfrac{\sin BCA}{AB}$

$\angle BCA = 180° - (\angle CAB + \angle ABC) = 180° - (31°20' + 130°20') = 18°20'$

$$BC = \frac{AB \sin CAB}{\sin BCA} = \frac{(200 \text{ m})(\sin 31°20')}{\sin 18°20'} = 331 \text{ m}$$

Then, in right triangle BCD, we have

$$\frac{h}{BC} = \sin CBD$$

$$h = BC \sin CBD = (331 \text{ m})(\sin 49°40') = 252 \text{ m}$$

50. We sketch a figure, labelling what we know.

In triangle OAB, we know AB = 34 cm, ∠AOB = 85°,

OA = OB = r. Thus, from the law of cosines:

$$AB^2 = OA^2 + OB^2 - 2(OA)(OB)\cos AOB$$

$$34^2 = r^2 + r^2 - 2r^2\cos 85° = 2r^2 - 2r^2\cos 85° = r^2(2 - 2\cos 85°)$$

$$r^2 = \frac{34^2}{2 - 2\cos 85°} \qquad r = \sqrt{\frac{34^2}{2 - 2\cos 85°}} \qquad \text{We discard the negative solution}$$
$$= 25 \text{ cm}$$

51. We will first find AD by applying the law of sines to triangle ABD, in which we know all angles
     and a side. We will then find AC by applying the law of sines to triangle ABC, in which we also
     know all angles and a side. We can then apply the law of cosines to triangle ADC to find DC.

In triangle ABD: $\dfrac{\sin ABD}{AD} = \dfrac{\sin ADB}{AB}$

$$AD = \frac{AB \sin ABD}{\sin ADB} = \frac{(450 \text{ ft})\sin 72°}{\sin 50°} = 558.7 \text{ ft}$$

In triangle ABC: $\dfrac{\sin ABC}{AC} = \dfrac{\sin ACB}{AB}$

$$AC = \frac{AB \sin ABC}{\sin ACB} = \frac{(450 \text{ ft})\sin(72° + 65°)}{\sin 20°} = 897.3 \text{ ft}$$

In triangle ACD: $CD^2 = AC^2 + AD^2 - 2(AC)(AD)\cos CAD$

$$= (558.7)^2 + (897.3)^2 - 2(558.7)(897.3)\cos 35°$$

$$= 295993.95 \ldots$$

$$CD = 540 \text{ ft}$$

52. Area of ABCD = Area of triangle ADC + Area of triangle ABC

$$= \frac{1}{2}(AD)(AC)\sin DAC + \frac{1}{2}(AC)(AB)\sin CAB$$

$$= \frac{1}{2}(558.7)(897.3)\sin 35° + \frac{1}{2}(897.3)(450)\sin 23° = 220{,}000 \text{ ft}^2$$

Chapter 6   Additional Triangle Topics;   Vectors

53. In triangle ABC, we are given two sides and
    the included angle; hence, we can apply the
    law of cosines to find side BC.

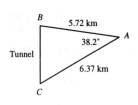

$$BC^2 = AB^2 + AC^2 - 2(AB)(AC)\cos BAC$$
$$= (5.72)^2 + (6.37)^2 - 2(5.72)(6.37)\cos 38.2°$$
$$= 16.027708 \ldots$$
$$BC = 4.00 \text{ km}$$

54. In triangle CST, we are given

    TS = 1,147 miles and TC = 3,964 miles.

    $\angle STC = \theta + 90° = 28.6° + 90° = 118.6°$.

    We are given two sides and the included angle;

    hence, we can apply the law of cosines to find

    side SC.

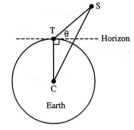

$$SC^2 = TS^2 + TC - 2(TS)(TC)\cos STC = (1,147)^2 + (3,964)^2 - 2(1,147)(3,964)\cos 118.6°$$
$$= 21,381,849 \ldots$$
$$SC = 4,624 \text{ mi}$$

Hence, the height of the satellite = SC – radius of the earth = 4,624 mi – 3,964 mi = 660 mi

55. Following the hint, we find all angles for triangle ACS first.

    $$\angle SAC = \theta + 90° = 21.7° + 90° = 111.7°$$

    For $\angle ACS$, we use the formula $s = \dfrac{\pi}{180} \theta R$ from Chapter 2, with

    $$s = 632 \text{ mi} \qquad \theta = \angle ACS \qquad R = 3,964 \text{ mi}$$

    Then, $\angle ACS = \dfrac{180°s}{\pi R} = \dfrac{180°(632 \text{ mi})}{\pi(3,964 \text{ mi})} = 9.13°$. Since $\angle SAC + \angle ACS + \angle ASC = 180°$, we have

    $$\angle ASC = 180° - (\angle ACS + \angle SAC) = 180° - (111.7° + 9.13°) = 59.2°$$

    Now, apply the law of sines to find side CS:

    $$\frac{\sin SAC}{CS} = \frac{\sin ASC}{AC}$$

    $$CS = \frac{AC \sin SAC}{\sin ASC} = \frac{(3,964 \text{ mi})\sin 111.7°}{\sin 59.2°} = 4,289 \text{ mi}$$

    Hence, the height of the satellite above B = BS = CS – BC.

    $$BS = 4,289 \text{ mi} - 3,964 \text{ mi} = 325 \text{ mi}$$

56.

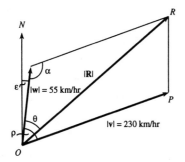

We are given $\varepsilon = 5°$ (wind heading), $\rho = 68°$ (plane heading). We want to find $|R|$, where R is the resultant sum of v, the plane's velocity, and w, the wind velocity. We also want θ, then $\theta + \varepsilon = \theta + 5°$ will be the plane's actual direction relative to north.

Solve for $|R|$: In triangle ORA, since $\angle AOP = \rho - \varepsilon = 68° - 5° = 63°$, α must equal

$180° - \angle AOP = 180° - 63° = 117°$ $|w| = 55$ km/hr $AR = |v| = 230$ km/hr

Now, apply the law of cosines.

$|R|^2 = |w|^2 + |v|^2 - 2\,|w|\,|v|\cos \alpha = 55^2 + 230^2 - 2(55)(230)\cos 117° = 67,410.96 \ldots$

$|R| = \sqrt{67,410.96 \ldots} \approx 260$ km/hr

Solve for θ: $\dfrac{\sin \theta}{AR} = \dfrac{\sin \alpha}{|R|}$ $\dfrac{\sin \theta}{230} = \dfrac{\sin 117°}{260}$

$\sin \theta = \dfrac{230}{260}\sin 117°$ $\theta = \sin^{-1}\left(\dfrac{230}{260}\sin 117°\right) \approx 52°$

The actual direction $= \theta + \varepsilon = 52° + 5° = 57°$.

57.

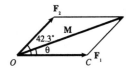 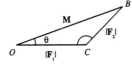

$\angle OCB = 180° - 42.3° = 137.7°$ $|F_1| = 352$ lb $|F_2| = 168$ lb

We can find the magnitude M of $F_1 + F_2$ using the law of cosines.

$M^2 = |F_1|^2 + |F_2|^2 - 2\,|F_1|\,|F_2|\cos OCB$

$= 352^2 + 168^2 - 2(352)(168)\cos 137.7° = 239,605.65 \ldots$

$M = \sqrt{239,605.65 \ldots} = 489$ lb

To find θ, we use the law of sines. $\dfrac{\sin \theta}{|F_2|} = \dfrac{\sin OCB}{M}$ $\dfrac{\sin \theta}{168} = \dfrac{\sin 137.7°}{489}$

$\sin \theta = \dfrac{168}{489}\sin 137.7°$ $\theta = \sin^{-1}\left(\dfrac{168}{489}\sin 137.7°\right) = 13.4°$ (relative to F_1)

58. $W = Fd = (489 \text{ lb})(22 \text{ ft}) = 10,800$ ft-lb

Chapter 6　　Additional Triangle Topics;　Vectors

59. First, form a force diagram with all force vectors

in standard position at the origin.

Let $\mathbf{F}_1$ = the force on the horizontal member AB

　$\mathbf{F}_2$ = the force on the supporting member BC

　W = the downward force (260 lb)

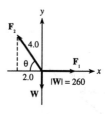

We note: $\cos\theta = \dfrac{2.0}{4.0}\ \theta = \cos^{-1}\dfrac{2.0}{4.0} = 60°$

Then write each force vector in terms of $\mathbf{i}$ and $\mathbf{j}$ unit vectors:

$$\mathbf{F}_1 = |\mathbf{F}_1|\,\mathbf{i} \qquad \mathbf{F}_2 = -|\mathbf{F}_2|\,(\cos 60°)\,\mathbf{i} + |\mathbf{F}_2|\,(\sin 60°)\,\mathbf{j} \qquad \mathbf{W} = -260\mathbf{j}$$

For the system to be in static equilibrium, we must have $\mathbf{F}_1 + \mathbf{F}_2 + \mathbf{W} = \mathbf{0}$ which becomes,

on addition,

$$[|\mathbf{F}_1| - |\mathbf{F}_2|\,(\cos 60°)]\,\mathbf{i} + [|\mathbf{F}_2|\,(\sin 60°) - 260]\,\mathbf{j} = 0\mathbf{i} + 0\mathbf{j}.$$

Since two vectors are equal if and only if their corresponding components are equal, we are led to the

following system of equations in $|\mathbf{F}_1|$ and $|\mathbf{F}_2|$:

$$|\mathbf{F}_1| - |\mathbf{F}_2|\,(\cos 60°) = 0 \qquad\qquad |\mathbf{F}_2|\,(\sin 60°) - 260 = 0$$

Solving,

$$|\mathbf{F}_2| = \dfrac{260}{\sin 60°} = 300\ \text{lb} \qquad\qquad |\mathbf{F}_1| = |\mathbf{F}_2|\cos 60° = 150\ \text{lb}$$

The force in the member AB is directed oppositely to the diagram, a compression of 150 lb.

The force in the member BC is directed oppositely to the diagram, a tension of 300 lb.

60. (A) We write each force vector in terms of $\mathbf{i}$ and $\mathbf{j}$ unit vectors.

Note: $\alpha = 90° - \theta$.

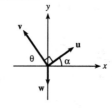

$$\begin{aligned}
\mathbf{u} &= |\mathbf{u}|\cos(90° - \theta)\,\mathbf{i} + |\mathbf{u}|\sin(90° - \theta)\,\mathbf{j} \\
&= |\mathbf{u}|\sin\theta\mathbf{i} + |\mathbf{u}|\cos\theta\mathbf{j} \\
\mathbf{v} &= -|\mathbf{v}|\cos\theta\mathbf{i} + |\mathbf{v}|\sin\theta\mathbf{j} \\
\mathbf{w} &= -|\mathbf{W}|\,\mathbf{j}
\end{aligned}$$

For the system to be in static equilibrium,

we must have $\mathbf{u} + \mathbf{v} + \mathbf{w} = \mathbf{0}$ which becomes, on addition,

$$[|\mathbf{u}|\sin\theta - |\mathbf{v}|\cos\theta]\,\mathbf{i} + [|\mathbf{u}|\cos\theta + |\mathbf{v}|\sin\theta - |\mathbf{w}|]\,\mathbf{j} = 0\mathbf{i} + 0\mathbf{j}$$

Since two vectors are equal if and only if their corresponding components are equal, we are led to

the following system of equations in $|\mathbf{u}|$ and $|\mathbf{v}|$:

(1)　$|\mathbf{u}|\sin\theta - |\mathbf{v}|\cos\theta = 0$

(2)　$|\mathbf{u}|\cos\theta + |\mathbf{v}|\sin\theta - |\mathbf{w}| = 0$

Solving, we have:

(3) $|\mathbf{u}| \sin^2 \theta - |\mathbf{v}| \sin \theta \cos \theta \qquad = 0$ Multiplying (1) by $\sin \theta$

(4) $|\mathbf{u}| \cos^2 \theta + |\mathbf{v}| \sin \theta \cos \theta - |\mathbf{w}| \cos \theta = 0$ Multiplying (2) by $\cos \theta$

$\overline{\qquad\qquad\qquad\qquad\qquad\qquad\qquad\qquad\qquad}$

$|\mathbf{u}| (\sin^2 \theta + \cos^2 \theta) \qquad - |\mathbf{w}| \cos \theta = 0$ Adding (3) and (4)

Since $\sin^2 \theta + \cos^2 \theta = 1$ by the Pythagorean identity, we have $|\mathbf{u}| = |\mathbf{w}| \cos \theta$.

Substituting this result into (1), we have

$\qquad |\mathbf{w}| \cos \theta \sin \theta - |\mathbf{v}| \cos \theta = 0$

Hence, $|\mathbf{v}| = \dfrac{|\mathbf{w}| \cos \theta \sin \theta}{\cos \theta} \qquad\qquad |\mathbf{v}| = |\mathbf{w}| \sin \theta$

(B) We are given $|\mathbf{w}| = 130$ lb and $\theta = 72°$. Hence,

$\qquad |\mathbf{u}| = |\mathbf{w}| \cos \theta = (130 \text{ lb}) \cos 72° = 40$ lb
and
$\qquad |\mathbf{v}| = |\mathbf{w}| \sin \theta = (130 \text{ lb}) \sin 72° = 124$ lb

(C) We are given $|\mathbf{u}| = \dfrac{1}{6} |\mathbf{w}|$. Hence

$\qquad \dfrac{1}{6} |\mathbf{w}| = |\mathbf{w}| \cos \theta \qquad\qquad \cos \theta = \dfrac{1}{6} \qquad\qquad \theta = \cos^{-1} \dfrac{1}{6} = 80°$

61. $W = \mathbf{F} \cdot \mathbf{d} = \langle -5, 8 \rangle \cdot \langle -8, 2 \rangle = 56$ ft-lb

CHAPTER 7   Polar Coordinates; Complex Numbers

EXERCISE 7.1 Polar and Rectangular Coordinates

1.

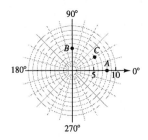

3.

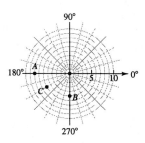

5.

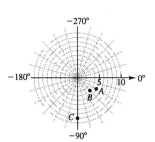

7.

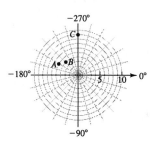

9.

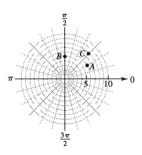

11.

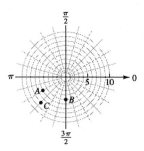

13.

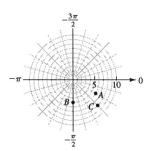

15.

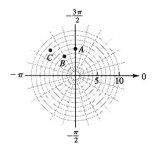

17.

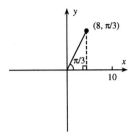

$$x = 8 \cos \frac{\pi}{3} = 8\left(\frac{1}{2}\right) = 4$$

$$y = 8 \sin \frac{\pi}{3} = 8\left(\frac{\sqrt{3}}{2}\right) = 4\sqrt{3}$$

Rectangular coordinates: $(4, 4\sqrt{3})$

19.

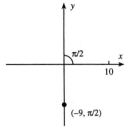

$$x = -9 \cos \frac{\pi}{2} = -9(0) = 0$$

$$x = -9 \sin \frac{\pi}{2} = -9(1) = -9$$

Rectangular coordinates: $(0, -9)$

21.

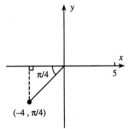

$$x = -4 \cos \frac{\pi}{4} = -4\left(\frac{\sqrt{2}}{2}\right) = -2\sqrt{2}$$

$$y = -4 \sin \frac{\pi}{4} = -4\left(\frac{\sqrt{2}}{2}\right) = -2\sqrt{2}$$

Rectangular coordinates: $(-2\sqrt{2}, -2\sqrt{2})$

23.

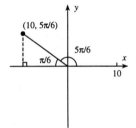

$$x = 10 \cos \frac{5\pi}{6} = 10\left(-\frac{\sqrt{3}}{2}\right) = -5\sqrt{3}$$

$$y = 10 \sin \frac{5\pi}{6} = 10\left(\frac{1}{2}\right) = 5$$

Rectangular coordinates: $(-5\sqrt{3}, 5)$

25.

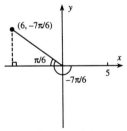

$$x = 6 \cos\left(-\frac{7\pi}{6}\right) = 6\left(-\frac{\sqrt{3}}{2}\right) = -3\sqrt{3}$$

$$y = 6 \sin\left(-\frac{7\pi}{6}\right) = 6\left(\frac{1}{2}\right) = 3$$

Rectangular coordinates: $(-3\sqrt{3}, 3)$

27.

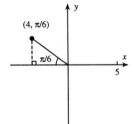

$$x = -4 \cos\left(-\frac{\pi}{6}\right) = -4\left(\frac{\sqrt{3}}{2}\right) = -2\sqrt{3}$$

$$y = -4 \sin\left(-\frac{\pi}{6}\right) = -4\left(-\frac{1}{2}\right) = 2$$

Rectangular coordinates: $(-2\sqrt{3}, 2)$

29. Use $r^2 = x^2 + y^2$ and $\tan\theta = \dfrac{y}{x}$

$r^2 = (2\sqrt{3})^2 + 2^2 = 16$ $r = 4$ $\tan\theta = \dfrac{2}{2\sqrt{3}} = \dfrac{1}{\sqrt{3}}$ $\theta = \dfrac{\pi}{6}$ since the point is in the first quadrant

Polar coordinates: $\left(4, \dfrac{\pi}{6}\right)$

31. Use $r^2 = x^2 + y^2$ and $\tan\theta = \dfrac{y}{x}$

$r^2 = (-4\sqrt{2})^2 + (4\sqrt{2})^2 = 64$ $r = 8$ $\tan\theta = \dfrac{4\sqrt{2}}{-4\sqrt{2}} = -1$ $\theta = \dfrac{3\pi}{4}$ since the point is in the second quadrant

Polar coordinates: $\left(8, \dfrac{3\pi}{4}\right)$

33. Use $r^2 = x^2 + y^2$ and $\tan\theta = \dfrac{y}{x}$

$r^2 = (-4)^2 + (-4\sqrt{3})^2 = 64$ $r = 8$ $\tan\theta = \dfrac{-4\sqrt{3}}{-4} = \sqrt{3}$ $\theta = \dfrac{4\pi}{3}$ since the point is in the third quadrant

Polar coordinates: $\left(8, \dfrac{4\pi}{3}\right)$

35. Use $r^2 = x^2 + y^2$ and $\tan\theta = \dfrac{y}{x}$

$r^2 = 0^2 + (-7)^2 = 49$ $r = 7$ $\tan\theta = \dfrac{-7}{0}$ is undefined $\theta = \dfrac{3\pi}{2}$ since the point is on the negative y axis

Polar coordinates: $\left(7, \dfrac{3\pi}{2}\right)$

37.

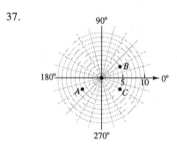

39.

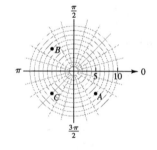

41. $6x - x^2 = y^2$

   $6x = x^2 + y^2$

   Use $x = r\cos\theta$ and $x^2 + y^2 = r^2$

   $6r\cos\theta = r^2$

   $0 = r^2 - 6r\cos\theta = r(r - 6\cos\theta)$

   $r = 0$ or $r - 6\cos\theta = 0$

The graph of $r = 0$ is the pole, and since the pole is included as a solution of

$r - 6\cos\theta = 0$ $\left(\text{let } \theta = \dfrac{\pi}{2}\right)$, we can

discard $r = 0$ and keep only

$r - 6\cos\theta = 0$ or $r = 6\cos\theta$

43. $2x + 3y = 5$

   Use $x = r \cos \theta$ and $y = r \sin \theta$

   $2r \cos \theta + 3r \sin \theta = 5$

   $r(2 \cos \theta + 3 \sin \theta) = 5$

45. $x^2 + y^2 = 9$

   Use $x^2 + y^2 = r^2$

   $r^2 = 9$ or $r = \pm 3$

47. $2xy = 1$

   Use $x = r \cos \theta$ and $y = r \sin \theta$

   $2r \cos \theta \, r \sin \theta = 1$

   $r^2(2 \sin \theta \cos \theta) = 1$

   $r^2 = \dfrac{1}{2 \sin \theta \cos \theta}$

   $= \dfrac{1}{\sin 2\theta}$

49. $4x^2 - y^2 = 4$

   Use $x = r \cos \theta$ and $y = r \sin \theta$

   $4(r \cos \theta)^2 - (r \sin \theta)^2 = 4$

   $4r^2 \cos^2 \theta - r^2 \sin^2 \theta = 4$

   $r^2(4 \cos^2 \theta - \sin^2 \theta) = 4$

   $r^2[4(1 - \sin^2 \theta) - \sin^2 \theta] = 4$

   $r^2(4 - 4 \sin^2 \theta - \sin^2 \theta) = 4$

   $r^2(4 - 5 \sin^2 \theta) = 4$

   $r^2 = \dfrac{4}{4 - 5 \sin^2 \theta}$

51. $r(2 \cos \theta + \sin \theta) = 4$

   $2r \cos \theta + r \sin \theta = 4$

   Use $x = r \cos \theta$ and $y = r \sin \theta$ $2x + y = 4$

53. $r = 8 \cos \theta$

   We multiply both sides by r, which adds the pole to the graph. But the pole is already part of the graph $\left(\text{let } \theta = \dfrac{\pi}{2} \right)$, so we have changed nothing. $r^2 = 8r \cos \theta$

   But $r^2 = x^2 + y^2$, $r \cos \theta = x$. Hence, $x^2 + y^2 = 8x$

55. $r^2 \cos 2\theta = 4$

   $r^2(\cos^2 \theta - \sin^2 \theta) = 4$

   $r^2 \cos^2 \theta - r^2 \sin^2 \theta = 4$

   $(r \cos \theta)^2 - (r \sin \theta)^2 = 4$

   Use $r \cos \theta = x$ and $r \sin \theta = y$

   $x^2 - y^2 = 4$

57. $r = 4$, so $r^2 = 16$

   Use $r^2 = x^2 + y^2$

   $x^2 + y^2 = 16$

59. $\theta = 30°$ $\tan \theta = \tan 30°$

   $\tan \theta = \dfrac{1}{\sqrt{3}}$ Use $\tan \theta = \dfrac{y}{x}$

   $\dfrac{y}{x} = \dfrac{1}{\sqrt{3}}$ $y = \dfrac{1}{\sqrt{3}} x$

61. $r = \dfrac{3}{\sin \theta - 2}$

   Multiply both sides by $\sin \theta - 2$, which is never 0 since $\sin \theta$ is never 2.

   $r(\sin \theta - 2) = 3$ $r \sin \theta - 2r = 3$

   Use $r \sin \theta = y$ and $r = -\sqrt{x^2 + y^2}$ $y + 2\sqrt{x^2 + y^2} = 3$

$y - 3 = -2\sqrt{x^2 + y^2}$ or $(y - 3)^2 = 4(x^2 + y^2)$

Note: The unusual choice of $r = -\sqrt{x^2 + y^2}$ is not a misprint. It is necessary to correspond with the fact that in the original polar equation, $\sin \theta < 2$; hence, $\sin \theta - 2$ is negative; hence, r is negative.

63. $d = \sqrt{r_1^2 + r_2^2 - 2r_1r_2 \cos(\theta_2 - \theta_1)}$ $(r_1, \theta_1) = (2, 30°)$ $(r_2, \theta_2) = (3, 60°)$

$\quad = \sqrt{2^2 + 3^2 - 2(2)(3)\cos(60° - 30°)}$

$\quad = \sqrt{2.6076952\ldots} \approx 1.615$ units

EXERCISE 7.2 Sketching Polar Equations

1.

θ	0	$\dfrac{\pi}{6}$	$\dfrac{\pi}{4}$	$\dfrac{\pi}{3}$	$\dfrac{\pi}{2}$
Exact value r	10	$5\sqrt{3}$	$5\sqrt{2}$	5	0
Calculator value r	10	8.7	7.1	5	0

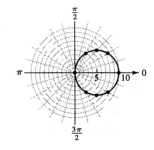

θ	$\dfrac{2\pi}{3}$	$\dfrac{3\pi}{4}$	$\dfrac{5\pi}{6}$	π
Exact value r	-5	$-5\sqrt{2}$	$-5\sqrt{3}$	-10
Calculator value r	-5	-7.1	-8.7	-10

3.

θ	0°	30°	60°	90°	120°	150°	180°	210°
Exact value r	$6\left(3 + \dfrac{3\sqrt{3}}{2}\right)$	$\dfrac{9}{2}$	3	$\dfrac{3}{2}\left(3 - \dfrac{3\sqrt{3}}{2}\right)$	0	$\left(3 - \dfrac{3\sqrt{3}}{2}\right)$		
Calculator value r	6	5.6	4.5	3	1.5	0.4	0	0.4

	θ	240°	270°	300°	330°	360°
Exact value	r	$\frac{3}{2}$	3	$\frac{9}{2}$	$\left(3 + \frac{3\sqrt{3}}{2}\right)$	6
Calculator value	r	1.5	3	4.5	5.6	6

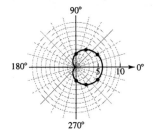

5.

θ	0	$\frac{\pi}{6}$	$\frac{\pi}{3}$	$\frac{\pi}{2}$	$\frac{2\pi}{3}$	$\frac{5\pi}{6}$	π
r	0	0.5	1.0	1.6	2.1	2.6	3.1

θ	$\frac{7\pi}{6}$	$\frac{4\pi}{3}$	$\frac{3\pi}{2}$	$\frac{5\pi}{3}$	$\frac{11\pi}{6}$	2π
r	3.7	4.2	4.7	5.2	5.8	6.3

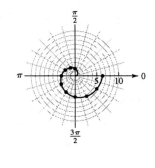

7.

θ	$-\frac{\pi}{2}$	$-\frac{\pi}{3}$	$-\frac{\pi}{6}$	0	$\frac{\pi}{6}$	$\frac{\pi}{3}$	$\frac{\pi}{2}$
r	Not defined	6	3.5	0	3.5	6	Not defined

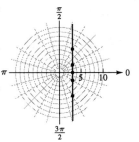

9. The graph consists of all points whose distance from the pole is 5, a circle with center at the pole, and radius 5.

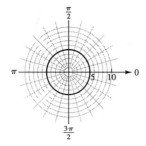

11. The graph consists of all points on a line that forms an angle of $\frac{\pi}{4}$ with the polar axis and passes through the pole.

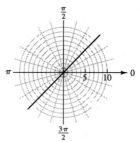

13. We start by graphing $r = 4\cos\theta$ in a rectangular coordinate system.

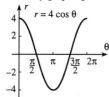

Rectangular coordinate system

From this, we observe how r varies as θ varies over particular intervals, summarize the information in a table, and sketch the polar curve from the information in the table.

θ	$4\cos\theta$	
0 to $\frac{\pi}{2}$	4 to 0	
$\frac{\pi}{2}$ to π	0 to -4	
π to $\frac{3\pi}{2}$	-4 to 0	Curve is traced out a second time in this region, although coordinate pairs are different
$\frac{3\pi}{2}$ to 2π	0 to 4	

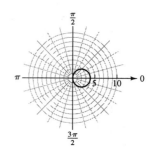

15. We start by graphing $r = 8 \cos 2\theta$ in a rectangular coordinate system.

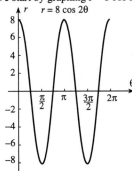

Rectangular coordinate system

From this, we observe how r varies as θ varies over particular intervals, summarize the information in a table, and sketch the polar curve from the information in the table.

θ	$8 \cos 2\theta$	θ	$8 \cos 2\theta$
0 to $\dfrac{\pi}{4}$	8 to 0	π to $\dfrac{5\pi}{4}$	8 to 0
$\dfrac{\pi}{4}$ to $\dfrac{\pi}{2}$	0 to -8	$\dfrac{5\pi}{4}$ to $\dfrac{3\pi}{2}$	0 to -8
$\dfrac{\pi}{2}$ to $\dfrac{3\pi}{4}$	-8 to 0	$\dfrac{3\pi}{2}$ to $\dfrac{7\pi}{4}$	-8 to 0
$\dfrac{3\pi}{4}$ to π	0 to 8	$\dfrac{7\pi}{4}$ to 2π	0 to 8

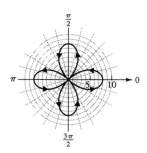

17. We start by graphing $r = 6 \sin 3\theta$ in a rectangular coordinate system.

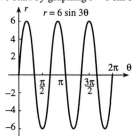

Rectangular coordinate system

From this, we observe how r varies as θ varies over particular intervals, summarize the information in a table, and sketch the polar curve from the information in the table.

θ	$6 \sin 3\theta$
0 to $\dfrac{\pi}{6}$	0 to 6
$\dfrac{\pi}{6}$ to $\dfrac{\pi}{3}$	6 to 0
$\dfrac{\pi}{3}$ to $\dfrac{\pi}{2}$	0 to -6
$\dfrac{\pi}{2}$ to $\dfrac{2\pi}{3}$	-6 to 0
$\dfrac{2\pi}{3}$ to $\dfrac{5\pi}{6}$	0 to 6
$\dfrac{5\pi}{6}$ to π	6 to 0

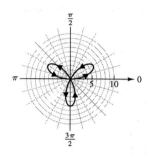

19. We start by graphing $r = 3 + 3 \cos \theta$ in a rectangular coordinate system.

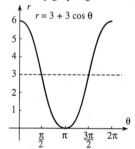

Rectangular coordinate system

From this, we observe how r varies as θ varies over particular intervals, summarize the information

in a table, and sketch the polar curve from the information in the table.

θ	$3 + 3 \cos \theta$
0 to $\dfrac{\pi}{2}$	6 to 3
$\dfrac{\pi}{2}$ to π	3 to 0
π to $\dfrac{3\pi}{2}$	0 to 3
$\dfrac{3\pi}{2}$ to 2π	3 to 6

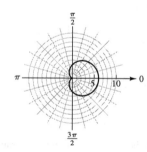

21. We start by graphing $r = 2 + 4 \cos \theta$ in a rectangular coordinate system.

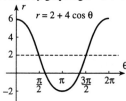

Rectangular coordinate system

From this, we observe how r varies as θ varies over particular intervals, summarize the information in a table, and sketch the polar curve from the information in the table. We note further that $r = 2 + 4 \cos \theta$ will equal 0 when $\cos \theta = -\dfrac{1}{2}$, that is, when $\theta = \dfrac{2\pi}{3}, \dfrac{4\pi}{3}$ (to list values between 0 and 2π only.)

θ	$2 + 4 \cos \theta$
0 to $\dfrac{\pi}{2}$	6 to 2
$\dfrac{\pi}{2}$ to $\dfrac{2\pi}{3}$	2 to 0
$\dfrac{2\pi}{3}$ to π	0 to –2
π to $\dfrac{4\pi}{3}$	–2 to 0
$\dfrac{4\pi}{3}$ to $\dfrac{3\pi}{2}$	0 to 2
$\dfrac{3\pi}{2}$ to 2π	2 to 6

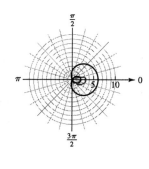

23. We start by graphing $r = 4 - 2 \sin \theta$ in a rectangular coordinate system.

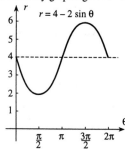

Rectangular coordinate system

From this, we observe how r varies as θ varies over particular intervals, summarize the information in a table, and sketch the polar curve from the information in the table.

θ	$4 - 2\sin\theta$
0 to $\dfrac{\pi}{2}$	4 to 2
$\dfrac{\pi}{2}$ to π	2 to 4
π to $\dfrac{3\pi}{2}$	4 to 6
$\dfrac{3\pi}{2}$ to 2π	6 to 4

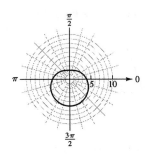

25. As θ increases, r increases at $\dfrac{1}{\pi}$ the rate that θ increases. The curve spirals leisurely out from the pole.

θ	r
0 to π	0 to 1
π to 2π	1 to 2
2π to 3π	2 to 3
3π to 4π	3 to 4
4π to 5π	4 to 5
5π to 6π	5 to 6
6π to 7π	6 to 7
7π to 8π	7 to 8
$> 8\pi$	r continues to increase

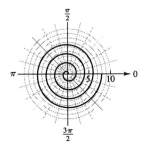

27. The following table can be used to investigate how r varies as θ varies over particular intervals. We then sketch the polar curve from the information in the table.

θ	$\dfrac{\theta}{2}$	$\cos\dfrac{\theta}{2}$	$5\cos\dfrac{\theta}{2}$	$5 + 5\cos\dfrac{\theta}{2}$
0 to $\dfrac{\pi}{2}$	0 to $\dfrac{\pi}{4}$	1 to $\dfrac{\sqrt{2}}{2}$	5 to $\dfrac{5\sqrt{2}}{2}$	10 to $5 + \dfrac{5\sqrt{2}}{2}$ (≈ 8.5)
$\dfrac{\pi}{2}$ to π	$\dfrac{\pi}{4}$ to $\dfrac{\pi}{2}$	$\dfrac{\sqrt{2}}{2}$ to 0	$\dfrac{5\sqrt{2}}{2}$ to 0	$5 + \dfrac{5\sqrt{2}}{2}$ to 5
π to $\dfrac{3\pi}{2}$	$\dfrac{\pi}{2}$ to $\dfrac{3\pi}{4}$	0 to $-\dfrac{\sqrt{2}}{2}$	0 to $-\dfrac{5\sqrt{2}}{2}$	5 to $5 - \dfrac{5\sqrt{2}}{2}$ (≈ 1.5)
$\dfrac{3\pi}{2}$ to 2π	$\dfrac{3\pi}{4}$ to π	$-\dfrac{\sqrt{2}}{2}$ to -1	$-\dfrac{5\sqrt{2}}{2}$ to -5	$5 - \dfrac{5\sqrt{2}}{2}$ to 0

θ	$\dfrac{\theta}{2}$	$\cos\dfrac{\theta}{2}$	$5\cos\dfrac{\theta}{2}$	$5+5\cos\dfrac{\theta}{2}$
2π to $\dfrac{5\pi}{2}$	π to $\dfrac{5\pi}{4}$	-1 to $-\dfrac{\sqrt{2}}{2}$	-5 to $-\dfrac{5\sqrt{2}}{2}$	0 to $5-\dfrac{5\sqrt{2}}{2}$
$\dfrac{5\pi}{2}$ to 3π	$\dfrac{5\pi}{4}$ to $\dfrac{3\pi}{2}$	$-\dfrac{\sqrt{2}}{2}$ to 0	$-\dfrac{5\sqrt{2}}{2}$ to 0	$5-\dfrac{5\sqrt{2}}{2}$ to 5
3π to $\dfrac{7\pi}{2}$	$\dfrac{3\pi}{2}$ to $\dfrac{7\pi}{4}$	0 to $\dfrac{\sqrt{2}}{2}$	0 to $\dfrac{5\sqrt{2}}{2}$	5 to $5+\dfrac{5\sqrt{2}}{2}$
$\dfrac{7\pi}{2}$ to 4π	$\dfrac{7\pi}{4}$ to 2π	$\dfrac{\sqrt{2}}{2}$ to 1	$\dfrac{5\sqrt{2}}{2}$ to 5	$5+\dfrac{5\sqrt{2}}{2}$ to 10

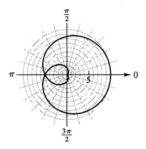

29. The following table can be used to investigate how r varies as θ varies over particular intervals. We then sketch the polar curve from the information in the table.

θ	2θ	$\cos 2\theta$	$64\cos 2\theta$	$r=\pm\sqrt{64\cos 2\theta}$	
0 to $\dfrac{\pi}{4}$	0 to $\dfrac{\pi}{2}$	1 to 0	64 to 0	$\begin{cases} 8 \text{ to } 0 \\ -8 \text{ to } 0 \end{cases}$	The two branches of the curve are reflections of each other in the x axis.
$\dfrac{\pi}{4}$ to $\dfrac{\pi}{2}$	$\dfrac{\pi}{2}$ to π	0 to -1	0 to -64	$\left.\begin{array}{l} \\ \end{array}\right\}$ No curve; no real square root of a negative no.	
$\dfrac{\pi}{2}$ to $\dfrac{3\pi}{4}$	π to $\dfrac{3\pi}{2}$	-1 to 0	-64 to 0		
$\dfrac{3\pi}{4}$ to π	$\dfrac{3\pi}{2}$ to 2π	0 to 1	0 to 64	$\begin{cases} 0 \text{ to } 8 \\ 0 \text{ to } -8 \end{cases}$	The two branches of the curve are reflections of each other in the x axis
π to $\dfrac{5\pi}{4}$	2π to $\dfrac{5\pi}{2}$	1 to 0	64 to 0	$\begin{cases} 8 \text{ to } 0 \\ -8 \text{ to } 0 \end{cases}$	
$\dfrac{5\pi}{4}$ to $\dfrac{3\pi}{2}$	$\dfrac{5\pi}{2}$ to 3π	0 to -1	0 to -64	$\left.\begin{array}{l} \\ \end{array}\right\}$ No curve	This repeats the curve already traced out
$\dfrac{3\pi}{2}$ to $\dfrac{7\pi}{4}$	3π to $\dfrac{7\pi}{2}$	-1 to 0	-64 to 0		
$\dfrac{7\pi}{4}$ to 2π	$\dfrac{7\pi}{2}$ to 4π	0 to 1	0 to 64	$\begin{cases} 0 \text{ to } 8 \\ -8 \text{ to } 0 \end{cases}$	

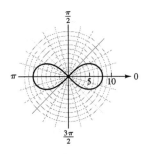

31. We sketch the two graphs using rapid sketching techniques. Tables of how r varies for each curve as θ varies over particular intervals can be readily constructed.

θ	$2\cos\theta$	$2\sin\theta$
0 to $\dfrac{\pi}{2}$	2 to 0	0 to 2
$\dfrac{\pi}{2}$ to π	0 to –2	2 to 0
π to $\dfrac{3\pi}{2}$	–2 to 0	0 to –2
$\dfrac{3\pi}{2}$ to π	0 to 2	–2 to 0

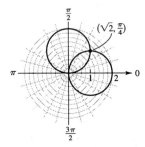

We solve the system, $r = 2\cos\theta$, $r = 2\sin\theta$, by equating the right sides: $2\cos\theta = 2\sin\theta$

$$\cos\theta = \sin\theta$$
$$1 = \tan\theta$$

The only solution of this equation, $0 \le \theta \le \pi$, is $\theta = \dfrac{\pi}{4}$. If we substitute this into either of the

original equations, we get $r = 2\cos\dfrac{\pi}{4} = 2\sin\dfrac{\pi}{4} = 2\left(\dfrac{\sqrt{2}}{2}\right) = \sqrt{2}.$

Solution: $\left(\sqrt{2}, \dfrac{\pi}{4}\right)$

The sketch shows that the pole is on both graphs; however, the pole has no ordered pairs of coordinates that simultaneously satisfy both equations. As $\left(0, \dfrac{\pi}{2}\right)$, it satisfies the first; as $(0, 0)$,

it satisfies the second; it is not a solution of the system.

33. We graph the system using rapid sketching techniques. See Problem 15 for a table for $r = 8 \cos 2\theta$;
a table for $r = 8 \sin \theta$ can be readily constructed.

θ	$8 \sin \theta$
0 to $\dfrac{\pi}{2}$	0 to 8
$\dfrac{\pi}{2}$ to π	8 to 0
π to $\dfrac{3\pi}{2}$	0 to -8
$\dfrac{3\pi}{2}$ to 2π	-8 to 0

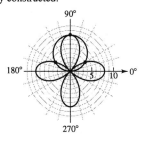

We solve the system: $r = 8 \sin \theta, r = 8 \cos 2\theta$, by equating the right-hand sides:

$$8 \sin \theta = 8 \cos 2\theta$$

$$\sin \theta = \cos 2\theta$$

$$\sin \theta = 1 - 2 \sin^2 \theta \qquad \text{Double-angle identity}$$

$$2 \sin^2 \theta + \sin \theta - 1 = 0$$

$$(2 \sin \theta - 1)(\sin \theta + 1) = 0 \qquad \text{Factoring}$$

Therefore, $2 \sin \theta - 1 = 0$ or $\sin \theta + 1 = 0$

$2 \sin \theta - 1 = 0$	$\sin \theta + 1 = 0$
$\sin \theta = \dfrac{1}{2}$	$\sin \theta = -1$
$\theta = 30°, 150°$	$\theta = 270°$

If we substitute these values into either of the original equations, we get (choosing $r = 8 \sin \theta$ for
ease of calculation):

$\theta = 30°$	$r = 8 \sin 30° = 8\left(\dfrac{1}{2}\right) = 4$	Solution: $(4, 30°)$
$\theta = 150°$	$r = 8 \sin 150° = 8\left(\dfrac{1}{2}\right) = 4$	Solution: $(4, 150°)$
$\theta = 270°$	$r = 8 \sin 270° = 8(-1) = -8$	Solution: $(-8, 270°)$

The sketch shows that all three solutions are on both graphs; it also shows that the pole is on both
graphs. However, the pole has no ordered pairs of coordinates that simultaneously satisfy both
equations. As $(0, 0°)$, it satisfies the first; as $(0, 45°)$, it satisfies the second; it is not a solution of
the system.

35. n = 2   n = 4   n = 6

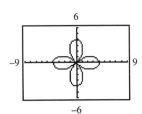 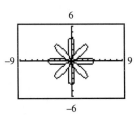 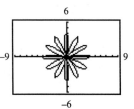

37. n = 3   n = 5   n = 7

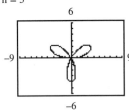 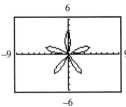 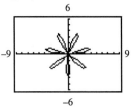

39.   41.

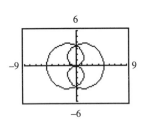 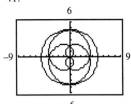

43.   45.

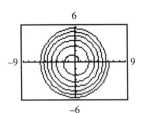 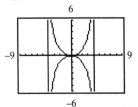

47. (A) The America's cup curve appears to pass through the point with polar coordinates (9.5, 30°).
         Speed = 9.5 knots.

     (B) The America's cup curve appears to pass through the point with polar coordinates (12.0, 60°).
         Speed = 12.0 knots.

     (C) The America's cup curve appears to pass through the point with polar coordinates (13.5, 90°).
         Speed = 13.5 knots.

     (D) The America's cup curve appears to pass through the point with polar coordinates (12.0, 120°).
         Speed = 12.0 knots.

49. (A) $r = \dfrac{8}{1 - 0.5 \cos \theta}$

$$\text{(graph, window } -18 \text{ to } 18, \; 12 \text{ to } -12)$$

The graph is an ellipse.

(B) $r = \dfrac{8}{1 - \cos \theta}$

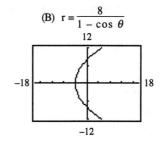

The graph is a parabola.

(C) $r = \dfrac{8}{1 - 2 \cos \theta}$

$$\text{(graph, window } -18 \text{ to } 18, \; 12 \text{ to } -12)$$

The graph is a hyperbola.

51. (A)   At aphelion, $\theta = 0°$, hence

$$r = \frac{3.44 \times 10^7 \text{ mi}}{1 - 0.206 \cos 0°} = 4.33 \times 10^7 \text{ mi}$$

At perihelion, $\theta = 180°$, hence

$$r = \frac{3.44 \times 10^7 \text{ mi}}{1 - 0.206 \cos 180°} = 2.85 \times 10^7 \text{ mi}$$

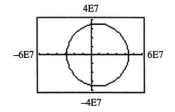

(B) The diagram in the text shows the areas swept out when the planet is near aphelion or perihelion as, approximately, triangles, whose areas would be proportional to their base (distance travelled by planet) and altitude (distance from planet to sun). Since at aphelion the distance from planet to sun is greater, the distance travelled by planet must be smaller in order to have equal area. Thus, the planet travels more slowly at aphelion, faster at perihelion.

Chapter 7   Polar Coordinates; Complex Numbers

EXERCISE 7.3 Complex Numbers in Rectangular and Polar Form

1.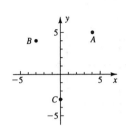

3.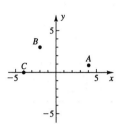

5.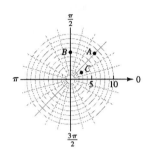

7.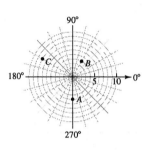

9. $z_1 z_2 = (10 \cdot 5)[\cos(45° + 32°) + i \sin(45° + 32°)] = 50(\cos 77° + i \sin 77°)$

$$\frac{z_1}{z_2} = \frac{10}{5}[\cos(45° - 32°) + i \sin(45° - 32°)] = 2(\cos 13° + i \sin 13°)$$

11. $z_1 z_2 = (7 \cdot 3)[\cos(163° + 102°) + i \sin(163° + 102°)] = 21(\cos 265° + i \sin 265°)$

$$\frac{z_1}{z_2} = \frac{7}{3}[\cos(163° - 102°) + i \sin(163° - 102°)] = \frac{7}{3}(\cos 61° + i \sin 61°)$$

13.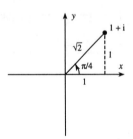

15.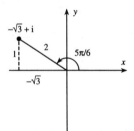

$r = \sqrt{2}$ $\theta = \dfrac{\pi}{4}$

$z = \sqrt{2}\left(\cos \dfrac{\pi}{4} + i \sin \dfrac{\pi}{4}\right)$

$r = 2$ $\theta = \dfrac{5\pi}{6}$

$z = 2\left(\cos \dfrac{5\pi}{6} + i \sin \dfrac{5\pi}{6}\right)$

17.

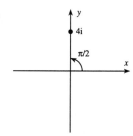

$r = 4$ $\qquad$ $\theta = \dfrac{\pi}{2}$

$z = 4\left(\cos\dfrac{\pi}{2} + i\,\sin\dfrac{\pi}{2}\right)$

19.

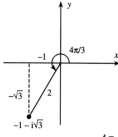

$r = 2$ $\qquad$ $\theta = \dfrac{4\pi}{3}$

$z = 2\left(\cos\dfrac{4\pi}{3} + i\,\sin\dfrac{4\pi}{3}\right)$

21.

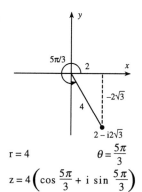

$r = 4$ $\qquad$ $\theta = \dfrac{5\pi}{3}$

$z = 4\left(\cos\dfrac{5\pi}{3} + i\,\sin\dfrac{5\pi}{3}\right)$

23.

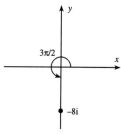

$r = 8$ $\qquad$ $\theta = \dfrac{3\pi}{2}$

$z = 8\left(\cos\dfrac{3\pi}{2} + i\,\sin\dfrac{3\pi}{2}\right)$

25. $\sqrt{2}(\cos 45° + i\,\sin 45°) = \sqrt{2}\left(\dfrac{1}{\sqrt{2}} + i\,\dfrac{1}{\sqrt{2}}\right) = 1 + i$

27. $\sqrt{2}(\cos 135° + i\,\sin 135°) = \sqrt{2}\left(-\dfrac{1}{\sqrt{2}} + i\,\dfrac{1}{\sqrt{2}}\right) = -1 + i$

29. $8(\cos \pi + i\,\sin \pi) = 8(-1 + i0) = -8$ $\qquad$ 31. $12(\cos 90° + i\,\sin 90°) = 12(0 + i1) = 12i$

33. $6\left(\cos\dfrac{4\pi}{3} + i\,\sin\dfrac{4\pi}{3}\right) = 6\left[-\dfrac{1}{2} + i\left(-\dfrac{\sqrt{3}}{2}\right)\right] = -3 - i3\sqrt{3}$

35. $4(\cos 330° + i\,\sin 330°) = 4\left[\dfrac{\sqrt{3}}{2} + i\left(-\dfrac{1}{2}\right)\right] = 2\sqrt{3} - 2i$

37. Directly: $(1 + i)^2 = (1)^2 + 2(1)\,i + i^2 = 1 + 2i + i^2 = 1 + 2i + (-1) = 2i$

Using polar form: $(1 + i)^2 = \left[\sqrt{2}\left(\cos\dfrac{\pi}{4} + i\,\sin\dfrac{\pi}{4}\right)\right]^2$ $\qquad$ (See Problem 13)

$$= \sqrt{2}\left(\cos\dfrac{\pi}{4} + i\,\sin\dfrac{\pi}{4}\right)\sqrt{2}\left(\cos\dfrac{\pi}{4} + i\,\sin\dfrac{\pi}{4}\right)$$

$$= \left(\sqrt{2} \cdot \sqrt{2}\right)\left[\cos\left(\frac{\pi}{4} + \frac{\pi}{4}\right) + i\,\sin\left(\frac{\pi}{4} + \frac{\pi}{4}\right)\right]$$

$$= 2\left(\cos\frac{\pi}{2} + i\,\sin\frac{\pi}{2}\right) \text{ or } 2(\cos 90° + i\,\sin 90°)$$

39. Directly: $(1 + \sqrt{3})(\sqrt{3} + i) = 1 \cdot \sqrt{3} + 1 \cdot i + i\sqrt{3} \cdot \sqrt{3} + i\sqrt{3} \cdot i$

$$= \sqrt{3} + i + 3i + i^2\sqrt{3}$$

$$= \sqrt{3} + 4i + (-1)\sqrt{3}$$

$$= \sqrt{3} + 4i - \sqrt{3} = 4i$$

Using polar form:

$$1 + i\sqrt{3} = 2\left(\cos\frac{\pi}{3} + i\,\sin\frac{\pi}{3}\right)$$

$$\sqrt{3} + i = 2\left(\cos\frac{\pi}{6} + i\,\sin\frac{\pi}{6}\right)$$

$$(1 + i\sqrt{3})(\sqrt{3} + i) = \left[2\left(\cos\frac{\pi}{3} + i\,\sin\frac{\pi}{3}\right)\right]\left[2\left(\cos\frac{\pi}{6} + i\,\sin\frac{\pi}{6}\right)\right]$$

$$= (2 \cdot 2)\left[\cos\left(\frac{\pi}{3} + \frac{\pi}{6}\right) + i\,\sin\left(\frac{\pi}{3} + \frac{\pi}{6}\right)\right] = 4\left(\cos\frac{\pi}{2} + i\,\sin\frac{\pi}{2}\right)$$

41. $r = \sqrt{3^2 + 5^2} = \sqrt{34} \qquad \tan\theta = \frac{5}{3} \qquad \theta = 59.0°$

$3 + 5i = \sqrt{34}(\cos 59.0° + \sin 59.0°)$

43. $r = \sqrt{(-7)^2 + 3^2} = \sqrt{58} \qquad \tan\theta = \frac{3}{-7} \qquad \cos\theta = -\frac{7}{\sqrt{58}}$

$\theta = 156.8°$ (second quadrant) $\qquad\qquad -7 + 3i = \sqrt{58}(\cos 156.8° + i\,\sin 156.8°)$

45. $r = \sqrt{6^2 + (-5)^2} = \sqrt{61} \qquad \tan\theta = -\frac{5}{6} \qquad \cos\theta = \frac{6}{\sqrt{61}}$

$\theta = 320.2°$ (fourth quadrant) $\qquad\qquad 6 - 5i = \sqrt{61}(\cos 320.2° + i\,\sin 320.2°)$

47. $9(\cos 37°20' + i\,\sin 37°20') = 9\cos 37°20' + 9i\sin 37°20' = 7.16 + 5.46i$

49. $5(\cos 197.2° + i\,\sin 197.2°) = 5\cos 197.2° + 5i\sin 197.2° = -4.78 - 1.48i$

51. $11(\cos 321°20' + i\,\sin 321°20') = 11\cos 321°20' + 11i\sin 321°20' = 8.59 - 6.87i$

53. If $z = r(\cos\theta + i\,\sin\theta)$, then

$$z^2 = zz = r(\cos + i\,\sin\theta)\,r(\cos\theta + i\,\sin\theta)$$

$$= (r \cdot r)[\cos(\theta + \theta) + i\,\sin(\theta + \theta)] = r^2(\cos 2\theta + i\,\sin 2\theta)$$

$$z^3 = zz^2 = r(\cos\theta + i\,\sin\theta)\,r^2(\cos 2\theta + i\,\sin 2\theta)$$

$$= (r \cdot r^2)[\cos(\theta + 2\theta) + i\,\sin(\theta + 2\theta)] = r^3(\cos 3\theta + i\,\sin 3\theta)$$

A logical (and correct) guess would be

$$z^n = r^n(\cos n\theta + i \sin n\theta) \qquad \text{(See Section 7.4 of the text)}$$

55. To show that $r^{1/n}\left[\cos \dfrac{\theta}{n} + i \sin \dfrac{\theta}{n}\right]$ is an nth root of $z = r(\cos \theta + i \sin \theta)$, we need only show that

$$\left\{ r^{1/n}\left[\cos \frac{\theta}{n} + i \sin \frac{\theta}{n}\right] \right\}^n = z$$

but, by the given assumption,

$$\left\{ r^{1/n}\left[\cos \frac{\theta}{n} + i \sin \frac{\theta}{n}\right] \right\}^n = (r^{1/n})^n\left[\cos n \frac{\theta}{n} + i \sin n \frac{\theta}{n}\right] = r(\cos \theta + i \sin \theta) = z$$

which is what was to be proved.

57. (A) $8(\cos 0° + i \sin 0°) = 8(1 + 0i) = 8 + 0i$

$6(\cos 30° + i \sin 30°) = 6\left(\dfrac{\sqrt{3}}{2} + i \dfrac{1}{2}\right) = 3\sqrt{3} + 3i$

$(8 + 0i) + (3\sqrt{3} + 3i) = (8 + 3\sqrt{3}) + 3i$

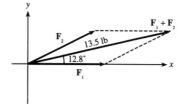

(B) $r = \sqrt{(8 + 3\sqrt{3})^2 + 3^2} \approx 13.5$

$\tan \theta = \dfrac{3}{8 + 3\sqrt{3}} \qquad \theta = 12.8°$

$(8 + 3\sqrt{3}) + 3i = 13.5(\cos 12.8° + i \sin 12.8°)$

(C) We can interpret $13.5(\cos 12.8° + i \sin 12.8°)$ as a force of 13.5 lb at an angle of 12.8° with respect to the force $\mathbf{F}_1$.

EXERCISE 7.4 De Moivre's Theorem and the nth Root Theorem

1. $[3(\cos 15° + i \sin 15°]^3 = 3^3[\cos 3(15°) + i \sin 3(15°)] = 27(\cos 45° + i \sin 45°)$

3. $[\sqrt{2}(\cos 45° + i \sin 45°)]^{10} = \left(\sqrt{2}\right)^{10}[\cos 10(45°) + i \sin 10(45°)]$

$$= (2^{1/2})^{10}(\cos 450° + i \sin 450°) = 2^5(\cos 450° + i \sin 450°)$$

$$= 32(\cos 450° + i \sin 450°) \ \text{ or } \ 32(\cos 90° + i \sin 90°)$$

5. $\left(\sqrt{3} + i\right)^6 = [2(\cos 30° + i \sin 30°)]^6 = 2^6(\cos 6 \cdot 30° + i \sin 6 \cdot 30°) = 64(\cos 180° + i \sin 180°)$

7. $(-1 + i)^4 = [\sqrt{2}(\cos 135° + i \sin 135°)]^4 = \left(\sqrt{2}\right)^4[\cos 4(135°) + i \sin 4(135°)]$

$$= (2^{1/2})^4 (\cos 540° + i \sin 540°) = 2^2(\cos 540° + i \sin 540°) = 4(\cos 180° + i \sin 180°)$$

$$= 4(-1 + i0) = -4$$

9. $\left(-\sqrt{3} + i\right)^5 = [2(\cos 150° + i \sin 150°)]^5 = 2^5(\cos 5 \cdot 150° + i \sin 5 \cdot 150°)$

Chapter 7   Polar Coordinates; Complex Numbers

$$= 32(\cos 750° + i \sin 750°) = 32(\cos 30° + i \sin 30°)$$

$$= 32\left(\frac{\sqrt{3}}{2} + i\ \frac{1}{2}\right) = 16\sqrt{3} + 16i$$

11. $\left(-\dfrac{1}{2} - \dfrac{\sqrt{3}}{2}\ i\right)^3 = [1(\cos 240° + i \sin 240°)]^3 = 1^3[\cos 3(240°) + i \sin 3(240°)]$

   $$= \cos 720° + i \sin 720° = \cos 0° + i \sin 0° = 1 + 0i = 1$$

13. The square roots of $4(\cos 30° + i \sin 30°)$ are given by

$$4^{1/2}\left(\cos \frac{30° + k \cdot 360°}{2} + i \sin \frac{30° + k \cdot 360°}{2}\right) \qquad k = 0, 1$$

Thus,

$$w_1 = 4^{1/2}\left(\cos \frac{30° + 0 \cdot 360°}{2} + i \sin \frac{30° + 0 \cdot 360°}{2}\right) = 2(\cos 15° + i \sin 15°)$$

$$w_2 = 4^{1/2}\left(\cos \frac{30° + 1 \cdot 360°}{2} + i \sin \frac{30° + 1 \cdot 360°}{2}\right) = 2(\cos 195° + i \sin 195°)$$

15. The cube roots of $8(\cos 90° + i \sin 90°)$ are given by

$$8^{1/3}\left(\cos \frac{90° + k \cdot 360°}{3} + i \sin \frac{90° + k \cdot 360°}{3}\right) \qquad k = 0, 1, 2$$

Thus,

$$w_1 = 8^{1/3}\left(\cos \frac{90° + 0 \cdot 360°}{3} + i \sin \frac{90° + 0 \cdot 360°}{3}\right) = 2(\cos 30° + i \sin 30°)$$

$$w_2 = 8^{1/3}\left(\cos \frac{90° + 1 \cdot 360°}{3} + i \sin \frac{90° + 1 \cdot 360°}{3}\right) = 2(\cos 150° + i \sin 150°)$$

$$w_3 = 8^{1/3}\left(\cos \frac{90° + 2 \cdot 360°}{3} + i \sin \frac{90° + 2 \cdot 360°}{3}\right) = 2(\cos 270° + i \sin 270°)$$

17. First, write $-1 + i$ in polar form: $-1 + i = \sqrt{2}(\cos 135° + i \sin 135°)$.

   Then, the fifth roots of $2^{1/2}(\cos 135° + i \sin 135°)$ are given by

$$(2^{1/2})^{1/5}\left(\cos \frac{135° + k \cdot 360°}{5} + i \sin \frac{135° + k \cdot 360°}{5}\right) \qquad k = 0, 1, 2, 3, 4$$

Thus,

$$w_1 = 2^{1/10}\left(\cos \frac{135° + 0 \cdot 360°}{5} + i \sin \frac{135° + 0 \cdot 360°}{5}\right) = 2^{1/10}(\cos 27° + i \sin 27°)$$

$$w_2 = 2^{1/10}\left(\cos \frac{135° + 1 \cdot 360°}{5} + i \sin \frac{135° + 1 \cdot 360°}{5}\right) = 2^{1/10}(\cos 99° + i \sin 99°)$$

$$w_3 = 2^{1/10}\left(\cos \frac{135° + 2 \cdot 360°}{5} + i \sin \frac{135° + 2 \cdot 360°}{5}\right) = 2^{1/10}(\cos 171° + i \sin 171°)$$

$$w_4 = 2^{1/10}\left(\cos \frac{135° + 3 \cdot 360°}{5} + i \sin \frac{135° + 3 \cdot 360°}{5}\right) = 2^{1/10}(\cos 243° + i \sin 243°)$$

$$w_5 = 2^{1/10}\left(\cos \frac{135° + 4 \cdot 360°}{5} + i \sin \frac{135° + 4 \cdot 360°}{5}\right) = 2^{1/10}(\cos 315° + i \sin 315°)$$

19. First, write 1 in polar form: $1 = 1(\cos 0° + i \sin 0°)$.

Then, the sixth roots of $1(\cos 0° + i \sin 0°)$ are given by

$$1^{1/6}\left(\cos \frac{0° + k \cdot 360°}{6} + i \sin \frac{0° + k \cdot 360°}{6}\right) \qquad k = 0, 1, 2, 3, 4, 5$$

Thus,

$$w_1 = 1^{1/6}\left(\cos \frac{0° + 0 \cdot 360°}{6} + i \sin \frac{0° + 0 \cdot 360°}{6}\right) = 1(\cos 0° + i \sin 0°)$$

$$w_2 = 1^{1/6}\left(\cos \frac{0° + 1 \cdot 360°}{6} + i \sin \frac{0° + 1 \cdot 360°}{6}\right) = 1(\cos 60° + i \sin 60°)$$

$$w_3 = 1^{1/6}\left(\cos \frac{0° + 2 \cdot 360°}{6} + i \sin \frac{0° + 2 \cdot 360°}{6}\right) = 1(\cos 120° + i \sin 120°)$$

$$w_4 = 1^{1/6}\left(\cos \frac{0° + 3 \cdot 360°}{6} + i \sin \frac{0° + 3 \cdot 360°}{6}\right) = 1(\cos 180° + i \sin 180°)$$

$$w_5 = 1^{1/6}\left(\cos \frac{0° + 4 \cdot 360°}{6} + i \sin \frac{0° + 4 \cdot 360°}{6}\right) = 1(\cos 240° + i \sin 240°)$$

$$w_6 = 1^{1/6}\left(\cos \frac{0° + 5 \cdot 360°}{6} + i \sin \frac{0° + 5 \cdot 360°}{6}\right) = 1(\cos 300° + i \sin 300°)$$

21. $x^3 - 8 = 0 \qquad x^3 = 8$

Therefore, x is a cube root of 8, and there are three of them. First, we write 8 in polar form:

$$8 = 8 + 0i = 8(\cos 0° + i \sin 0°)$$

All three cube roots of $8(\cos 0° + i \sin 0°)$ are given by

$$8^{1/3}\left(\cos \frac{0° + k \cdot 360°}{3} + i \sin \frac{0° + k \cdot 360°}{3}\right) \qquad k = 0, 1, 2$$

Thus,

$$w_1 = 8^{1/3}\left(\cos \frac{0° + 0 \cdot 360°}{3} + i \sin \frac{0° + 0 \cdot 360°}{3}\right)$$

$$= 2(\cos 0° + i \sin 0°) = 2(1 + 0i) = 2 + 0i \text{ or } 2$$

$$w_2 = 8^{1/3}\left(\cos \frac{0° + 1 \cdot 360°}{3} + i \sin \frac{0° + 1 \cdot 360°}{3}\right)$$

$$= 2(\cos 120° + i \sin 120°) = 2\left(-\frac{1}{2} + i \frac{\sqrt{3}}{2}\right) = -1 + i\sqrt{3}$$

$$w_3 = 8^{1/3}\left(\cos \frac{0° + 2 \cdot 360°}{3} + i \sin \frac{0° + 2 \cdot 360°}{3}\right)$$

$$= 2(\cos 240° + i \sin 240°) = 2\left(-\frac{1}{2} - i \frac{\sqrt{3}}{2}\right) = -1 - i\sqrt{3}$$

23. $x^3 + 27 = 0 \qquad x^3 = -27$

Therefore, x is a cube root of –27, and there are three of them. First, we write –27 in polar form:

$$-27 = -27 + 0i = 27(\cos 180° + i \sin 180°)$$

All three cube roots of 27(cos 180° + i sin 180°) are given by

$$27^{1/3}\left(\cos\frac{180° + k\cdot 360°}{3} + i\sin\frac{180° + k\cdot 360°}{3}\right) \quad k = 0, 1, 2$$

Thus,

$$w_1 = 27^{1/3}\left(\cos\frac{180° + 0\cdot 360°}{3} + i\sin\frac{180° + 0\cdot 360°}{3}\right)$$

$$= 3(\cos 60° + i\sin 60°) = 3\left(\frac{1}{2} + i\,\frac{\sqrt{3}}{2}\right) = \frac{3}{2} + \frac{3\sqrt{3}}{2}\,i$$

$$w_2 = 27^{1/3}\left(\cos\frac{180° + 1\cdot 360°}{3} + i\sin\frac{180° + 1\cdot 360°}{3}\right)$$

$$= 3(\cos 180° + i\sin 180°) = 3(-1 + 0i) = -3 + 0i \text{ or } -3$$

$$w_3 = 27^{1/3}\left(\cos\frac{180° + 2\cdot 360°}{3} + i\sin\frac{180° + 2\cdot 360°}{3}\right)$$

$$= 3(\cos 300° + i\sin 300°) = 3\left(\frac{1}{2} - i\,\frac{\sqrt{3}}{2}\right) = \frac{3}{2} - \frac{3\sqrt{3}}{2}$$

25. $$\left[r^{1/n}\left(\cos\frac{\theta + k\cdot 360°}{n} + i\sin\frac{\theta + k\cdot 360°}{n}\right)\right]^n$$

$$= (r^{1/n})^n\left[\cos\left(n\cdot\frac{\theta + k\cdot 360°}{n}\right) + i\sin\left(n\cdot\frac{\theta + k\cdot 360°}{n}\right)\right] \quad \text{De Moivre's theorem}$$

$$= r[\cos(\theta + k\cdot 360°) + i\sin(\theta + k\cdot 360°)] \quad \text{Algebra}$$

$$= r(\cos\theta + i\sin\theta) \quad\quad\quad\quad \text{Periodic property of cosine and sine functions}$$

27. $x^5 - 1 = 0 \quad\quad x^5 = 1$

Therefore, x is a fifth root of 1, and there are five of them. First, we write 1 in polar form:

$$1 = 1 + 0i = 1(\cos 0° + i\sin 0°)$$

All five fifth roots of 1 are given by

$$1^{1/5}\left(\cos\frac{0° + k\cdot 360°}{5} + i\sin\frac{0° + k\cdot 360°}{5}\right) \quad k = 0, 1, 2, 3, 4$$

Thus,

$$w_1 = 1^{1/5}\left(\cos\frac{0° + 0\cdot 360°}{5} + i\sin\frac{0° + 0\cdot 360°}{5}\right) = 1(\cos 0° + i\sin 0°) = 1$$

$$w_2 = 1^{1/5}\left(\cos\frac{0° + 1\cdot 360°}{5} + i\sin\frac{0° + 1\cdot 360°}{5}\right)$$

$$= 1(\cos 72° + i\sin 72°) = 0.309 + 0.951i$$

$$w_3 = 1^{1/5}\left(\cos\frac{0° + 2\cdot 360°}{5} + i\sin\frac{0° + 2\cdot 360°}{5}\right)$$

$$= 1(\cos 144° + i\sin 144°) = -0.809 + 0.588i$$

$$w_4 = 1^{1/5}\left(\cos\frac{0° + 3\cdot 360°}{5} + i\sin\frac{0° + 3\cdot 360°}{5}\right)$$

$$= 1(\cos 216° + i \sin 216°) = -0.809 - 0.588i$$

$$w_5 = 1^{1/5}\left(\cos \frac{0° + 4 \cdot 360°}{5} + i \sin \frac{0° + 4 \cdot 360°}{5}\right)$$

$$= 1(\cos 288° + i \sin 288°) = 0.309 - 0.951i$$

29. $x^3 + 5 = 0$ $x^3 = -5$

Therefore, x is a cube root of –5, and there are three of them. First, we write –5 in polar form:

$$-5 = -5 + 0i = 5(-1 + 0i) = 5(\cos 180° + i \sin 180°)$$

All three cube roots of 5(cos 180° + i sin 180°) are given by

$$5^{1/3}\left(\cos \frac{180° + k \cdot 360°}{3} + i \sin \frac{180° + k \cdot 360°}{3}\right)$$

Thus,

$$w_1 = 5^{1/3}\left(\cos \frac{180° + 0 \cdot 360°}{3} + i \sin \frac{180° + 0 \cdot 360°}{3}\right)$$

$$= 5^{1/3}(\cos 60° + i \sin 60°) = 0.855 + 1.481i$$

$$w_2 = 5^{1/3}\left(\cos \frac{180° + 1 \cdot 360°}{3} + i \sin \frac{180° + 1 \cdot 360°}{3}\right)$$

$$= 5^{1/3}(\cos 180° + i \sin 180°) = -1.710$$

$$w_3 = 5^{1/3}\left(\cos \frac{180° + 2 \cdot 360°}{3} + i \sin \frac{180° + 2 \cdot 360°}{3}\right)$$

$$= 5^{1/3}(\cos 240° + i \sin 240°) = 0.855 - 1.481i$$

CHAPTER 7 REVIEW EXERCISE

1.

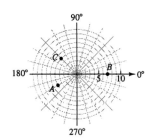

Chapter 7   Polar Coordinates; Complex Numbers

2. We graph r = 5 sin θ using rapid sketching techniques. First, we graph r = 5 sin θ in a rectangular coordinate system.

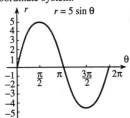

Rectangular coordinate system

From this, we observe how r varies as θ varies over particular intervals, summarize the information in a table, and sketch the polar curve from the information in the table.

θ	5 sin θ	
0 to $\dfrac{\pi}{2}$	0 to 5	
$\dfrac{\pi}{2}$ to π	5 to 0	
π to $\dfrac{3\pi}{2}$	0 to −5	Curve is traced out a second time in this region, although coordinate pairs are different
$\dfrac{3\pi}{2}$ to 2π	−5 to 0	

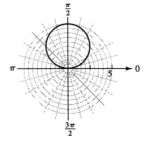

3. We graph r = 4 + 4 cos θ using rapid sketching techniques. First, we graph r = 4 + 4 cos θ in a rectangular coordinate system.

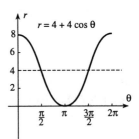

From this, we observe how r varies as θ varies over particular intervals, summarize the information in a table, and sketch the polar curve from the information in the table.

θ	$4 + 4 \cos \theta$
0 to $\dfrac{\pi}{2}$	8 to 4
$\dfrac{\pi}{2}$ to π	4 to 0
π to $\dfrac{3\pi}{2}$	0 to 4
$\dfrac{3\pi}{2}$ to 2π	4 to 8

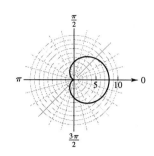

4. The graph consists of all points whose distance from the pole is 8, a circle with center at the pole, and radius 8.

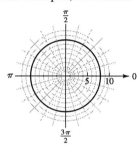

5.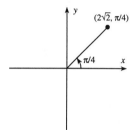

$$x = 2\sqrt{2} \cos \frac{\pi}{4} = 2\sqrt{2}\left(\frac{1}{\sqrt{2}}\right) = 2$$

$$y = 2\sqrt{2} \sin \frac{\pi}{4} = 2\sqrt{2}\left(\frac{1}{\sqrt{2}}\right) = 2$$

6. Use $r^2 = x^2 + y^2$ and $\tan \theta = \dfrac{y}{x}$.

$r^2 = (-\sqrt{3})^2 + 1^2 = 4$ $r = 2$ $\tan \theta = \dfrac{1}{-\sqrt{3}} = -\dfrac{1}{\sqrt{3}}$

$\theta = \dfrac{5\pi}{6}$ since the point is in the second quadrant.   Polar coordinates: $\left(2, \dfrac{5\pi}{6}\right)$

7.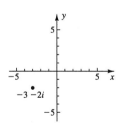

8.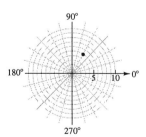

9. $z_1 z_2 = 9(\cos 42° + i \sin 42°) \cdot 3(\cos 37° + i \sin 37°)$

$= (9 \cdot 3)[\cos(42° + 37°) + i \sin(42° + 37°)] = 27(\cos 79° + i \sin 79°)$

$\dfrac{z_1}{z_2} = \dfrac{9(\cos 42° + i \sin 42°)}{3(\cos 37° + i \sin 37°)} = 3[\cos(42° - 37°) + i \sin(42° - 37°)] = 3(\cos 5° + i \sin 5°)$

10. $[2(\cos 10° + i \sin 10°)]^4 = 2^4(\cos 4 \cdot 10° + i \sin 4 \cdot 10°) = 16(\cos 40° + i \sin 40°)$

11.

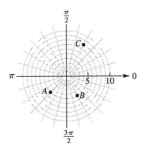

12. We graph $r = 8 \sin 3\theta$ using rapid sketching techniques. First, we graph $r = 8 \sin 3\theta$ in a rectangular coordinate system.

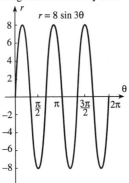

Rectangular coordinate system

From this, we observe how r varies as θ varies over particular intervals, summarize the information in a table, and sketch the polar curve from the information in the table.

θ	$8 \sin 3\theta$
0 to $\dfrac{\pi}{6}$	0 to 8
$\dfrac{\pi}{6}$ to $\dfrac{\pi}{3}$	8 to 0
$\dfrac{\pi}{3}$ to $\dfrac{\pi}{2}$	0 to −8
$\dfrac{\pi}{2}$ to $\dfrac{2\pi}{3}$	−8 to 0
$\dfrac{2\pi}{3}$ to $\dfrac{5\pi}{6}$	0 to 8
$\dfrac{5\pi}{6}$ to π	8 to 0

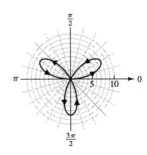

13. We graph r = 4 sin 2θ using rapid sketching techniques.

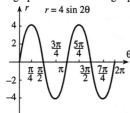

Rectangular coordinate system

From this, we observe how r varies as θ varies over particular intervals, summarize the information in a table, and sketch the polar curve from the information in the table.

θ	4 sin 2θ
0 to $\frac{\pi}{4}$	0 to 4
$\frac{\pi}{4}$ to $\frac{\pi}{2}$	4 to 0
$\frac{\pi}{2}$ to $\frac{3\pi}{4}$	0 to −4
$\frac{3\pi}{4}$ to π	−4 to 0

θ	4 sin 2θ
π to $\frac{5\pi}{4}$	0 to 4
$\frac{5\pi}{4}$ to $\frac{3\pi}{2}$	4 to 0
$\frac{3\pi}{2}$ to $\frac{7\pi}{4}$	0 to −4
$\frac{7\pi}{4}$ to 2π	−4 to 0

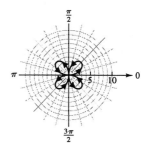

14. The graph consists of all points on a line that forms an angle of $\frac{\pi}{6}$ with the polar axis and passes through the pole.

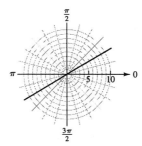

15. $8x - y^2 = x^2$

   $8x = x^2 + y^2$

   Use $x = r \cos \theta$ and $x^2 + y^2 = r^2$
   $8r \cos \theta = r^2$

   $0 = r^2 - 8r \cos \theta = r(r - 8 \cos \theta)$

   $r = 0$ or $r - 8 \cos \theta = 0$

   The graph of r = 0 is the pole, and since the pole is included as a solution of r − 8 cos θ = 0 $\left(\text{let } \theta = \frac{\pi}{2}\right)$, we can discard r = 0 and keep only r − 8 cos θ = 0 or r = 8 cos θ

16. $r(3 \cos \theta - 2 \sin \theta) = -2$

   $3r \cos \theta - 2r \sin \theta = -2$

   Use $x = r \cos \theta$ and $y = r \sin \theta$ $3x - 2y = -2$

Chapter 7   Polar Coordinates; Complex Numbers

17. $r = -3 \cos \theta$

We multiply both sides by r, which adds the pole to the graph. But the pole is already part of the graph $\left(\text{let } \theta = \dfrac{\pi}{2} \right)$, so we have changed nothing. $r^2 = -3r \cos \theta$

But $r^2 = x^2 + y^2$, $r \cos \theta = x$. Hence, $x^2 + y^2 = -3x$

18.

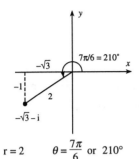

$r = 2 \qquad \theta = \dfrac{7\pi}{6}$ or $210°$

$z = 2(\cos 210° + i \sin 210°)$

19. $3\sqrt{2}\left(\cos \dfrac{3\pi}{4} + i \sin \dfrac{3\pi}{4} \right)$

$= 3\sqrt{2}\left(-\dfrac{1}{\sqrt{2}} + i \dfrac{1}{\sqrt{2}} \right)$

$= -3 + 3i$

20.   In polar form:

$2 + i\, 2\sqrt{3} = 4(\cos 60° + i \sin 60°)$

$-\sqrt{2} + i\, \sqrt{2} = 2(\cos 135° + i \sin 135°)$

$(2 + i\, 2\sqrt{3})(-\sqrt{2} + i\sqrt{2})$

$= 4(\cos 60° + i \sin 60°) \cdot 2(\cos 135° + i \sin 135°)$

$= (4 \cdot 2)[\cos(60° + 135°) + i \sin(60° + 135°)]$

$= 8(\cos 195° + i \sin 195°)$

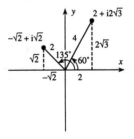

21.   From Problem 20, we have

$2 + i\, 2\sqrt{3} = 4 \cos(60° + i \sin 60°)$ and $-\sqrt{2} + i\, \sqrt{2} = 2(\cos 135° + i \sin 135°)$

in polar form. Thus,

$\dfrac{-\sqrt{2} + i\sqrt{2}}{2 + i\, 2\sqrt{3}} = \dfrac{2(\cos 135° + i \sin 135°)}{4(\cos 60° + i \sin 60°)}$

$= \dfrac{2}{4}[\cos(135° - 60°) + i \sin(135° - 60°)] = \dfrac{1}{2}(\cos 75° + i \sin 75°)$

Chapter 7   Polar Coordinates; Complex Numbers

22. First convert $-1-i$ to polar form, then apply

DeMoivre's theorem. $-1-i = \sqrt{2}(\cos 225° + i \sin 225°)$

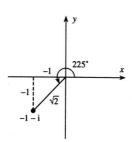

$$(-1-i)^4 = [\sqrt{2}(\cos 225° + i \sin 225°)]^4$$
$$= \left(\sqrt{2}\right)^4(\cos 4 \cdot 225° + i \sin 4 \cdot 225°)$$
$$= (2^{1/2})^4(\cos 900° + i \sin 900°)$$
$$= 2^2 \cos(180° + i \sin 180°)$$
$$= 4(-1 + i0) = -4$$

23. $x^3 - 64 = 0 \qquad x^3 = 64$

Therefore, x is a cube root of 64, and there are three of them. First, we write 64 in polar form:

$$64 = 64 + 0i = 64(\cos 0° + i \sin 0°)$$

All three cube roots of $64(\cos 0° + i \sin 0°)$ are given by

$$64^{1/3}\left(\cos \frac{0° + k \cdot 360°}{3} + i \sin \frac{0° + k \cdot 360°}{3}\right) \qquad k = 0, 1, 2$$

Thus,

$$w_1 = 64^{1/3}\left(\cos \frac{0° + 0 \cdot 360°}{3} + i \sin \frac{0° + 0 \cdot 360°}{3}\right)$$
$$= 4(\cos 0° + i \sin 0°) = 4(1 + 0i) = 4 + 0i \ \text{ or } \ 4$$

$$w_2 = 64^{1/3}\left(\cos \frac{0° + 1 \cdot 360°}{3} + i \sin \frac{0° + 1 \cdot 360°}{3}\right)$$
$$= 4(\cos 120° + i \sin 120°) = 4\left(-\frac{1}{2} + i \frac{\sqrt{3}}{2}\right) = -2 + 2i\sqrt{3}$$

$$w_3 = 64^{1/3}\left(\cos \frac{0° + 2 \cdot 360°}{3} + i \sin \frac{0° + 2 \cdot 360°}{3}\right)$$
$$= 4(\cos 240° + i \sin 240°) = 4\left(-\frac{1}{2} - i \frac{\sqrt{3}}{2}\right) = -2 - 2i\sqrt{3}$$

24. First convert $-4\sqrt{3} - 4i$ to polar form:

$$-4\sqrt{3} - 4i = 8(\cos 210° + i \sin 210°)$$

All cube roots of $8(\cos 210° + i \sin 210°)$ are

given by

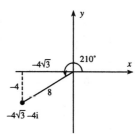

$$8^{1/3}\left(\cos \frac{210° + k \cdot 360°}{3} + i \sin \frac{210° + k \cdot 360°}{3}\right) \qquad k = 0, 1, 2$$

Thus,

$$w_1 = 8^{1/3}\left(\cos \frac{210° + 0 \cdot 360°}{3} + i \sin \frac{210° + 0 \cdot 360°}{3}\right) = 2(\cos 70° + i \sin 70°)\Bigg)$$

$$w_2 = 8^{1/3}\left(\cos \frac{210° + 1 \cdot 360°}{3} + i \sin \frac{210° + 1 \cdot 360°}{3}\right) = 2(\cos 190° + i \sin 190°)$$

$$w_3 = 8^{1/3}\left(\cos \frac{210° + 2 \cdot 360°}{3} + i \sin \frac{210° + 2 \cdot 360°}{3}\right) = 2(\cos 310° + i \sin 310°)$$

25. To show that $2(\cos 30° + i \sin 30°)$ is a square root of $2 + i \, 2\sqrt{3}$, we need only show that

$$[2(\cos 30° + i \sin 30°)]^2 = 2 + i \, 2\sqrt{3}.$$

But, by De Moivre's theorem,

$$[2(\cos 30° + i \sin 30°)]^2 = 2^2(\cos 2 \cdot 30° + i \sin 2 \cdot 30°) = 4(\cos 60° + i \sin 60°)$$

$$= 4\left(\frac{1}{2} + i \, \frac{\sqrt{3}}{2}\right) = 2 + i \, 2\sqrt{3}$$

26.

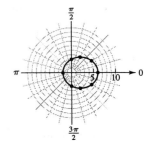

θ	0	$\frac{\pi}{6}$	$\frac{\pi}{3}$	$\frac{\pi}{2}$	$\frac{2\pi}{3}$	$\frac{5\pi}{6}$	π
r	6.0	5.3	4.0	3.0	2.4	2.1	2.0

θ	$\frac{7\pi}{6}$	$\frac{4\pi}{3}$	$\frac{3\pi}{2}$	$\frac{5\pi}{3}$	$\frac{11\pi}{6}$	2π
r	2.1	2.4	3.0	4.0	5.3	6.0

27. We graph $r = 4 + 4 \cos \dfrac{\theta}{2}$ using rapid sketching techniques. The following table can be used to investigate how r varies as θ varies over particular intervals. We then sketch the polar curve from the information in the table.

θ	$\frac{\theta}{2}$	$\cos \frac{\theta}{2}$	$4 \cos \frac{\theta}{2}$	$4 + 4 \cos \frac{\theta}{2}$
0 to $\frac{\pi}{2}$	0 to $\frac{\pi}{4}$	1 to $\frac{\sqrt{2}}{2}$	4 to $2\sqrt{2}$	8 to $4 + 2\sqrt{2}$ (≈ 6.8)
$\frac{\pi}{2}$ to π	$\frac{\pi}{4}$ to $\frac{\pi}{2}$	$\frac{\sqrt{2}}{2}$ to 0	$2\sqrt{2}$ to 0	$4 + 2\sqrt{2}$ to 4
π to $\frac{3\pi}{2}$	$\frac{\pi}{2}$ to $\frac{3\pi}{4}$	0 to $-\frac{\sqrt{2}}{2}$	0 to $-2\sqrt{2}$	4 to $4 - 2\sqrt{2}$ (≈ 1.2)
$\frac{3\pi}{2}$ to 2π	$\frac{3\pi}{4}$ to π	$-\frac{\sqrt{2}}{2}$ to -1	$-2\sqrt{2}$ to -4	$4 - 2\sqrt{2}$ to 0

θ	$\dfrac{\theta}{2}$	$\cos\dfrac{\theta}{2}$	$4\cos\dfrac{\theta}{2}$	$4+4\cos\dfrac{\theta}{2}$
2π to $\dfrac{5\pi}{2}$	π to $\dfrac{5\pi}{4}$	-1 to $-\dfrac{\sqrt{2}}{2}$	-4 to $-2\sqrt{2}$	0 to $4-2\sqrt{2}$
$\dfrac{5\pi}{2}$ to 3π	$\dfrac{5\pi}{4}$ to $\dfrac{3\pi}{2}$	$-\dfrac{\sqrt{2}}{2}$ to 0	$-2\sqrt{2}$ to 0	$4-2\sqrt{2}$ to 4
3π to $\dfrac{7\pi}{2}$	$\dfrac{3\pi}{2}$ to $\dfrac{7\pi}{4}$	0 to $\dfrac{\sqrt{2}}{2}$	0 to $2\sqrt{2}$	4 to $4+2\sqrt{2}$
$\dfrac{7\pi}{2}$ to 4π	$\dfrac{7\pi}{4}$ to 2π	$\dfrac{\sqrt{2}}{2}$ to 1	$2\sqrt{2}$ to 4	$4+2\sqrt{2}$ to 8

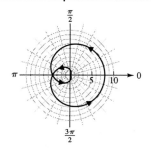

28. $r(\sin\theta - 2) = 3$ $r\sin\theta - 2r = 3$

Use $r\sin\theta = y$ $r = -\sqrt{x^2 + y^2}$ $y + 2\sqrt{x^2 + y^2} = 3$

$y - 3 = -2\sqrt{x^2 + y^2}$ or $(y - 3)^2 = 4(x^2 + y^2)$

Note: See comment, Exercise 7.1, Problem 61.

29. $x^3 - 12 = 0$ $x^3 = 12$

Therefore, x is a cube root of 12 and there are three of them. First, we write 12 in polar form:

$12 = 12(1 + 0i) = 12(\cos 0° + i\sin 0°)$

All three cube roots of $12(\cos 0° + i\sin 0°)$ are given by

$$12^{1/3}\left(\cos\frac{0° + k\cdot 360°}{3} + i\sin\frac{0° + k\cdot 360°}{3}\right)\quad k = 0, 1, 2$$

Thus,

$$w_1 = 12^{1/3}\left(\cos\frac{0° + 0\cdot 360°}{3} + i\sin\frac{0° + 0\cdot 360°}{3}\right) = 2.289$$

$$w_2 = 12^{1/3}\left(\cos\frac{0° + 1\cdot 360°}{3} + i\sin\frac{0° + 1\cdot 360°}{3}\right) = -1.145 + 1.983i$$

$$w_3 = 12^{1/3}\left(\cos\frac{0° + 2\cdot 360°}{3} + i\sin\frac{0° + 2\cdot 360°}{3}\right) = -1.145 - 1.983i$$

30. $\left[r^{1/3}\left(\cos \dfrac{\theta + k \cdot 360°}{3} + i \sin \dfrac{\theta + k \cdot 360°}{3} \right) \right]^3$

$= (r^{1/3})^3 \left[\cos\left(3 \cdot \dfrac{\theta + k \cdot 360°}{3} \right) + i \sin\left(3 \cdot \dfrac{\theta + k \cdot 360°}{3} \right) \right]$ DeMoivre's theorem

$= r[\cos(\theta + k \cdot 360°) + i \sin(\theta + k \cdot 360°)]$ Algebra

$= r(\cos \theta + i \sin \theta)$ Periodic property of sine and cosine functions

31.
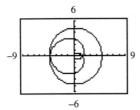

32.
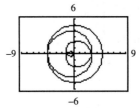

33. $n = 1$ $n = 2$ $n = 3$

$r = 5(\sin \theta)^2$ $r = 5(\sin \theta)^4$ $r = 5(\sin \theta)^6$

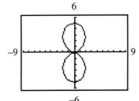   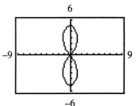

Generalizing from these graphs, we expect that the graph always has two leaves.

34. (A) $r = \dfrac{2}{1 - 1.6 \sin \theta}$ (B) $r = \dfrac{2}{1 - \sin \theta}$ (C) $r = \dfrac{2}{1 - 0.4 \sin \theta}$

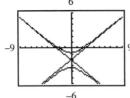   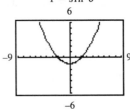   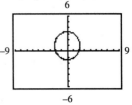

The graph is a hyperbola.   The graph is a parabola.   The graph is an ellipse.

CUMULATIVE REVIEW EXERCISE   Chapters 1–7

1. We compare α and β by changing to decimal degrees.

Since $\dfrac{\theta_{\text{deg}}}{180°} = \dfrac{\theta_{\text{rad}}}{\pi \text{ rad}}$, $\alpha_{\text{deg}} = \dfrac{180°}{\pi}\,\alpha_{\text{rad}} = \dfrac{180°}{\pi}\cdot\dfrac{2\pi}{7} = 51.42857°\ldots$

Since $25' = \dfrac{25°}{60}$ and $40'' = \dfrac{40°}{3600}$, then, $\beta = 51°25'40'' = 51.427777°\ldots$ Thus, $\alpha > \beta$.

2. *Solve for the hypotenuse c:* $c^2 = a^2 + b^2$

$$c = \sqrt{a^2 + b^2} = \sqrt{(1.27\text{ cm})^2 + (4.65\text{ cm})^2} = 4.82\text{ cm}$$

Solve for θ: We use the tangent. $\tan\theta = \dfrac{1.27}{4.65}$ $\theta = \tan^{-1}\dfrac{1.27}{4.65} = 15.3°$

Solve for the complementary angle: $90° - \theta = 90° - 15.3° = 74.7°$

3. $P(a, b) = (7, -24)$

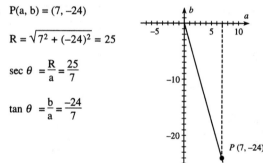

$R = \sqrt{7^2 + (-24)^2} = 25$

$\sec\theta = \dfrac{R}{a} = \dfrac{25}{7}$

$\tan\theta = \dfrac{b}{a} = \dfrac{-24}{7}$

4. (A) $\cot x \sec x \sin x = \dfrac{\cos x}{\sin x}\cdot\dfrac{1}{\cos x}\cdot\sin x$ Quotient and reciprocal identities

$= 1$ Algebra

(B) $\tan\theta + \cot\theta = \dfrac{\sin\theta}{\cos\theta} + \dfrac{\cos\theta}{\sin\theta}$ Quotient identities

$= \dfrac{\sin^2\theta}{\sin\theta\cos\theta} + \dfrac{\cos^2\theta}{\sin\theta\cos\theta}$ Algebra

$= \dfrac{\sin^2\theta + \cos^2\theta}{\sin\theta\cos\theta}$ Algebra

$= \dfrac{1}{\sin\theta\cos\theta}$ Pythagorean identity

$= \dfrac{1}{\sin\theta}\cdot\dfrac{1}{\cos\theta}$ Algebra

$= \csc\theta\sec\theta$ Reciprocal identity

$= \sec\theta\csc\theta$ Algebra

Chapter 7　　　Polar Coordinates; Complex Numbers

5.　Locate the 30° – 60° reference triangle, determine (a, b) and R, then evaluate.

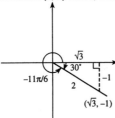

$$\sin \frac{11\pi}{6} = \frac{-1}{2} = -\frac{1}{2}$$

6.　Locate the 30° – 60° reference triangle, determine (a, b) and R, then evaluate.

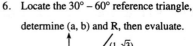
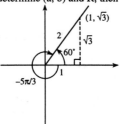

$$\tan \frac{-5\pi}{3} = \frac{\sqrt{3}}{1} = \sqrt{3}$$

7.　$y = \cos^{-1}(-0.5)$ is equivalent to $\cos y = -0.5$. What y between 0 and π has cosine equal to –0.5? y must be associated with a reference triangle in the second quadrant. Reference triangle is a special 30° – 60° triangle.

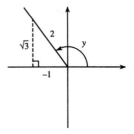

$$y = \frac{2\pi}{3} \qquad \cos^{-1}(-0.5) = \frac{2\pi}{3}$$

8.　$y = \csc^{-1}(\sqrt{2})$ is equivalent to $\csc y = \sqrt{2}$ and $-\frac{\pi}{2} \le y \le \frac{\pi}{2}$, $y \ne 0$. What number between $-\frac{\pi}{2}$ and $\frac{\pi}{2}$ has cosecant equal to $\sqrt{2}$? y must be in the first quadrant.

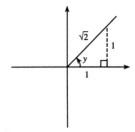

$$\csc y = \sqrt{2} = \frac{\sqrt{2}}{1} \qquad y = \frac{\pi}{4}$$

Thus, $\csc^{-1}(\sqrt{2}) = \frac{\pi}{4}$

9.　Calculator in degree mode:　$\sin 43°22' = \sin(43.366\ldots°)$　　　　Convert to decimal degrees

$$= 0.6867$$

10.　Use the reciprocal relationship $\cot \theta = \frac{1}{\tan \theta}$

Calculator in radian mode:　$\cot \frac{2\pi}{5} = \dfrac{1}{\tan \dfrac{2\pi}{5}} = 0.3249$

Chapter 7   Polar Coordinates; Complex Numbers

11. Calculator in radian mode: $\sin^{-1}(0.8) = 0.9273$

12. Calculator in radian mode: $\sec^{-1}(4.5) = \cos^{-1}\dfrac{1}{4.5} = 1.347$

13. $\sin x + \sin y = 2 \sin\dfrac{x + y}{2}\cos\dfrac{x - y}{2}$

   $\sin 3t + \sin t = 2 \sin\dfrac{3t + t}{2}\cos\dfrac{3t - t}{2} = 2\sin 2t\cos t$

14. We are given two angles and the included side (ASA).

   We use the law of sines.

   Solve for γ: $\alpha + \beta + \gamma = 180°$

   $\qquad\qquad\gamma = 180° - (52° + 47°) = 81°$

   Solve for a: $\dfrac{\sin\alpha}{a} = \dfrac{\sin\gamma}{c}$ $a = \dfrac{c\sin\alpha}{\sin\gamma} = \dfrac{(28\text{ cm})\sin 52°}{\sin 81°} = 22\text{ cm}$

   Solve for b: $\dfrac{\sin\beta}{b} = \dfrac{\sin\gamma}{c}$ $b = \dfrac{c\sin\beta}{\sin\gamma} = \dfrac{(28\text{ cm})\sin 47°}{\sin 81°} = 21\text{ cm}$

15. We are given two angles and a non-included side (AAS).

   We use the law of sines.

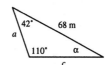

   Solve for α: $\alpha + \beta + \gamma = 180°$

   $\qquad\qquad\alpha = 180° - (42° + 110°) = 28°$

   Solve for a: $\dfrac{\sin\alpha}{a} = \dfrac{\sin\beta}{b}$ $a = \dfrac{b\sin\alpha}{\sin\beta} = \dfrac{(68\text{ m})\sin 28°}{\sin 110°} = 34\text{ m}$

   Solve for c: $\dfrac{\sin\gamma}{c} = \dfrac{\sin\beta}{b}$ $c = \dfrac{b\sin\gamma}{\sin\beta} = \dfrac{(68\text{ m})\sin 42°}{\sin 110°} = 48\text{ m}$

16. We are given two sides and the included angle (SAS).

   We use the law of cosines to find the third side, then

   the law of sines to find a second angle.

   Solve for b: $b^2 = a^2 + c^2 - 2ac\cos\beta$

   $\qquad\qquad = 16^2 + 24^2 - 2(16)(24)\cos 34° = 195.2991\ldots$

   $\qquad b = \sqrt{195.2991\ldots} = 14\text{ in.}$

   Solve for γ: $\dfrac{\sin\gamma}{c} = \dfrac{\sin\beta}{b}$ $\sin\gamma = \dfrac{c\sin\beta}{b} = \dfrac{(24\text{ in.})\sin 34°}{(14\text{ in.})} = 0.9604$

   $\gamma = \sin^{-1} 0.9604 = 74°$ or $\gamma = 180° - \sin^{-1} 0.9604 = 106°$

To determine which of the two possible values for γ is correct, we examine the third angle, α.

Solve for α: $\alpha + \beta + \gamma = 180°$

$$\alpha = 180° - (\beta + \gamma) = 180° - (34° + 74°) = 72°$$

or $$\alpha' = 180° - (\beta + \gamma') = 180° - (34° + 106°) = 40°$$

Substituting in the law of sines:

$$\frac{\sin \alpha}{a} = \frac{\sin \gamma}{c} \qquad\qquad \frac{\sin \alpha'}{a} = \frac{\sin \gamma'}{c}$$

$$\frac{\sin 72°}{16} = \frac{\sin 74°}{24} \qquad\qquad \frac{\sin 40°}{16} = \frac{\sin 106°}{24}$$

$$0.0594 \neq 0.0400 \qquad\qquad 0.401 \approx 0.400$$

Thus, we conclude that the choices γ' and α' are correct. $\gamma' = 106°$, $\alpha' = 40°$.

17. We are given three sides (SSS). We solve for the largest angle, γ (largest because it is opposite the largest side, c), using the law of cosines. We then solve for a second angle using the law of sines, because it involves simpler calculations.

Solve for γ: $c^2 = a^2 + b^2 - 2ab \cos\gamma$ $\cos \gamma = \dfrac{a^2 + b^2 - c^2}{2ab}$

$$\gamma = \cos^{-1}\frac{a^2 + b^2 - c^2}{2ab} = \cos^{-1}\frac{18^2 + 23^2 - 32^2}{2 \cdot 18 \cdot 23} = 102°$$

Solve for β: $\dfrac{\sin \beta}{b} = \dfrac{\sin \gamma}{c}$ $\sin \beta = \dfrac{b \sin \gamma}{c}$

$$\beta = \sin^{-1}\frac{b \sin \gamma}{c} \quad \begin{cases} \beta \text{ is acute, because there} \\ \text{is room for only one} \\ \text{obtuse angle in a triangle.} \end{cases}$$

$$= \sin^{-1}\frac{(23 \text{ ft}) \sin 102°}{32 \text{ ft}} = 45°$$

Solve for α: $\alpha + \beta + \gamma = 180°$

$$\alpha = 180° - (\beta + \gamma) = 180° - (45° + 102°) = 33°$$

18. The given information consists of two sides and the included angle; hence, we use the formula

$$A = \frac{ab}{2} \sin \theta \quad \text{in the form} \quad A = \frac{1}{2} ac \sin \beta = \frac{1}{2}(16 \text{ in.})(24 \text{ in.}) \sin 34° = 110 \text{ in}^2$$

19. The given information consists of three sides; hence, we use Heron's formula. First, find the semiperimeter s:

$$s = \frac{a + b + c}{2} = \frac{18 + 23 + 32}{2} = 36.5 \text{ ft}$$

Then, $s - a = 36.5 - 18 = 18.5$

$s - b = 36.5 - 23 = 13.5$

$s - c = 36.5 - 32 = 4.5$

Thus, $A = \sqrt{s(s-a)(s-b)(s-c)} = \sqrt{(36.5)(18.5)(13.5)(4.5)} = 200 \text{ ft}^2$

20. Horizontal component H: $\cos 25° = \dfrac{H}{13}$ $H = 13 \cos 25° = 12$

Vertical component V: $\sin 25° = \dfrac{V}{13}$ $V = 13 \sin 25° = 5.5$

21. To find $|\mathbf{u} + \mathbf{v}|$: Apply the Pythagorean theorem to triangle OCB.

$|\mathbf{u} + \mathbf{v}|^2 = OB^2 = OC^2 + BC^2 = 6.4^2 + 3.9^2 = 56.17$

$|\mathbf{u} + \mathbf{v}| = \sqrt{56.17} = 7.5$

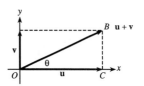

Solve triangle OCB for θ: $\tan \theta = \dfrac{BC}{OC} = \dfrac{|\mathbf{v}|}{|\mathbf{u}|}$

$\theta = \tan^{-1}\dfrac{|\mathbf{v}|}{|\mathbf{u}|}$ θ is acute

$= \tan^{-1}\dfrac{3.9}{6.4} = 31°$

22. The algebraic vector $\langle a, b \rangle$ has coordinates given by

$a = x_b - x_a = (-3) - 4 = -7$ $b = y_b - y_a = 7 - (-2) = 9$

Hence, $\langle a, b \rangle = \langle -7, 9 \rangle$

Magnitude of $\langle a, b \rangle = |\langle -7, 9 \rangle| = \sqrt{a^2 + b^2} = \sqrt{(-7)^2 + 9^2} = \sqrt{130}$

23. $|\mathbf{u}| = \sqrt{2^2 + (-7)^2} = \sqrt{53}$ $|\mathbf{v}| = \sqrt{3^2 + 8^2} = \sqrt{73}$

$\cos \theta = \dfrac{\mathbf{u} \cdot \mathbf{v}}{|\mathbf{u}|\,|\mathbf{v}|} = \dfrac{(2\mathbf{i} - 7\mathbf{j}) \cdot (3\mathbf{i} + 8\mathbf{j})}{\sqrt{53}\,\sqrt{73}} = \dfrac{2 \cdot 3 + (-7) \cdot 8}{\sqrt{53}\,\sqrt{73}}$ $\theta = \cos^{-1}\dfrac{-50}{\sqrt{53}\,\sqrt{73}} = 143.5°$

24.

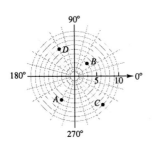

Chapter 7   Polar Coordinates; Complex Numbers

25. We graph $r = 5 + 5 \sin \theta$ using rapid sketching techniques.
First, we graph $r = 5 + 5 \sin \theta$ in a rectangular coordinate
system.
From this, we observe how r varies as θ varies over
particular intervals, summarize the information in a table,
and sketch the polar curve from the information in the table.

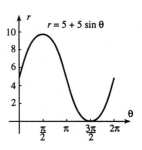

θ	$5 + 5 \sin \theta$
0 to $\dfrac{\pi}{2}$	5 to 10
$\dfrac{\pi}{2}$ to π	10 to 5
π to $\dfrac{3\pi}{2}$	5 to 0
$\dfrac{3\pi}{2}$ to 2π	0 to 5

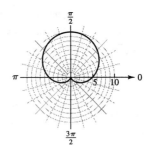

26.

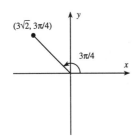

$$x = 3\sqrt{2} \cos \frac{3\pi}{4} = 3\sqrt{2}\left(-\frac{1}{\sqrt{2}}\right) = -3$$

$$y = 3\sqrt{2} \sin \frac{3\pi}{4} = 3\sqrt{2}\left(\frac{1}{\sqrt{2}}\right) = 3$$

27. Use $r^2 = x^2 + y^2$ and $\tan \theta = \dfrac{y}{x}$

$$r^2 = (-2\sqrt{3})^2 + 2^2 = 16$$

$$r = 4$$

$$\tan \theta = \frac{2}{-2\sqrt{3}} = -\frac{1}{\sqrt{3}}$$

$\theta = \dfrac{5\pi}{6}$ since the point is in
the second quadrant

Polar coordinates: $\left(4, \dfrac{5\pi}{6}\right)$

28.

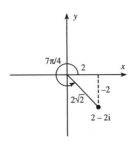

29. $3\left(\cos \dfrac{3\pi}{2} + i \sin \dfrac{3\pi}{2}\right)$

$= 3[0 + i(-1)] = -3i$

$r = 2\sqrt{2}$ $\theta = \dfrac{7\pi}{4}$

$z = 2\sqrt{2}\left(\cos \dfrac{7\pi}{4} + i \sin \dfrac{7\pi}{4}\right)$

30. $z_1 z_2 = (3 \cdot 5)[\cos(50° + 15°) + i \sin(50° + 15°)] = 15(\cos 65° + i \sin 65°)$

$\dfrac{z_1}{z_2} = \dfrac{3}{5}[\cos(50° - 15°) + i \sin(50° - 15°)] = 0.6(\cos 35° + i \sin 35°)$

31. $[3(\cos 25° + i \sin 25°)]^4 = 3^4(\cos 4 \cdot 25° + i \sin 4 \cdot 25°) = 81(\cos 100° + i \sin 100°)$

32.   Since the two right triangles are similar, we can write

$$\dfrac{9}{x} = \dfrac{9 + x}{10}$$

$$10x \cdot \dfrac{9}{x} = 10x \cdot \dfrac{9 + x}{10}$$

$$90 = x(9 + x) = 9x + x^2$$

$$0 = x^2 + 9x - 90 = (x + 15)(x - 6)$$

$x - 6 = 0$ or $x + 15 = 0$

$x = 6$ $x = -15$

We discard the negative answer. Since $\tan \theta = \dfrac{x}{9}$, we can write

$$\tan \theta = \dfrac{6}{9} \qquad \theta = \tan^{-1} \dfrac{6}{9} = 33°40'$$

33.   Since $\sin \theta < 0$ and $\cot \theta > 0$, the terminal side
of θ lies in quadrant III. We sketch a reference
triangle and label what we know. ·
Since $\cot \theta = 4 = \dfrac{4}{1} = \dfrac{-4}{-1}$, we know that $a = -4$
and $b = -1$. We use the Pythagorean theorem to
find R.

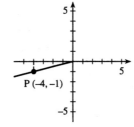

$$(-4)^2 + (-1)^2 = R^2$$

$$R = \sqrt{17} \ \ (\text{R is never negative})$$

Therefore, $\csc \theta = \dfrac{R}{b} = \dfrac{\sqrt{17}}{-1} = -\sqrt{17} \qquad \cos \theta = \dfrac{a}{R} = \dfrac{-4}{\sqrt{17}}$

34.   This graph is the graph of $y = \sin(2x + \pi)$ moved up one unit. We first find the period and phase
shift by solving

$2x + \pi = 0,$ and $2x + \pi = 2\pi,$

$x = -\dfrac{\pi}{2}$ $x = -\dfrac{\pi}{2} + \pi$

Period $= \pi$ Phase Shift $= -\dfrac{\pi}{2}$ Frequency $= \dfrac{1}{\text{Period}} = \dfrac{1}{\pi}$

We then sketch one period of the graph starting at $x = -\dfrac{\pi}{2}$ (the phase shift) and ending at

$x = -\dfrac{\pi}{2} + \pi = \dfrac{\pi}{2}$ (the phase shift plus one period).

The graph is a basic sine curve relative to the horizontal line $y = 1$ (shown as a broken line) and the
y axis. We then extend the graph from $-\pi$ to 2π.

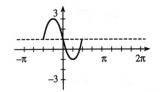

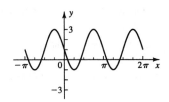

35. $\dfrac{\cos x}{1 - \sin x} + \dfrac{\cos x}{1 + \sin x} = \dfrac{\cos x(1 + \sin x)}{(1 - \sin x)(1 + \sin x)} + \dfrac{\cos x(1 - \sin x)}{(1 - \sin x)(1 + \sin x)}$ Algebra

$= \dfrac{\cos x(1 + \sin x) + \cos x(1 - \sin x)}{(1 - \sin x)(1 + \sin x)}$ Algebra

$= \dfrac{\cos x + \sin x \cos x + \cos x - \sin x \cos x}{1 - \sin^2 x}$ Algebra

$= \dfrac{2 \cos x}{1 - \sin^2 x}$ Algebra

$= \dfrac{2 \cos x}{\cos^2 x}$ Pythagorean identity

$= \dfrac{2}{\cos x}$ Algebra

$= 2 \sec x$ Reciprocal identity

36. $\tan \dfrac{\theta}{2} = \dfrac{\sin \theta}{1 + \cos \theta}$ Half-angle identity

$= \dfrac{\dfrac{\sin \theta}{\sin \theta}}{\dfrac{1 + \cos \theta}{\sin \theta}}$ Algebra

$= \dfrac{1}{\dfrac{1}{\sin \theta} + \dfrac{\cos \theta}{\sin \theta}}$ Algebra

$= \dfrac{1}{\csc \theta + \cot \theta}$ Reciprocal and Quotient identities

37. $\dfrac{\cos x - \sin x}{\cos x + \sin x} = \dfrac{(\cos x - \sin x)(\cos x - \sin x)}{(\cos x + \sin x)(\cos x - \sin x)}$ Algebra

$= \dfrac{\cos^2 x - 2 \cos x \sin x + \sin^2 x}{\cos^2 x - \sin^2 x}$ Algebra

$= \dfrac{1 - 2 \cos x \sin x}{\cos^2 x - \sin^2 x}$ Pythagorean identity

$= \dfrac{1 - \sin 2x}{\cos 2x}$ Double-angle identities

$= \dfrac{1}{\cos 2x} - \dfrac{\sin 2x}{\cos 2x}$ Algebra

$= \sec 2x - \tan 2x$ Reciprocal and Quotient identities

366

Chapter 7   Polar Coordinates; Complex Numbers

38. First draw a reference triangle in the first quadrant and find sin x.

$$b = \sqrt{25^2 - 24^2} = 7 \qquad \sin x = \frac{b}{R} = \frac{7}{25}$$

We can now find $\tan \frac{x}{2}$ from the half-angle identity.

$$\tan \frac{x}{2} = \frac{1 - \cos x}{\sin x} = \frac{1 - \frac{24}{25}}{\frac{7}{25}} = \frac{25 - 24}{7} = \frac{1}{7}$$

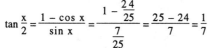

To find sin 2x, we use a double-angle identity.

$$\sin 2x = 2 \sin x \cos x = 2 \cdot \frac{7}{25} \cdot \frac{24}{25} = \frac{336}{625}$$

39. $M = -1$ and $N = 1$

Locate $P(M, N) = P(-1, 1)$ to determine C.

$$R = \sqrt{(-1)^2 + 1^2} = \sqrt{2}$$

$$\sin C = \frac{1}{\sqrt{2}} \qquad \tan C = -1$$

$$C = \frac{3\pi}{4} \qquad |C| \text{ is minimum for this choice.}$$

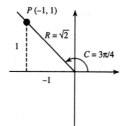

Thus, $y = -\sin 2\pi t + \cos 2\pi t = \sqrt{2} \sin\left(2\pi t + \frac{3\pi}{4}\right)$

Amplitude $= |\sqrt{2}| = \sqrt{2}$

Period and Phase Shift:

$$2\pi t + \frac{3\pi}{4} = 0 \qquad\qquad 2\pi t + \frac{3\pi}{4} = 2\pi$$

$$2\pi t = -\frac{3\pi}{4} \qquad\qquad 2\pi t = -\frac{3\pi}{4} + 2\pi$$

$$t = -\frac{3}{8} \qquad\qquad t = -\frac{3}{8} + 1$$

Period $= 1$ \qquad Phase Shift $= -\frac{3}{8}$

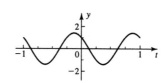

Frequency $= \dfrac{1}{\text{Period}} = \dfrac{1}{1} = 1$

40. Let $y = \sin^{-1} \frac{3}{4}$, then $\sin y = \frac{3}{4}$, $-\frac{\pi}{2} \le y \le \frac{\pi}{2}$.

Sketch the reference triangle associated with y, then $\sec y = \sec\left(\sin^{-1} \frac{3}{4}\right)$, can be determined directly from the triangle.

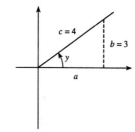

$$a^2 + b^2 = c^2 \qquad a = \sqrt{4^2 - 3^2} = \sqrt{7}$$

$$\sec\left(\sin^{-1}\frac{3}{4}\right) = \sec y = \frac{4}{\sqrt{7}}$$

41. (A) Calculator in radian mode: $\sec(\sin^{-1} 0.25) = \dfrac{1}{\cos(\sin^{-1} 0.25)} = 1.033$

(B) Calculator in radian mode: $\tan^{-1}(\csc 3.75) = \tan^{-1}\left(\dfrac{1}{\sin 3.75}\right) = -1.052$

(C) Calculator in radian mode: $\tan(\text{arccsc } 3.75) = \tan\left(\arcsin \dfrac{1}{3.75}\right) = 0.2767$

42.
$$\sin 2x + \sin x = 0$$
$$2 \sin x \cos x + \sin x = 0$$
$$\sin x(2 \cos x + 1) = 0$$

$$\sin x = 0 \qquad \text{or} \qquad 2 \cos x + 1 = 0$$
$$x = 0, \pi \qquad\qquad \cos x = -\frac{1}{2}$$
$$x = \frac{2\pi}{3}, \frac{4\pi}{3}$$

Thus, the solutions over one period, $0 \le x < 2\pi$, are $0, \pi, \dfrac{2\pi}{3}, \dfrac{4\pi}{3}$. Thus, if x can range over all real numbers,

$$x = \begin{cases} 0 + 2k\pi \\ \pi + 2k\pi \end{cases} \text{or } k\pi$$
$$\frac{2\pi}{3} + 2k\pi$$
$$\frac{4\pi}{3} + 2k\pi$$

k any integer

43. $2 \cos 2x = 5 \sin x - 4$

$$2(1 - 2 \sin^2 x) = 5 \sin x - 4$$
$$2 - 4 \sin^2 x = 5 \sin x - 4$$
$$0 = 4 \sin^2 x + 5 \sin x - 6$$
$$0 = (4 \sin x - 3)(\sin x + 2)$$

$$4 \sin x - 3 = 0 \qquad\qquad \sin x + 2 = 0$$
$$\sin x = \frac{3}{4} \qquad\qquad \sin x = -2$$
$$\qquad\qquad\qquad \text{No solution}$$

$$x = \begin{cases} \sin^{-1}\dfrac{3}{4} \\ \pi - \sin^{-1}\dfrac{3}{4} \end{cases} = \begin{cases} 0.8481 \\ 2.294 \end{cases} \qquad \text{solutions over one period } 0 \le x < 2\pi.$$

If x can range over all real numbers

$$x = \begin{cases} 0.8481 + 2k\pi \\ 2.294 \ + 2k\pi \end{cases} \quad k \text{ any integer}$$

44. We are given two sides and a non-included angle (SSA).

    (A) a = 11.5 cm.

    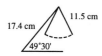

    Solve for β: $\quad \dfrac{\sin \beta}{b} = \dfrac{\sin \alpha}{a}$

    $\qquad\qquad \sin \beta = \dfrac{b \sin \alpha}{a} = \dfrac{17.4 \sin 49°30'}{11.5}$

    $\qquad\qquad\qquad = 1.151$

    Since $\sin \beta = 1.151$ has no solution, no triangle exists with the given measurements.

    No solution.

    (B)

    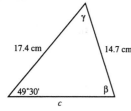 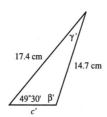

    Solve for β: $\quad \dfrac{\sin \beta}{b} = \dfrac{\sin \alpha}{a} \qquad \sin \beta = \dfrac{b \sin \alpha}{a} = \dfrac{(17.4 \text{ cm}) \sin 49°30'}{14.7 \text{ cm}} = 0.9001$

    Two triangles are possible; angle β can be either acute or obtuse.

    $\qquad \beta = \sin^{-1} 0.9001 = 64°10' \qquad\qquad\qquad \beta' = 180° - \sin^{-1} 0.9001 = 115°50'$

    Solve for γ *and* γ':

    $\qquad \alpha + \beta + \gamma = 180° \qquad\qquad\qquad\qquad \alpha + \beta' + \gamma' = 180°$

    $\qquad\qquad \gamma = 180° - (49°30' + 64°10') \qquad\qquad\quad \gamma' = 180° - (49°30' + 115°50')$

    $\qquad\qquad\quad = 66°20' \qquad\qquad\qquad\qquad\qquad\qquad = 14°40'$

    Solve for c *and* c':

    $\qquad \dfrac{\sin \alpha}{a} = \dfrac{\sin \gamma}{c} \qquad\qquad\qquad\qquad\qquad \dfrac{\sin \alpha}{a} = \dfrac{\sin \gamma'}{c'}$

    $\qquad\qquad c = \dfrac{a \sin \gamma}{\sin \alpha} \qquad\qquad\qquad\qquad\qquad\quad c = \dfrac{a \sin \gamma'}{\sin \alpha}$

    $\qquad\qquad = \dfrac{(14.7 \text{ cm}) \sin 66°20'}{\sin 49°30'} \qquad\qquad\qquad = \dfrac{(14.7 \text{ cm}) \sin 14°40'}{\sin 49°30'}$

    $\qquad\qquad = 17.7 \text{ cm} \qquad\qquad\qquad\qquad\qquad\quad = 4.89 \text{ cm}$

(C) *Solve for β:*

$$\frac{\sin \beta}{b} = \frac{\sin \alpha}{a}$$

$$\sin \beta = \frac{b \sin \alpha}{a} = \frac{(17.4 \text{ cm}) \sin 49°30'}{21.1}$$

$$= 0.6271$$

$$\beta = \sin^{-1} 0.6271 = 38°50'$$

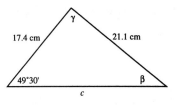

(There is another solution of sin β = 0.6271 that deserves brief consideration:

$$\beta' = 180° - \sin^{-1} 0.6271 = 141°10'.$$

However, there is not enough room in a triangle for an angle of 141°10' and an angle of 49°30', since their sum is greater than 180°.)

Solve for γ. $\alpha + \beta + \gamma = 180°$

$$\gamma = 180° - (49°30' + 38°50') = 91°40'$$

Solve for c: $\dfrac{\sin \alpha}{a} = \dfrac{\sin \gamma}{c}$

$$c = \frac{a \sin \gamma}{\sin \alpha} = \frac{(21.1 \text{ cm}) \sin 91°40'}{\sin 49°30'} = 27.7 \text{ cm}$$

45.

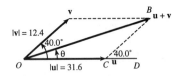

$\angle BCD = 40.0°$. Hence, $\angle OCB = 180° - 40.0° = 140.0°$

We can find $|\mathbf{u} + \mathbf{v}|$ using the law of cosines:

$$|\mathbf{u} + \mathbf{v}|^2 = |\mathbf{u}|^2 + |\mathbf{v}|^2 - 2|\mathbf{u}||\mathbf{v}|\cos(\text{OCB}) = 31.6^2 + 12.4^2 - 2(31.6)(12.4) \cos 140.0°$$

$$= 1752.65370 \ldots$$

$$|\mathbf{u} + \mathbf{v}| = \sqrt{1752.65370 \ldots} = 41.9$$

To find θ, we use the law of sines:

$$\frac{\sin \theta}{|\mathbf{v}|} = \frac{\sin \text{OCB}}{|\mathbf{u} + \mathbf{v}|}$$

$$\frac{\sin \theta}{12.4} = \frac{\sin 140.0°}{41.9}$$

$$\sin \theta = \frac{12.4}{41.9} \sin 140.0°$$

$$\theta = \sin^{-1}\left(\frac{12.4}{41.9} \sin 140.0°\right) = 11.0°$$

46. (A) $3\mathbf{u} - 4\mathbf{v} = 3\langle 1, -2 \rangle - 4\langle 0, 3 \rangle = \langle 3, -6 \rangle + \langle 0, -12 \rangle = \langle 3, -18 \rangle$

    (B) $3\mathbf{u} - 4\mathbf{v} = 3(2\mathbf{i} + 3\mathbf{j}) - 4(-\mathbf{i} + 5\mathbf{j}) = 6\mathbf{i} + 9\mathbf{j} + 4\mathbf{i} - 20\mathbf{j} = 10\mathbf{i} - 11\mathbf{j}$

47. $|\mathbf{v}| = \sqrt{7^2 + (-24)^2} = 25$

    $\mathbf{u} = \dfrac{1}{|\mathbf{v}|}\mathbf{v} = \dfrac{1}{25}\langle 7, -24 \rangle = \left\langle \dfrac{7}{25}, -\dfrac{24}{25} \right\rangle$ or $\langle 0.28, -0.96 \rangle$

48. The algebraic vector $\langle a, b \rangle$ has coordinates given by

    $a = x_b - x_a = (-1) - (-3) = 2$ $b = y_b - y_a = 5 - 2 = 3$

    Hence, $\langle a, b \rangle = \langle 2, 3 \rangle = \langle 2, 0 \rangle + \langle 0, 3 \rangle = 2\langle 1, 0 \rangle + 3\langle 0, 1 \rangle = 2\mathbf{i} + 3\mathbf{j}$

49. (A) $\mathbf{u} \cdot \mathbf{v} = \langle 4, 0 \rangle \cdot \langle 0, -5 \rangle = 0 + 0 = 0$. Thus, $\mathbf{u}$ and $\mathbf{v}$ are orthogonal.

    (B) $\mathbf{u} \cdot \mathbf{v} = \langle 3, 2 \rangle \cdot \langle -3, 4 \rangle = -9 + 8 = -1 \neq 0$. Thus, $\mathbf{u}$ and $\mathbf{v}$ are not orthogonal.

    (C) $\mathbf{u} \cdot \mathbf{v} = (\mathbf{i} - 2\mathbf{j}) \cdot (6\mathbf{i} + 3\mathbf{j}) = 6 - 6 = 0$. Thus, $\mathbf{u}$ and $\mathbf{v}$ are orthogonal.

50. We start by graphing $r = 8 \cos 2\theta$ in a rectangular coordinate system.

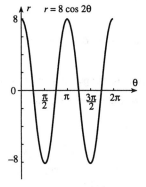

Rectangular coordinate system

From this we observe how r varies as θ varies over particular intervals, summarize the information in a table, and sketch the polar curve from the information in the table.

θ	$8\cos 2\theta$	θ	$8\cos 2\theta$
0 to $\dfrac{\pi}{4}$	8 to 0	π to $\dfrac{5\pi}{4}$	8 to 0
$\dfrac{\pi}{4}$ to $\dfrac{\pi}{2}$	0 to -8	$\dfrac{5\pi}{4}$ to $\dfrac{3\pi}{2}$	0 to -8
$\dfrac{\pi}{2}$ to $\dfrac{3\pi}{4}$	-8 to 0	$\dfrac{3\pi}{2}$ to $\dfrac{7\pi}{4}$	-8 to 0
$\dfrac{3\pi}{4}$ to π	0 to 8	$\dfrac{7\pi}{4}$ to 2π	0 to 8

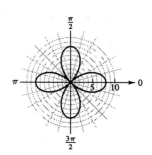

51. $x^2 = 6y$

Use $x = r\cos\theta$ and $y = r\sin\theta$

$$(r\cos\theta)^2 = 6(r\sin\theta)$$
$$r^2\cos^2\theta = 6r\sin\theta$$

$r^2\cos^2\theta - 6r\sin\theta = 0$

$r(r\cos^2\theta - 6\sin\theta) = 0$

$r = 0$ or $r\cos^2\theta - 6\sin\theta = 0$

$$r\cos^2\theta = 6\sin\theta$$

The graph of $r = 0$ is the pole, and since the pole is included as a solution of $r = 6\tan\theta\sec\theta$ (let $\theta = 0$), we can discard $r = 0$ and keep only $r = 6\tan\theta\sec\theta$.

$$r = \frac{6\sin\theta}{\cos^2\theta} = 6\,\frac{\sin\theta}{\cos\theta}\,\frac{1}{\cos\theta} = 6\tan\theta\sec\theta$$

52. $r = 4\sin\theta$

We multiply both sides by r, which adds the pole to the graph. But the pole is already part of the graph (let $\theta = 0$), so we have changed nothing. $r^2 = 4r\sin\theta$. But $r^2 = x^2 + y^2$, $r\sin\theta = y$. Hence, $x^2 + y^2 = 4y$.

53. Converting to polar form:

$$3 + 3i = 3\sqrt{2}(\cos 45° + i\sin 45°)$$
$$-1 + i\sqrt{3} = 2(\cos 120° + i\sin 120°)$$

$(3 + 3i)(-1 + i\sqrt{3})$

$\quad = [3\sqrt{2}(\cos 45° + i\sin 45°)][2(\cos 120° + i\sin 120°)]$

$\quad = (3\sqrt{2})\cdot 2[\cos(45° + 120°) + i\sin(45° + 120°)]$

$\quad\quad = 6\sqrt{2}(\cos 165° + i\sin 165°)$

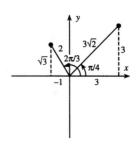

54. $\dfrac{-1 + i\sqrt{3}}{3 + 3i} = \dfrac{2(\cos 120° + i\sin 120°)}{3\sqrt{2}(\cos 45° + i\sin 45°)} = \dfrac{2}{3\sqrt{2}}[\cos(120° - 45°) + i\sin(120° - 45°)]$

$\qquad\qquad = \dfrac{\sqrt{2}}{3}(\cos 75° + i\sin 75°)$

Chapter 7   Polar Coordinates; Complex Numbers

55. First, write $1 - i$ in polar form:

$$1 - i = \sqrt{2}\left(\cos \frac{7\pi}{4} + i \sin \frac{7\pi}{4}\right)$$

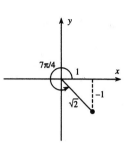

$$(1 - i)^6 = \left[\sqrt{2}\left(\cos \frac{7\pi}{4} + i \sin \frac{7\pi}{4}\right)\right]^6$$

$$= \left(\sqrt{2}\right)^6\left[\cos 6\left(\frac{7\pi}{4}\right) + i \sin 6\left(\frac{7\pi}{4}\right)\right]$$

$$= (2^{1/2})^6\left(\cos \frac{21\pi}{2} + i \sin \frac{21\pi}{2}\right)$$

$$= 2^3\left(\cos \frac{\pi}{2} + i \sin \frac{\pi}{2}\right) = 8(0 + i1) = 8i$$

56. First, write $-i$ in polar form: $-i = 1(\cos 270° + i \sin 270°)$. Then, the cube roots of $1(\cos 270° + i \sin 270°)$ are given by

$$1^{1/3}\left(\cos \frac{270° + k \cdot 360°}{3} + i \sin \frac{270° + k \cdot 360°}{3}\right) \qquad k = 0, 1, 2$$

Thus,

$$w_1 = 1^{1/3}\cos\left(\frac{270° + 0 \cdot 360°}{3} + i \sin \frac{270° + 0 \cdot 360°}{3}\right)$$

$$= \cos 90° + i \sin 90° = i$$

$$w_2 = 1^{1/3}\cos\left(\frac{270° + 1 \cdot 360°}{3} + i \sin \frac{270° + 1 \cdot 360°}{3}\right)$$

$$= \cos 210° + i \sin 210° = -\frac{\sqrt{3}}{2} - \frac{1}{2}i$$

$$w_3 = 1^{1/3}\cos\left(\frac{270° + 2 \cdot 360°}{3} + i \sin \frac{270° + 2 \cdot 360°}{3}\right)$$

$$= \cos 330° + i \sin 330° = \frac{\sqrt{3}}{2} - \frac{1}{2}i$$

57. Since $\theta = \frac{s}{R}$, and $s = 8$, and $R = 2$, we have $\theta = \frac{8}{2} = 4$ rad.

Since $\cos \theta = \frac{a}{R}$ and $\sin \theta = \frac{b}{R}$, we have
$$a = R \cos \theta = 2 \cos 4 \quad \text{and} \quad b = R \sin \theta = 2 \sin 4$$

Thus, $(a, b) = (2 \cos 4, 2 \sin 4) = (-1.307, -1.514)$.

58. Since $\tan \theta = \frac{b}{a}$, we have $\tan \theta = \frac{-1.2}{-1.6} = 0.75$. Since (a, b) is in Quadrant III,

$$\theta = \tan^{-1} \theta + \pi = \tan^{-1} 0.75 + \pi = 3.785 \text{ rad}$$

Since $\theta = \frac{s}{R}$, we can write $3.785 = \frac{s}{2}$, $s = 2(3.785) = 7.570$ units

373

59. We first find the period and phase shift by solving

$$\pi x + \frac{\pi}{4} = 0 \qquad \text{and} \qquad \pi x + \frac{\pi}{4} = 2\pi$$

$$x = -\frac{1}{4} \qquad\qquad\qquad x = -\frac{1}{4} + 2$$

Period = 2   Phase shift $= -\frac{1}{4}$

Now, since $2\sec\left(\pi x + \frac{\pi}{4}\right) = \dfrac{1}{\frac{1}{2}\cos\left(\pi x + \frac{\pi}{4}\right)}$, we graph $y = \frac{1}{2}\cos\left(\pi x + \frac{\pi}{4}\right)$ for one cycle

from $-\frac{1}{4}$ to $-\frac{1}{4} + 2 = \frac{7}{4}$ with a broken line graph, then take reciprocals. We also place asymptotes

through the x intercepts of the cosine graph to guide us when we sketch the secant function.

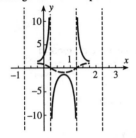

We then extend the one cycle over the required interval from –1 to 3.

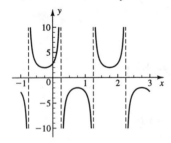

60. Let $y = \tan^{-1} x$ $-\frac{\pi}{2} < y < \frac{\pi}{2}$ or, equivalently, $x = \tan y$ $-\frac{\pi}{2} < y < \frac{\pi}{2}$

Geometrically,

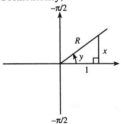   or   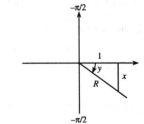

In either case, $R = \sqrt{x^2 + 1}$

$$\sec(2 \tan^{-1} x) \; = \sec(2y) = \frac{1}{\cos 2y} = \frac{1}{\cos^2 y - \sin^2 y} = 1 \div (\cos^2 y - \sin^2 y)$$

$$= 1 \div \left[\left(\frac{1}{\sqrt{x^2 + 1}} \right)^2 - \left(\frac{x}{\sqrt{x^2 + 1}} \right)^2 \right] = 1 \div \left[\frac{1}{x^2 + 1} - \frac{x^2}{x^2 + 1} \right]$$

$$= 1 \div \frac{1 - x^2}{x^2 + 1} = \frac{x^2 + 1}{1 - x^2} \; \text{ or } \; \frac{1 + x^2}{1 - x^2}$$

61. $\tan 3x = \tan(x + 2x)$ Algebra

$$= \frac{\tan x + \tan 2x}{1 - \tan x \tan 2x}$$ Sum identity

$$= \frac{\tan x + \dfrac{2 \tan x}{1 - \tan^2 x}}{1 - \tan x \cdot \dfrac{2 \tan x}{1 - \tan^2 x}}$$ Double-angle identity

$$= \frac{(1 - \tan^2 x)}{(1 - \tan^2 x)} \cdot \frac{\tan x + \dfrac{2 \tan x}{1 - \tan^2 x}}{1 - \dfrac{2 \tan^2 x}{1 - \tan^2 x}}$$ Algebra

$$= \frac{(1 - \tan^2 x)\tan x + 2 \tan x}{1 - \tan^2 x - 2 \tan^2 x}$$ Algebra

$$= \frac{\tan x - \tan^3 x + 2 \tan x}{1 - 3 \tan^2 x}$$ Algebra

$$= \frac{3 \tan x - \tan^3 x}{1 - 3 \tan^2 x}$$ Algebra

$$= \frac{\tan x(3 - \tan^2 x)}{1 - 3 \tan^2 x}$$ Algebra

$$= \frac{1}{\cot x} \frac{3 - \tan^2 x}{1 - 3 \tan^2 x}$$ Reciprocal identity

$$= \frac{3 - \tan^2 x}{\cot x - 3 \tan^2 x \cot x}$$ Algebra

$$= \frac{3 - \tan^2 x}{\cot x - 3 \tan x \cdot (\tan x \cot x)}$$ Algebra

$$= \frac{3 - \tan^2 x}{\cot x - 3 \tan x \cdot 1}$$ Reciprocal identity

$$= \frac{3 - \tan^2 x}{\cot x - 3 \tan x}$$ Algebra

62. Use $r = \sqrt{x^2 + y^2}$ and $r \cos \theta = x$

$$r(\cos \theta + 1) = 1$$

$$r \cos \theta + r = 1$$

$$x + \sqrt{x^2 + y^2} = 1$$

$$x = 1 - \sqrt{x^2 + y^2}$$

$$x - 1 = -\sqrt{x^2 + y^2} \text{ or, squaring both sides}$$

$$(x - 1)^2 = \left(-\sqrt{x^2 + y^2}\right)^2 = x^2 + y^2$$

63. We graph $r = 5 - 3 \sin \dfrac{\theta}{2}$ using rapid sketching techniques. The following table can be used to investigate how r varies as θ varies over particular intervals. We then sketch the polar curve from the information in the table.

θ	$\dfrac{\theta}{2}$	$\sin \dfrac{\theta}{2}$	$-3 \sin \dfrac{\theta}{2}$	$5 - 3 \sin \dfrac{\theta}{2}$	
0 to $\dfrac{\pi}{2}$	0 to $\dfrac{\pi}{4}$	0 to $\dfrac{\sqrt{2}}{2}$	0 to $-\dfrac{3\sqrt{2}}{2}$	5 to $5 - \dfrac{3\sqrt{2}}{2}$	(≈ 2.9)
$\dfrac{\pi}{2}$ to π	$\dfrac{\pi}{4}$ to $\dfrac{\pi}{2}$	$\dfrac{\sqrt{2}}{2}$ to 1	$-\dfrac{3\sqrt{2}}{2}$ to -3	$5 - \dfrac{3\sqrt{2}}{2}$ to 2	
π to $\dfrac{3\pi}{2}$	$\dfrac{\pi}{2}$ to $\dfrac{3\pi}{4}$	1 to $\dfrac{\sqrt{2}}{2}$	-3 to $-\dfrac{3\sqrt{2}}{2}$	2 to $5 - \dfrac{3\sqrt{2}}{2}$	
$\dfrac{3\pi}{2}$ to 2π	$\dfrac{3\pi}{4}$ to π	$\dfrac{\sqrt{2}}{2}$ to 0	$-\dfrac{3\sqrt{2}}{2}$ to 0	$5 - \dfrac{3\sqrt{2}}{2}$ to 5	

θ	$\dfrac{\theta}{2}$	$\sin \dfrac{\theta}{2}$	$-3 \sin \dfrac{\theta}{2}$	$5 - 3 \sin \dfrac{\theta}{2}$	
2π to $\dfrac{5\pi}{2}$	π to $\dfrac{5\pi}{4}$	0 to $\dfrac{-\sqrt{2}}{2}$	0 to $\dfrac{3\sqrt{2}}{2}$	5 to $5 + \dfrac{3\sqrt{2}}{2}$	(≈ 7.1)
$\dfrac{5\pi}{2}$ to 3π	$\dfrac{5\pi}{4}$ to $\dfrac{3\pi}{2}$	$\dfrac{-\sqrt{2}}{2}$ to -1	$\dfrac{3\sqrt{2}}{2}$ to 3	$5 + \dfrac{3\sqrt{2}}{2}$ to 8	
3π to $\dfrac{7\pi}{2}$	$\dfrac{3\pi}{2}$ to $\dfrac{7\pi}{4}$	-1 to $\dfrac{-\sqrt{2}}{2}$	3 to $\dfrac{3\sqrt{2}}{2}$	8 to $5 + \dfrac{3\sqrt{2}}{2}$	
$\dfrac{7\pi}{2}$ to 4π	$\dfrac{7\pi}{4}$ to 2π	$\dfrac{-\sqrt{2}}{2}$ to 0	$\dfrac{3\sqrt{2}}{2}$ to 0	$5 + \dfrac{3\sqrt{2}}{2}$ to 5	

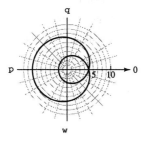

64. $x^3 - 4 = 0$

$x^3 = 4$

Therefore, x is a cube root of 4 and there are three of them. First, we write 4 in polar form:

$4 = 4(1 + 0i) = 4(\cos 0° + i \sin 0°)$

All three cube roots of $4(\cos 0° + i \sin 0°)$ are given by

$$4^{1/3}\left(\cos \frac{0° + k \cdot 360°}{3} + i \sin \frac{0° + k \cdot 360°}{3}\right) \qquad k = 0, 1, 2$$

Thus,

$$w_1 = 4^{1/3}\left(\cos \frac{0° + 0 \cdot 360°}{3} + i \sin \frac{0° + 0 \cdot 360°}{3}\right) = 1.587$$

$$w_2 = 4^{1/3}\left(\cos \frac{0° + 1 \cdot 360°}{3} + i \sin \frac{0° + 1 \cdot 360°}{3}\right) = -0.794 + 1.375i$$

$$w_3 = 4^{1/3}\left(\cos \frac{0° + 2 \cdot 360°}{3} + i \sin \frac{0° + 2 \cdot 360°}{3}\right) = -0.794 - 1.375i$$

65. (A) By De Moivre's theorem, $(\cos \theta + i \sin \theta)^3 = \cos 3\theta + i \sin 3\theta$

By the binomial theorem,

$(\cos \theta + i \sin \theta)^3 = \cos^3 \theta + 3 \cos^2 \theta (i \sin \theta) + 3 \cos \theta (i \sin \theta)^2 + (i \sin \theta)^3$

$\qquad\qquad\qquad\quad = \cos^3 \theta + 3i \cos^2 \theta \sin \theta - 3 \cos \theta \sin^2 \theta - i \sin^3 \theta$

Thus,

$\cos 3\theta + i \sin 3\theta = \cos^3 \theta - 3 \cos \theta \sin^2 \theta + i(3 \cos^2 \theta \sin \theta - \sin^3 \theta)$

Equating the real and imaginary parts of the left and right sides, we obtain

$\cos 3\theta = \cos^3 \theta - 3 \cos \theta \sin^2 \theta$ and $\sin 3\theta = 3 \cos^2 \theta \sin \theta - \sin^3 \theta$

(B) $\cos 3\theta = \cos(\theta + 2\theta)$ Algebra

$\qquad\quad = \cos \theta \cos 2\theta - \sin \theta \sin 2\theta$ Sum identity

$\qquad\quad = \cos \theta(\cos^2 \theta - \sin^2 \theta) - \sin \theta(2 \sin \theta \cos \theta)$ Double-angle identities

$\qquad\quad = \cos^3 \theta - \cos \theta \sin^2 \theta - 2 \cos \theta \sin^2 \theta$ Algebra

$\qquad\quad = \cos^3 \theta - 3 \cos \theta \sin^2 \theta$ Algebra

$$
\begin{aligned}
\sin 3\theta &= \sin(\theta + 2\theta) && \text{Algebra} \\
&= \sin\theta \cos 2\theta + \cos\theta \sin 2\theta && \text{Sum identity} \\
&= \sin\theta(\cos^2\theta - \sin^2\theta) + \cos\theta(2\sin\theta\cos\theta) && \text{Double-angle identities} \\
&= \sin\theta\cos^2\theta - \sin^3\theta + 2\sin\theta\cos^2\theta && \text{Algebra} \\
&= 3\sin\theta\cos^2\theta - \sin^3\theta && \text{Algebra}
\end{aligned}
$$

66. The graph of f(x) is shown in the figure.

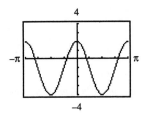

The graph appears to be a basic cosine curve with period π,
amplitude $= \dfrac{1}{2}$ (y max $-$ y min) $= \dfrac{1}{2}$ [2 $-$ ($-$4)] = 3, displaced downward by k = 1 unit.
It appears that g(x) = $-1 + 3\cos 2x$ would be an appropriate choice. We verify f(x) = g(x) as
follows:

$$
\begin{aligned}
f(x) &= 2\cos^2 x - 4\sin^2 x \\
&= 2\cos^2 x - 4(1 - \cos^2 x) && \text{Pythagorean identity} \\
&= 2\cos^2 x - 4 + 4\cos^2 x && \text{Algebra} \\
&= 6\cos^2 x - 4 && \text{Algebra} \\
&= 3(2\cos^2 x - 1) - 1 && \text{Algebra} \\
&= 3\cos 2x - 1 && \text{Double-angle identity} \\
&= -1 + 3\cos 2x = g(x) && \text{Algebra}
\end{aligned}
$$

67. The graph of f(x) is shown in the figure.

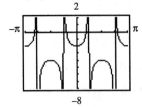

The graph appears to have vertical asymptotes $x = -\dfrac{3\pi}{4}$, $-\dfrac{\pi}{4}$, $\dfrac{\pi}{4}$, and $\dfrac{3\pi}{4}$ and period π.
It appears to have high and low points with y coordinates -4 and -2, respectively.

It appears that g(x) = sec 2x $-$ 3 would be an appropriate choice. We verify f(x) = g(x) as follows:

$$f(x) = \frac{6\sin^2 x - 2}{2\cos^2 x - 1}$$

$$= \frac{3(2\sin^2 x - 1) + 1}{2\cos^2 x - 1} \qquad \text{Algebra}$$

$$= \frac{1 - 3(1 - 2\sin^2 x)}{2\cos^2 x - 1} \qquad \text{Algebra}$$

$$= \frac{1 - 3\cos 2x}{\cos 2x} \qquad \text{Double-angle identities}$$

$$= \frac{1}{\cos 2x} - 3 \qquad \text{Algebra}$$

$$= \sec 2x - 3 = g(x) \qquad \text{Reciprocal identity}$$

68. The graph of f(x) is shown in the figure.

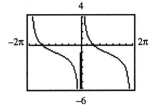

The graph appears to have vertical asymptotes $x = -2\pi$, $x = 0$, and $x = -2\pi$, and period 2π.

It appears to have x intercepts $-\frac{3\pi}{2}$ and $\frac{\pi}{2}$, and symmetry with respect to points where the curve crosses the line $y = -1$.

It appears to be a cotangent curve displaced downward by $|k| = 1$ unit.

It appears that $g(x) = -1 + \cot\frac{x}{2}$ would be an appropriate choice. We verify f(x) = g(x) as follows:

$$f(x) = \frac{\sin x + \cos x - 1}{1 - \cos x}$$

$$= \frac{\sin x - (1 - \cos x)}{1 - \cos x} \qquad \text{Algebra}$$

$$= \frac{\sin x}{1 - \cos x} - \frac{1 - \cos x}{1 - \cos x} \qquad \text{Algebra}$$

$$= 1 + \frac{1 - \cos x}{\sin x} - 1 \qquad \text{Algebra}$$

$$= 1 + \tan\frac{x}{2} - 1 \qquad \text{Half-angle identity}$$

$$= \cot\frac{x}{2} - 1 = g(x) \qquad \text{Reciprocal identity}$$

69.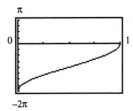

70.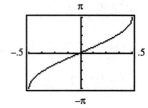

71.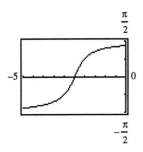

72. The solutions of the equation tan x = 5 are the x coordinates of the points of intersection of the graphs of y1 = tan x and y2 = 5 in the graphing calculator figure.

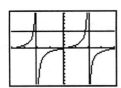

Examining the graph, we see that there are two points of intersection on the indicated interval. Using the zoom and trace procedures of the calculator, we obtain the following approximations to the x coordinates of these intersection points: $x = \begin{cases} -1.768 \\ 1.373 \end{cases}$

73. From the figure, we see that y1 = cos x and y2 = $\sqrt[3]{x}$ intersect only once on the interval $-2\pi \le x \le 2\pi$.

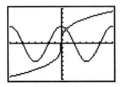

Using zoom and trace procedures, the solution is found to be 0.582. Since $|\cos x| \leq 1$, while $|\sqrt[3]{x}| > 1$ for real x not shown, there can be no other solutions.

74. From the figure, we see that $y1 = 3 \sin 2x \cos 3x$ and $y2 = 2$ intersect four times on the indicated interval.

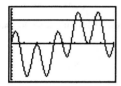

Using zoom and trace procedures, the solutions are found to be 3.909, 4.313, 5.111, and 5.516.

75.

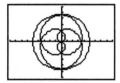

76. (A) $r = \dfrac{2}{1 - 0.7 \sin(\theta + 0.6)}$ (B) $r = \dfrac{2}{1 - \sin(\theta + 0.6)}$

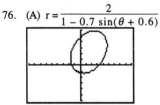

The graph is an ellipse.   The graph is a parabola.

(C) $r = \dfrac{2}{1 - 1.5 \sin(\theta + 0.6)}$

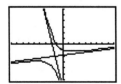

The graph is a hyperbola.

77. We are given two angles and the included side (ASA). We find the third angle, then apply the law of sines to find side BC. $\angle ABC + \angle BCA + \angle CAB = 180°$

$$\angle ABC = 180° - (52° + 77°) = 51°$$

$$\frac{\sin CAB}{BC} = \frac{\sin ABC}{AC}$$

$$BC = \frac{AC \sin CAB}{\sin ABC} = \frac{(520 \text{ ft}) \sin 77°}{\sin 51°} = 650 \text{ ft}$$

78. Here we are given two sides and the included angle, hence we can use the law of cosines to find side BC.

$$BC^2 = AB^2 + AC^2 - 2(AB)(AC) \cos CAB$$
$$= (580)^2 + (430)^2 - 2(580)(530) \cos 64° = 302,640.4 \ldots$$
$$BC = 550 \text{ ft}$$

79. (A) Triangle ABC is a right triangle.

$$\tan BAC = \frac{BC}{AC}$$

$$BC = AC \tan BAC$$
$$= (35 \text{ ft}) \tan 54° = 48 \text{ ft.}$$

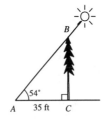

(B) Here triangle ABC is an oblique triangle.

$$\angle BAC = 54° - 11° = 43°$$
$$\angle BCA = 90° + 11° = 101°.$$

We are given two angles and the included side.

We find the third angle, then apply the law of

sines to find side BC.

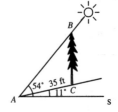

$$\angle ABC + \angle ACB + \angle BAC = 180°$$

$$\angle ABC = 180° - (43° + 101°) = 36°$$

$$\frac{\sin ABC}{AC} = \frac{\sin BAC}{BC}$$

$$BC = \frac{AC \sin BAC}{\sin ABC} = \frac{(35 \text{ ft}) \sin 43°}{\sin 36°} = 41 \text{ ft.}$$

80. In previous exercises, we have solved similar problems using right triangle methods. (See Chapter 1, Review Exercise, Problem 30, for example.)

   For comparison, we solve this problem using oblique triangle methods. We are given two angles, $\angle ABC = 180° - 67° = 113°$ and $\angle CAB = 42°$, and the included side, hence, we can find the third angle, then use the law of sines to find the other two sides.

$$\angle ABC + \angle CAB + \angle BCA = 180°$$

$$\angle BCA = 180° - (113° + 42°) = 25°$$

Solve for BC: $\dfrac{\sin CAB}{BC} = \dfrac{\sin BCA}{AB}$

$$BC = \frac{AB \sin CAB}{\sin BCA}$$

$$= \frac{(4.0 \text{ mi}) \sin 42°}{\sin 25°}$$
$$= 6.3 \text{ mi from Station B}$$

Solve for AC: $\dfrac{\sin ABC}{AC} = \dfrac{\sin BCA}{AB}$

$$AC = \frac{AB \sin ABC}{\sin BCA}$$

$$= \frac{(4.0 \text{ mi}) \sin 113°}{\sin 25°} = 8.7 \text{ mi from Station A}$$

81. (A) $\text{Period} = \dfrac{1}{\text{Frequency}} = \dfrac{1}{70 \text{ Hz}} = \dfrac{1}{70} \text{ sec.}$ Since $\text{Period} = \dfrac{2\pi}{B}$, $B = \dfrac{2\pi}{\text{Period}} = \dfrac{2\pi}{\frac{1}{70}} = 140\pi$

   (B) $\text{Frequency} = \dfrac{1}{\text{Period}} = \dfrac{1}{0.0125 \text{ sec}} = 80 \text{ Hz.}$ Since $\text{Period} = \dfrac{2\pi}{B}$, $B = \dfrac{2\pi}{\text{Period}} = \dfrac{2\pi}{0.0125} = 160\pi$

   (C) $\text{Period} = \dfrac{2\pi}{B} = \dfrac{2\pi}{100\pi} = \dfrac{1}{50} \text{ sec.}$ $\text{Frequency} = \dfrac{1}{\text{Period}} = \dfrac{1}{\frac{1}{50} \text{ sec}} = 50 \text{ Hz}$

82. The height of the wave from trough to crest is the difference in height between the crest (height A) and the trough (height –A). In this case, A = 2 ft.

   $$A - (-A) = 2A = 2(2 \text{ ft}) = 4 \text{ ft.}$$

To find the wavelength λ, we note: $\lambda = 5.12\ T^2$ $T = 4$ sec $\lambda = 5.12(4)^2 \approx 82$ ft

To find the speed S, we use

$$S = \sqrt{\frac{g\ \lambda}{2\pi}} \qquad g = 32\ \text{ft/sec}^2 \qquad S = \sqrt{\frac{32(82)}{2\pi}} \approx 20\ \text{ft/sec}$$

83. Area OCBA = Area of Sector OCB + Area of triangle OAB.

Area of Sector OCB $= \dfrac{1}{2} r^2 \theta = \dfrac{1}{2} \cdot 1^2 \cdot \theta = \dfrac{1}{2} \theta$

Area of right triangle OAB $= \dfrac{1}{2} (\text{base})(\text{height}) = \dfrac{1}{2} xy$

Since OAB is a right triangle, we have

$$\sin \theta = \frac{x}{1} \quad x = \sin \theta \qquad\qquad \cos \theta = \frac{y}{1} \quad y = \cos \theta$$

Hence, area of triangle OAB $= \dfrac{1}{2} xy = \dfrac{1}{2} \sin \theta \cos \theta.$

Thus, Area of OCBA $= \dfrac{1}{2} \theta + \dfrac{1}{2} \sin \theta \cos \theta$

84. Since $x = \sin \theta$, $\theta = \sin^{-1} x$ (θ is acute)

Since OAB is a right triangle, applying the Pythagorean theorem, we have

$$x^2 + y^2 = 1^2$$
$$y^2 = 1 - x^2$$
$$y = \sqrt{1 - x^2}$$

Thus, Area of OCBA $= \dfrac{1}{2} \theta + \dfrac{1}{2} xy$ \qquad (see previous problem)

$$= \dfrac{1}{2} \sin^{-1} x + \dfrac{1}{2} x\sqrt{1 - x^2}$$

85. We are to solve $\dfrac{1}{2} \theta + \dfrac{1}{2} \sin \theta \cos \theta = 0.5.$

We graph $y1 = \dfrac{1}{2} \theta + \dfrac{1}{2} \sin \theta \cos \theta$ and

$y2 = 0.5$ on the interval from 0 to $\dfrac{\pi}{2}$.

From the figure, we see that y1 and y2

intersect once on the interval.

Using zoom and trace procedures, the

solution is found to be $\theta = 0.553.$

86. We are to solve $\frac{1}{2}\sin^{-1}x + \frac{1}{2}x\sqrt{1-x^2} = 0.4$.

    We graph $y1 = \frac{1}{2}\sin^{-1}x + \frac{1}{2}x\sqrt{1-x^2}$ and
    $y2 = 0.4$ on the interval from 0 to 1. From the
    figure, we see that y1 and y2 intersect once on
    the interval.

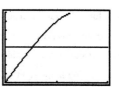

    Using zoom and trace procedures, the solution
    is found to be x = 0.412.

87. (A) Since the right triangles in the figure are similar, we can write $\frac{r}{h} = \frac{R}{H}$.

    Since $\tan\alpha = \frac{r}{h} = \frac{R}{H}$, we can write $\quad \frac{r}{h} = \tan\alpha \qquad r = h\tan\alpha$

    $$\frac{R}{H} = \tan\alpha \qquad R = H\tan\alpha$$

    Then, $R = H\tan\alpha = (H - h + h)\tan\alpha = (H - h)\tan\alpha + h\tan\alpha$

    $$R = (H - h)\tan\alpha + r$$

    (B) Solving the previous equation for $\tan\alpha$, we can write

    $$R - r = (H - h)\tan\alpha \qquad \tan\alpha = \frac{R - r}{H - h}$$

    Since α and β are complementary angles, we can write

    $$\tan\beta = \cot\alpha = \frac{1}{\tan\alpha} = 1 + \tan\alpha = 1 + \frac{R - r}{H - h} = \frac{H - h}{R - r}$$

    Thus, $\beta = \tan^{-1}\left(\frac{H - h}{R - r}\right)$.

88. We require θ such that the actual velocity **R** will be the resultant of the
    apparent velocity **v** and the wind velocity **w**.

    From the diagram it should be clear that

    $$\sin\theta = \frac{|\mathbf{w}|}{|\mathbf{v}|} = \frac{81.5}{265} \qquad\qquad \theta = \sin^{-1}\frac{81.5}{265} = 18°$$

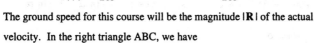

    The ground speed for this course will be the magnitude $|\mathbf{R}|$ of the actual
    velocity. In the right triangle ABC, we have

    $$\cos\theta = \frac{|\mathbf{R}|}{|\mathbf{v}|} \qquad\qquad |\mathbf{R}| = |\mathbf{v}|\cos\theta = 265\cos\left(\sin^{-1}\frac{81.5}{265}\right)$$

    $$= 265\sqrt{1 - \left(\frac{81.5}{265}\right)^2} = 252 \text{ mph}$$

89. (A) First, form a force diagram with all force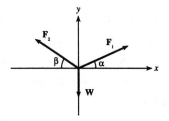
vectors in standard position at the origin.

Let $\mathbf{F}_1$ = the tension in the left side

$\mathbf{F}_2$ = the tension in the right side

$|\mathbf{F}_1| = T_L$ $|\mathbf{F}_2| = T_R$ $|\mathbf{W}| = w$

Write each force vector in terms of $\mathbf{i}$ and $\mathbf{j}$
unit vectors.

$$\mathbf{F}_1 = T_R \cos\alpha\,\mathbf{i} + T_R \sin\alpha\,\mathbf{j} \qquad \mathbf{F}_2 = T_L(-\cos\beta)\mathbf{i} + T_L \sin\beta\,\mathbf{j} \qquad \mathbf{W} = -w\mathbf{j}$$

For the system to be in static equilibrium, we must have $\mathbf{F}_1 + \mathbf{F}_2 + \mathbf{W} = \mathbf{0}$ which becomes, on
addition,

$$(T_R \cos\alpha - T_L \cos\beta)\mathbf{i} + (T_R \sin\alpha + T_L \sin\beta - w)\mathbf{j} = 0\mathbf{i} + 0\mathbf{j}$$

Since two vectors are equal if and only if their corresponding components are equal, we are led to
the following system of equations in T_L and T_R:

$$T_R \cos\alpha - T_L \cos\beta = 0 \qquad\qquad T_R \sin\alpha + T_L \sin\beta - w = 0$$

$$\text{Solving, } T_R = T_L \frac{\cos\beta}{\cos\alpha} \qquad\qquad T_L \sin\alpha \frac{\cos\beta}{\cos\alpha} + T_L \sin\beta = w$$

$$T_L\left(\frac{\sin\alpha\cos\beta}{\cos\alpha} + \sin\beta\right) = w$$

$$T_L\left(\frac{\sin\alpha\cos\beta + \cos\alpha\sin\beta}{\cos\alpha}\right) = w$$

$$T_L \frac{\sin(\alpha + \beta)}{\cos\alpha} = w$$

Thus, $T_L = \dfrac{w\cos\alpha}{\sin(\alpha + \beta)}$. Hence, $T_R = \dfrac{w\cos\alpha}{\sin(\alpha + \beta)} \dfrac{\cos\beta}{\cos\alpha} = \dfrac{w\cos\beta}{\sin(\alpha + \beta)}$

(B) If $\alpha = \beta$, then

$$T_L = T_R = \frac{w\cos\alpha}{\sin(\alpha + \alpha)} = \frac{w\cos\alpha}{\sin 2\alpha} = \frac{w\cos\alpha}{2\sin\alpha\cos\alpha} = \frac{w}{2\sin\alpha} = \frac{w}{2}\frac{1}{\sin\alpha}$$

$$= \frac{1}{2}\,w\,\csc\alpha$$

90. We can apply the law of cosines to the triangle shown in the figure. Then,

$$100^2 = r^2 + r^2 - 2r\cdot r\cdot \cos\theta$$

$$10{,}000 = 2r^2 - 2r^2 \cos\theta = 2r^2(1 - \cos\theta)$$

$$5000 = r^2(1 - \cos\theta)$$

(A) Given $\theta = 10°$, then $5000 = r^2(1 - \cos 10°)$

$$r^2 = \frac{5000}{1 - \cos 10°}$$

$$r = \sqrt{\frac{5000}{1 - \cos 10°}} = 574 \text{ ft}$$

(B) Given r = 2000, then $5000 = (2000)^2(1 - \cos \theta)$

$$\frac{5000}{(2000)^2} = 1 - \cos \theta$$

$$\cos \theta = 1 - \frac{5000}{(2000)^2}$$

$$\theta = \cos^{-1}\left[1 - \frac{5000}{(2000)^2}\right] = 2.9°$$

91. (A)

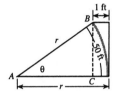

Since s = rθ, we can write rθ = 50.

To determine θ, we note that triangle

ABC is a right triangle, with side

AC = r – 1. Then,

$$\cos \theta = \frac{AC}{AB} = \frac{r - 1}{r}$$

$$\theta = \cos^{-1}\left(\frac{r - 1}{r}\right)$$

Thus, $r \cos^{-1}\left(\frac{r-1}{r}\right) = 50$. To solve this,

we graph $y1 = r \cos^{-1}\left(\frac{r-1}{r}\right)$ and $y2 = 50$

on the interval from 1000 to 2000. From the

figure, we see that y1 and y2 intersect once on

the interval.

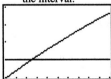

Using zoom and trace procedures, the solution

is found to be r = 1250 ft.

(B) From Problem 90, we have $5000 = r^2(1 - \cos \theta)$. If $r = 1250$, then

$$5000 = (1250)^2(1 - \cos \theta)$$

$$\frac{5000}{(1250)^2} = 1 - \cos \theta$$

$$\cos \theta = 1 - \frac{5000}{(1250)^2}$$

$$\theta = \cos^{-1}\left[1 - \frac{5000}{(1250)^2} \right] = 4.6°$$

92. The areas of the end pieces are given by the formula from Section 5.4, Problem 69 to be
$\frac{1}{2}r^2(\theta - \sin \theta)$. Subtracting this twice from the area of the total cross-section, πr^2, we obtain

$$A = \pi r^2 - \frac{1}{2}r^2(\theta - \sin \theta) - \frac{1}{2}r^2(\theta - \sin \theta) = \pi r^2 - r^2(\theta - \sin \theta)$$

$$= \pi r^2 - r^2\theta + r^2 \sin \theta = r^2(\pi - \theta + \sin \theta)$$

93. If all three pieces of the log have the same
cross-sectional area, then each of these areas
is one-third of the entire area, that is

$$r^2(\pi - \theta + \sin \theta) = \frac{1}{3} \pi r^2$$

Thus, $\pi - \theta + \sin \theta = \frac{1}{3} \pi$. To solve this, we graph $y1 = \pi - \theta + \sin \theta$ and $y2 = \frac{1}{3} \pi$ on the interval
from 0 to π. From the figure, we see that y1 and y2 intersect once on the interval.

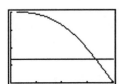

Using zoom and trace procedures, the solution is found to be $\theta = 2.6053$ rad.

94. (A)

x (months)	1, 13	2, 14	3, 15	4, 16	5, 17	6, 18
$y\left(\genfrac{}{}{0pt}{}{\text{twilight}}{\text{duration}}\right)$	1.62	1.82	2.35	2.98	3.55	4.12

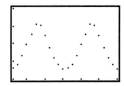

x (months)	7, 19	8, 20	9, 21	10, 22	11, 23	12, 24
y $\left(\begin{array}{c}\text{twilight}\\ \text{duration}\end{array}\right)$	4.05	3.50	2.80	2.22	1.80	1.57

(B) From the table, Max y = 4.12 and Min y = 1.57. Then,

$$A = \frac{\text{Max y} - \text{Min y}}{2} = \frac{4.12 - 1.57}{2} = 1.275$$

$$B = \frac{2\pi}{\text{Period}} = \frac{2\pi}{12} = \frac{\pi}{6}$$

$$k = \text{Min y} + A = 1.57 + 1.275 = 2.845$$

From the plot in (A) or the table, we estimate the smallest value of x for which y = k = 2.845 to be approximately 3.4. Then, this is the phase shift for the graph. Substitute $B = \frac{\pi}{6}$ and x = 3.4 into the phase-shift equation

$$x = -\frac{C}{B} \qquad 3.4 = -\frac{C}{\frac{\pi}{6}} \qquad C = \frac{-3.4\pi}{6} = -1.8$$

Thus, the equation required is $y = 2.845 + 1.275 \sin\left(\frac{\pi x}{6} - 1.8\right)$

(C)

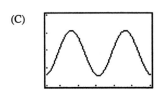

APPENDICES

APPENDIX A Comments on Numbers

EXERCISE A.1 Real numbers

1. There are infinitely many negative integers. Examples include –3.

    The only integer that is neither positive nor negative is 0.

    There are infinitely many positive integers. Examples include 5.

3. There are infinitely many rational numbers that are not integers. Examples include $\frac{2}{3}$.

5. (A) True.   (B) False (6 is a real number, but it is not irrational).   (C) True.

7. (A) $0.\overline{363636}$, rational   (B) $0.7\overline{777}$, rational

    (C) $2.64575131\ldots$, irrational   (D) $1.6\overline{2500}$, rational

9. (A) Since $\frac{26}{9} = 2.\overline{888}$, it lies between 2 and 3

    (B) Since $-\frac{19}{5} = -3.8$, it lies between -4 and -3.

    (C) Since $-\sqrt{23} = -4.7958\ldots$, it lies between -5 and -4.

11. $y + 3$ 13. $(3 \cdot 2)\,x$ 15. $7x$ 17. $(5 + 7)\,x$

19. $3m$ 21. $u + v$ 23. $(2 + 3)\,x$

25. (A) True. (This is the commutative property for addition of real numbers.)

    (B) False. For example, $4 - 2 \neq 2 - 4$.

    (C) True. (This is the commutative property for multiplication of real numbers.)

    (D) False. For example, $8 \div 2 \neq 2 \div 8$

EXERCISE A.2 Complex Numbers

1. $(3 - 2i) + (4 - 7i) = 3 - 2i + 4 + 7i = 3 + 4 - 2i + 7i = 7 + 5i$

3. $(3 - 2i) - (4 + 7i) = 3 - 2i - 4 - 7i = 3 - 4 - 2i + 7i = -1 - 9i$

5. $(6i)(3i) = 18i^2 = 18(-1) = -18$ 7. $(2i)(3 - 4i) = 6i - 8i^2 = 6i - 8(-1) = 8 + 6i$

9. $(3 - 4i)(1 - 2i) = 3 - 10i + 8i^2 = 3 - 10i + 8(-1) = 3 - 10i - 8 = -5 - 10i$

11. $(3 + 5i)(3 - 5i) = 9 - 25i^2 = 9 - 25(-1) = 9 + 25 = 34$

13. $\dfrac{1}{2 + i} = \dfrac{1}{2 + i} \dfrac{2 - i}{2 - i} = \dfrac{2 - i}{4 - i^2} = \dfrac{2 - i}{4 + 1} = \dfrac{2 - i}{5} = \dfrac{2}{5} - \dfrac{1}{5}i$

15. $\dfrac{2 - i}{3 + 2i} = \dfrac{2 - i}{3 + 2i} \dfrac{3 - 2i}{3 - 2i} = \dfrac{6 - 7i + 2i^2}{9 - 4i^2} = \dfrac{6 - 7i - 2}{9 + 4} = \dfrac{4 - 7i}{13} = \dfrac{4}{13} - \dfrac{7}{13}i$

17. $\dfrac{-1 + 2i}{4 + 3i} = \dfrac{-1 + 2i}{4 + 3i} \dfrac{4 - 3i}{4 - 3i} = \dfrac{-4 + 11i - 6i^2}{16 - 9i^2} = \dfrac{-4 + 11i + 6}{16 + 9} = \dfrac{2 + 11i}{25} = \dfrac{2}{25} + \dfrac{11}{25}i$

19. $(3 + \sqrt{-4}) + (2 - \sqrt{-16}) = (3 + i\sqrt{4}) + (2 - i\sqrt{16}) = 3 + 2i + 2 - 4i = 3 + 2 + 2i - 4i = 5 - 2i$

21. $(5 - \sqrt{-1}) - (2 - \sqrt{-36}) = (5 - i) - (2 - i\sqrt{36}) = (5 - i) - (2 - 6i)$

$\qquad\qquad = 5 - i - 2 + 6i = 5 - 2 - i + 6i = 3 + 5i$

23. $(-3 - \sqrt{-1})(-2 + \sqrt{-49}) = (-3 - i)(-2 + i\sqrt{49}) = (-3 - i)(-2 + 7i) = 6 - 19i - 7i^2$

$\qquad\qquad = 6 - 19i + 7 = 13 - 19i$

25. $\dfrac{5 - \sqrt{-1}}{2 + \sqrt{-4}} = \dfrac{5 - i}{2 + i\sqrt{-4}} = \dfrac{5 - i}{2 + 2i} = \dfrac{5 - i}{2 + 2i} \dfrac{2 - 2i}{2 - 2i} = \dfrac{10 - 12i + 2i^2}{4 - 4i^2}$

$\qquad\qquad = \dfrac{10 - 12i - 2}{4 + 4} = \dfrac{8 - 12i}{8} = 1 - \dfrac{3}{2}i$

27. $(1 - i)^2 - 2(1 - i) + 2 = 1 - 2i + i^2 - 2 + 2i + 2 = 1 - 2i - 1 - 2 + 2i + 2 = 0 + 0i = 0$

29. $\left(\dfrac{-1}{2} + \dfrac{\sqrt{3}}{2}i\right)^3 = \left(-\dfrac{1}{2} + \dfrac{\sqrt{3}}{2}i\right)\left(-\dfrac{1}{2} + \dfrac{\sqrt{3}}{2}i\right)^2$

$\qquad = \left(-\dfrac{1}{2} + \dfrac{\sqrt{3}}{2}i\right)\left[\left(-\dfrac{1}{2}\right)^2 + 2\left(-\dfrac{1}{2}\right)\left(\dfrac{\sqrt{3}}{2}i\right) + \left(\dfrac{\sqrt{3}}{2}i\right)^2\right]$

$\qquad = \left(-\dfrac{1}{2} + \dfrac{\sqrt{3}}{2}i\right)\left[\dfrac{1}{4} - \dfrac{2\sqrt{3}}{4}i + \dfrac{3}{4}i^2\right] = \left(-\dfrac{1}{2} + \dfrac{\sqrt{3}}{2}i\right)\left(\dfrac{1}{4} - \dfrac{2\sqrt{3}}{4}i - \dfrac{3}{4}\right)$

$\qquad = \left(-\dfrac{1}{2} + \dfrac{\sqrt{3}}{2}i\right)\left(-\dfrac{1}{2} - \dfrac{2\sqrt{3}}{4}i\right)$

$\qquad = \left(-\dfrac{1}{2}\right)\left(-\dfrac{1}{2}\right) + \left(-\dfrac{1}{2}\right)\left(-\dfrac{2\sqrt{3}}{4}i\right) + \left(\dfrac{\sqrt{3}}{2}i\right)\left(-\dfrac{1}{2}\right) + \left(\dfrac{\sqrt{3}}{2}i\right)\left(-\dfrac{2\sqrt{3}}{4}i\right)$

$\qquad = \dfrac{1}{4} + \dfrac{\sqrt{3}}{4}i - \dfrac{\sqrt{3}}{4}i - \dfrac{3}{4}i^2 = \dfrac{1}{4} + \dfrac{3}{4} = 1$

APPENDICES

EXERCISE A.3 Significant Digits

1. $640 = 6.40. \times 10^2 = 6.4 \times 10^2$
    ↑ |

    2 places left
    |
    positive exponent

3. $5,460,000,000 = 5.460,000,000. \times 10^9 = 5.46 \times 10^9$
    ↑ |

    9 places left
    |
    positive exponent

5. $0.73 = 0.7.3 \times 10^{-1} = 7.3 \times 10^{-1}$
    | ↑

    1 place right
    |
    negative exponent

7. $0.00000032 = 0.0000003.2 \times 10^{-7} = 3.2 \times 10^{-7}$
    | ↑

    7 places right
    |
    negative exponent

9. $0.0000491 = 0.00004.91 \times 10^{-5} = 4.91 \times 10^{-5}$
    | ↑

    5 places right
    |
    negative exponent

11. $67,000,000,000 = 6.7,000,000,000. \times 10^{10}$
    ↑ |

    10 places left
    |
    positive exponent

13. $5.6 \times 10^4 = 5.6 \times 10,000 = 56,000$

15. $9.7 \times 10^{-3} = 9.7 \times 0.001 = 0.0097$

17. $4.61 \times 10^{12} = 4.61 \times 1,000,000,000,000 = 4,610,000,000,000$

19. $1.08 \times 10^{-1} = 1.08 \times 0.1 = 0.108$

21. 12.3 has a decimal point. From the first nonzero digit (1) to the last digit (3), there are 3 digits. 3 significant digits.

23. $12 \cdot 300$ has a decimal point. From the first nonzero digit (1) to the last digit (0), there are 5 digits. 5 significant digits.

25. 0.01230 has a decimal point. From the first nonzero digit (1) to the last digit (0), there are 4 digits. 4 significant digits.

27. 6.7×10^{-1} is in scientific notation. There are 2 digits in 6.7. 2 significant digits.

29. 6.700×10^{-1} is in scientific notation. There are 4 digits in 6.700. 4 significant digits.

31. 7.090×10^5 is in scientific notation. There are 4 digits in 7.090. 4 significant digits.

33. 635,000

35. 86.8 (convention of leaving the digit before the 5 alone, if it is even)

37. 0.00465

39. $734 = 7.34 \times 10^2 \approx 7.3 \times 10^2$

41. $0.040 = 4.0 \times 10^{-2}$

43. $0.000435 = 4.35 \times 10^{-4} \approx 4.4 \times 20^{-4}$ (convention of rounding the digit before the 5 up, if it is odd)

45. 3, since there are three significant digits in the number (32.8) with the least number of significant digits in the calculation.

47. 2, since there are two signficant digits in the numbers (360 and 1,200) with the least number of significant digits in the calculation.

49. 1, since there is one significant digit in the number (6×10^4) with the least number of significant digits in the calculation.

51. $\dfrac{6.07}{0.5057}$　　　　6.07 has the least number of significant digits (3).

　　 = 12.0　　　　Answer must have 3 significant digits.

53. $(6.14 \times 10^9)(3.154 \times 10^{-1})$　　　6.14×10^9 has the least number of significant digits (3).

　　 = 1.94×10^9　　　Answer must have 3 significant digits.

55. $\dfrac{6,730}{(2.30)(0.0551)}$　　　All numbers in the calculation have 3 significant digits.

　　 = 53,100　　　　Answer must have 3 significant digits.

57. C = 2π(25.31 cm)　　　There are 4 significant digits in 25.31 cm.

　　 = 159.0 cm　　　Answer must have 4 significant digits.

59. A = $\dfrac{1}{2}$(22.4 ft)(8.6 ft)　　　8.6 has the least number of significant digits (2)

　　 = 96 ft^2　　　Answer must have 2 significant digits

61. s = 4π (1.5 mm)^2　　　There are 2 significant digits in 1.5 mm

　　 = 28 mm^2　　　Answer must have 2 significant digits

63. V = ℓwh

　　 h = $\dfrac{V}{\ell w}$

　　 h = $\dfrac{24.2 \text{ cm}^3}{(3.25 \text{ cm})(4.50 \text{ cm})}$　　　All numbers in the calculation have 3 significant digits

　　 = 1.65 cm　　　Answer must have 3 significant digits

65. V = $\dfrac{1}{3}\pi r^2 h$

　　 3V = $\pi r^2 h$

　　 r^2 = $\dfrac{3V}{\pi h}$

　　 r = $\sqrt{\dfrac{3V}{\pi h}} = \sqrt{\dfrac{3(1200 \text{ in}^3)}{\pi(6.55 \text{ in.})}}$　　　1200 in^3 has the least number of significant digits (2)

　　 = 13 in.　　　Answer must have 2 significant digits

APPENDICES

APPENDIX B Functions and Inverse Functions

EXERCISE B.1 Functions

1. $f(x) = 4x - 1$
    $f(1) = 4(1) - 1$
    $= 3$

3. $f(x) = 4x - 1$
    $f(-1) = 4(-1) - 1$
    $= -5$

5. $f(x) = 4x - 1$
    $f(0) = 4(0) - 1$
    $= -1$

7. $g(x) = x - x^2$
    $g(1) = 1 - 1^2$
    $= 0$

9. $g(x) = x - x^2$
    $g(5) = 5 - 5^2$
    $= -20$

11. $g(x) = x - x^2$
    $g(-2) = -2 - (-2)^2$
    $= -2 - 4 = -6$

13. $f(0) + g(0) = 1 - 2 \cdot 0 + 4 - 0^2 = 1 + 4 = 5$

15. $\dfrac{f(3)}{g(1)} = \dfrac{1 - 2 \cdot 3}{4 - 1^2} = \dfrac{-5}{3} = -\dfrac{5}{3}$

17. $2f(-1) = 2[1 - 2(-1)] = 2(1 + 2) = 6$

19. $f(2 + h) = 1 - 2(2 + h) = 1 - 4 - 2h = -3 - 2h$

21. $\dfrac{f(2 + h) - f(2)}{h} = \dfrac{[1 - 2(2 + h)] - (1 - 2 \cdot 2)}{h} = \dfrac{-3 - 2h - (-3)}{h} = \dfrac{-2h}{h} = -2$

23. $g[f(2)] = g(1 - 2 \cdot 2) = g(-3) = 4 - (-3)^2 = 4 - 9 = -5$

25. $x^2 + y^2 = 25$ does not specify a function, since both (3, 4) and (3, -4) are solutions, in which the same domain value corresponds to more than one range value.

27. $2x - 3y = 6$ specifies a function, since each domain value x corresponds to exactly one range value $\left(\dfrac{2x - 6}{3}\right)$.

29. $y^2 = x$ does not specify a function, since both (9, 3) and (9, -3) are solutions, in which the same domain value corresponds to more than one range value.

31. $y = |x|$ specifies a function, since each domain value x corresponds to exactly one range value ($|x|$).

33. $f(x) = x^2 - x + 1$ Domain $X = \{-2, -1, 0, 1, 2\}$
    $f(-2) = (-2)^2 - (-2) + 1 = 7$
    $f(-1) = (-1)^2 - (-1) + 1 = 3$
    $f(0) = 0^2 - 0 + 1 = 1$
    $f(1) = 1^2 - 1 + 1 = 1$
    $f(2) = 2^2 - 2 + 1 = 3$ Range $Y = \{1, 3, 7\}$

35. G does not specify a function since the domain value -4 corresponds to more than one range value (3 and 0). F specifies a function.

    Domain of F = Set of all first components = $\{-2, -1, 0\} = X$
    Range of F = Set of all second components = $\{0, 1\} = Y$

37. $s(t) = 4.88t^2$

    $s(0) = 4.88(0)^2 = 0$

    $s(1) = 4.88(1)^2 = 4.88$ m

    $s(2) = 4.88(2)^2 = 19.52$ m

    $s(3) = 4.88(3)^2 = 43.92$ m

39. $\dfrac{s(2+h) - s(2)}{h} = \dfrac{4.88(2+h)^2 - 4.88(2)^2}{h} = \dfrac{4.88(4 + 4h + h^2) - 19.52}{h}$

    $= \dfrac{19.52 + 19.52h + 4.88h^2 - 19.52}{h} = \dfrac{19.52h + 4.88h^2}{h} = 19.52 + 4.88h$

As h gets closer to 0, this gets closer to 19.52; the average speed $\dfrac{s(2+h) - s(2)}{h}$ tends to a quantity called the speed at $t = 2$.

APPENDICES

EXERCISE B.2 Inverse Functions

1. This function is one-to-one since each range element corresponds to exactly one domain element.

3. This function is not one-to-one since the range element 9 corresponds to several domain elements.

5. This function passes the horizontal line test; it is one-to-one.

7. This function fails the horizontal line test; it is not one-to-one.

9. This function passes the horizontal line test; it is one-to-one.

11. The function fails the horizontal line test; it is not one-to-one.

13. First, we note that f is not one-to-one, since the range element 0 corresponds to two domain elements, -1 and 2. The function g is one-to-one, since each range element corresponds to exactly one domain element. Reversing the ordered pairs in the function g produces the inverse function.

$$g^{-1} = \{(-8, -2), (1, 1), (8, 2)\}$$

Its domain is $\{-8, 1, 8\}$. Its range is $\{-2, 1, 2\}$.

15.

17. Replace f(x) with y:

Interchange the variables x and y to form f^{-1}:

Solve for y in terms of x: $x + 7 = 2y$

$$y = \frac{x + 7}{2}$$

Replace y with $f^{-1}(x)$: $f^{-1}(x) = \frac{x + 7}{2}$

f: $y = 2x - 7$

f^{-1}: $x = 2y - 7$

19. Replace h(x) with y: h: $y = \dfrac{x + 3}{3}$

    Interchange the variables x and y to form h^{-1}: h^{-1}: $x = \dfrac{y + 3}{3}$

    Solve for y in terms of x: $3x = y + 3$

    $\qquad\qquad\qquad\qquad\qquad y = 3x - 3$

    Replace y with $h^{-1}(x)$: $h^{-1}(x) = 3x - 3$

21.

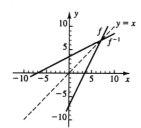

23. Replace f(x) with y: f: $y = 2x - 7$

    Interchange the variables x and y to form f^{-1}: f^{-1}: $x = 2y - 7$

    Solve for y in terms of x: $x + 7 = 2y$

    $\qquad\qquad\qquad\qquad\qquad y = \dfrac{x + 7}{2}$

    Replace y with $f^{-1}(x)$: $f^{-1}(x) = \dfrac{x + 7}{2}$

    Thus, $f^{-1}(3) = \dfrac{3 + 7}{2}$ $f^{-1}(3) = 5$

25. Replace h(x) with y: h: $y = \dfrac{x}{3} + 1$

    Interchange the variables x and y to form h^{-1}: h^{-1}: $x = \dfrac{y}{3} + 1$

    Solve for y in terms of x: $x - 1 = \dfrac{y}{3}$

    $\qquad\qquad\qquad\qquad\qquad y = 3(x - 1) = 3x - 3$

    Replace y with $h^{-1}(x)$: $h^{-1}(x) = 3x - 3$

    Thus, $h^{-1}(2) = 3 \cdot 2 - 3$ $h^{-1}(2) = 3$

27. $f[f^{-1}(4)] = 2[f^{-1}(4)] - 7 = 2\left(\dfrac{4 + 7}{2}\right) - 7 = 4 + 7 - 7 = 4$

29. $h^{-1}[h(x)] = 3h(x) - 3 = 3\left(\dfrac{x}{3} + 1\right) - 3 = x + 3 - 3 = x$

31. $h[h^{-1}(x)] = \dfrac{h^{-1}(x)}{3} + 1 = \dfrac{3x - 3}{3} + 1 = x - 1 + 1 = x$